ANIMAL AGGREGATIONS

AMS PRESS
NEW YORK

ANIMAL AGGREGATIONS

A Study in General Sociology

By W. C. ALLEE

The University of Chicago

THE UNIVERSITY OF CHICAGO PRESS
CHICAGO · ILLINOIS

Library of Congress Cataloging in Publication Data

Allee, Warder Clyde, 1885-1955.
Animal aggregations.

Reprint of the ed. published by University of Chicago Press, Chicago.
Bibliography: p.
Includes index.
1. Animal societies. I. Title.
QL775.A44 1978 591.5′2′4 75-41007
ISBN 0-404-14501-9

Reprinted from an original in the collections
of the Ohio State University Libraries

From the edition of 1931, Chicago
First AMS edition published in 1978

Manufactured in the United States of America

AMS PRESS, INC.
NEW YORK, N.Y.

IN MEMORY OF MY SON

WARDER ALLEE

1913–23

IN APPRECIATION OF HIS BOYISH ENTHUSIASM
OVER THE PARTS OF THIS WORK
WHICH HE UNDERSTOOD

PREFACE

The attempt to summarize knowledge concerning the relations within and between different sorts of animal societies is not new. Espinas in 1878 undertook such an effort, and shows in his introduction that Aristotle, Spinoza, Leibnitz, Montesquieu, Kant, Hegel, A. Comte, Herbert Spencer, and others had preceded him in the consideration of certain aspects of this problem. Since the time of Espinas, knowledge concerning the social insects and the development of insect societies has greatly increased. Wheeler, Forel, Buttel-Reepen, and many others have contributed both personal observations and generalizing summaries of value; and I have no desire to enter into a field so ably covered. There does remain, however, a field of social, or perhaps subsocial, life almost entirely untouched by these students. They have been concerned with the fascinating problems and intricate relationships presented by fairly well-developed societies. Here, I propose to investigate the relationships existing among the more loosely integrated collections of animals, which may rightly be designated as "animal aggregations," with regard to their ecological and behavioristic physiology, as well as with regard to their strictly social implications.

This book is built about a phenomenon or a series of phenomena, rather than about a philosophy. In the present form it may even be designated as notes on an unsolved problem; but since a presentation of a problem is necessary for its ultimate solution, and since an inquiry into the universality of a given problem is imperative before undertaking laborious experimentation directed toward finding a solution, no apology is offered for summarizing our growing knowledge on the subject of animal aggregations at the present stage of inquiry into the problems involved.

My own experimental work within the field covered by the present book began in 1911 and has continued intermittently to date. The investigation of animal aggregations has been at the center of my research program for the last twelve years, during which time work has been actively carried on with the aid of a number of graduate students, with facilities furnished by the University of Chicago, the Marine Biological Laboratory, and, more recently, with financial aid from a grant from the Rockefeller Foundation to aid investigations in the biological sciences at this university. The preparation of the manuscript of this book has been, in part, supported by aid from this latter source.

In addition to the loyal co-operation of students and colleagues in the accumulation of experimental data, of citations, and of criticisms, aid with the scattered literature has come from friends and acquaintances from five continents. The extended literature list is incomplete, but the labor of gathering and selecting the references used has been appreciably decreased by this cordial co-operation.

Certain specific acknowledgments I have made in the text. I am also indebted to Drs. Marie A Hinrichs, A. M. Holmquist, Walburga A. Petersen, J. M. Shaver, and O. Park; to Messrs. M. R. Garner, J. R. Fowler, J. F. Schuett, W. A. Dreyer, Carl Welty, W. H. Johnson, E. O. Deere, Ralph Buchsbaum, D. A. D. Boyer, J. F. W. Pearson, and T. Park; to Mrs. Frances Church van Pelt and Mrs. Gretchen Shaw Rudnick; and to Miss Edith Bowen, for citations to pertinent literature, for permission to give results before publication, or for criticism of parts of the manuscript, or for all three; to Professor F. R. Lillie, who read critically chapters xvi and xvii; to Professor A. E. Emerson for similar service with chapters xix and xx; to Mr. K. Toda, who drew or copied the text figures; and to Marjorie Hill Allee for editorial assistance with the manuscript. Acknowledgment of courtesy in permitting reproduction of figures will be given elsewhere.

W. C. Allee

Whitman Laboratory of Experimental Zoölogy
University of Chicago
August, 1930

CONTENTS

INTRODUCTION

HARMFUL EFFECTS OF AGGREGATIONS

BENEFICIAL EFFECTS OF AGGREGATIONS

GENERAL EFFECTS OF AGGREGATIONS

CONCLUSION

INTRODUCTION

CHAPTER I

THE GENERAL BACKGROUND

INTRODUCTION

This study of animal aggregations is concerned with some of the physiological effects of crowding upon the individuals composing the crowd, and is offered as a contribution toward the development of general sociology upon a physiological basis. A few years ago it would have been possible to summarize the knowledge then existing on the subject with the statement that, except in hibernation or at breeding time, the physiological effects of crowding are uniformly harmful, whether attention is given to the effect upon rate of reproduction, rate of individual growth, or longevity. Data on these harmful results will be presented later, but they are no longer accepted as a complete picture; in this study they are needed to obtain a correct perspective for the recent discoveries of beneficial effects of relatively unorganized crowds of animals.

Much attention has deservedly been given to the study of organized societies, particularly those of mammals, birds, and insects—sometimes with relation to the light they may throw upon the social relations of man, but frequently on account of their own inherent interest. In the main, consideration of these highly organized social groups falls outside the interests of the present discussion, which will be limited, so far as possible, to the physiological effects of crowding upon organisms whose interrelations have not reached the level of development usually called "social."

The general physiologists contend with justice that one cannot understand the physiology of man without a knowledge of the general physiology of all animals and much of that of plants as well. The comparative psychologists conclude similarly that one cannot understand the working of the human nervous system without knowing how other nervous systems function. Similarly, an increasing number of investigators are convinced that without a knowledge of general sociology we are likely to regard the social traits exhibited by man or by the ants as being peculiarly human or peculiarly formicine, when many of them are merely human or ant variations of social traits common to animals in general. Again, it is difficult to evaluate properly the origin and function of many of these general social traits without a proper understanding of their physiological antecedents among animals not usually regarded as having reached the social level. It is quite easy to consider certain bits of behavior as definitely social in origin and inherent in the social type of organization which may be merely specialized developments of general behavior common to most animals when crowded.

One fallacy may be suggested at the beginning. All too frequently one gains the impression that sex forms the main, if not the only, physiological connecting link between the infrasocial and the social animals. I believe that a consideration of the facts to be presented will allow us to place this important social factor more nearly in its proper relation to other factors equally important.

The problems dealt with in the present study are of interest also to the large group of students of animal ecology. It is generally known that ecology deals with the relations between the organism and its environment. This environment is roughly divided into two parts—the non-living and the living—which are commonly referred to as the "physical" and the "biotic" elements of the environment. Of these, the former has received particular attention at the hands of modern animal ecologists, since such factors as light, hydrogen-ion concentration, humidity, wind velocity, and temperature are more or less readily and definitely measured, and since others, such as soil type or the chemical composition of the waters of a lake or river, though less readily analyzed, are still capable of being studied on a quantitative basis. Meantime the analysis of the biotic relations of the environment has lagged, probably on account of the greater difficulties involved in the quantitative treatment of this exceedingly complex part of the environment. Even so, marked progress has been made in this analysis by recent students of the ecological relations within animal and plant communities (Smith, 1928; Shackleford, 1929). In the present studies we shall find ourselves concerned with animal communities which, from their concentrated nature, necessarily make the biotic elements an important aspect in the environment of any particular individual, while the physical elements of the environment act mainly through their influence on the entire aggregation or crowd. Such a situation must frequently obtain in assemblages of sea anemones, of *Mytilus*, of ascidians, or of crabs. In working at the aggregation level here considered, we find the ratio of importance of the physical and the biotic environment in a transition stage between that present in definitely social groups and that occurring in the more typical animal community, or biocoenose of the ecologist.

We must emphasize the fact that all studies dealing with the biotic elements of the environment are likely to be less definitely quantitative than those dealing with the non-living environment. This is no reason for their neglect, but it is a reason why we may not expect their treatment to be precise and final. The present summary, gained from pioneering in this relatively new field, must be regarded as tentative in many respects. My own research program dealing with various aspects of the subject is only well under way; the present statements furnish a point of departure, rather than a gathering-in of conclusions. With the accumulation of evidence now being actively collected, the conclusions tentatively advanced here may be further confirmed, or they may soon be modified or entirely abandoned. This must always be the case, even in the well-developed fields of physics and chemistry; and does not prevent summaries of knowledge to date having definite value, if they stimulate further research or give point to researches already in progress.

TERMINOLOGY

The general terminology causes unexpected difficulty. One usually thinks that such words as "society," "association," and "community" have a relatively stable meaning and that "biocoenosis," for example, might be expected to be a quite exact term; but this is unfortunately not the case.

According to writers on human sociology (Park and Burgess,

1921), "the terms society, community and social group are now used by students with a certain difference of emphasis, but with very little difference in meaning. Society is the more abstract and inclusive term, and society is made up of social groups, each possessing its own specific type of organization but having at the same time all the general characteristics of society in the abstract. Community is the term applied to societies and the social groups where they are considered from the point of view of the geographical distribution of the individuals and institutions of which they are composed. It follows that every community is a society, but not every society is a community. An individual may belong to many social groups but he will not ordinarily belong to more than one community, except in so far as a smaller community of which he is a member is included in a larger of which he is also a member. However, an individual is not, at least from a sociological point of view, a member of a community because he lives in it but rather because, and to the extent that, he participates in the common life of the community."

The same authors evidently do not consider "association" as being a sufficiently significant term to be given formal definition. In general sociology the contrast between what is ordinarily called an "association" and a "society" is important. Students differ concerning the proper criteria to use in making this distinction. Espinas (1878) recognized that there was a difference, and called "associations" accidental societies between animals of different species. According to this pioneer in the field of general sociology, the characteristic trait of social life is to be found in habitual reciprocity between activities which are more or less independent. He recognized certain similarities between associations and societies but regarded the former as less necessary for their constituent elements. Associations, according to Espinas, are groups of convenience, not of necessity. Deegener sharpened this distinction (1918) on the basis of usefulness of the animal group to the individual members. He designated an "association" as a collection of similar or dissimilar animals, which does not have value for the individuals composing the group; and a "society" as one in which the collection does have distinct value for the individuals of which it is composed.

Deegener's criteria for the social value of his categories were far less sensitive than those which were shortly developed by other workers in this field, and which will be summarized in the body of the present discussion. Application of such distinctions, even in their present incomplete form, would necessitate a marked revision in Deegener's scheme.

Later, Deegener recognized that certain groups of animals are held together by a social force or instinct of which we know at present relatively little. The arrangement of such groups in his original system is obviously difficult. One may think of the satisfaction of the so-called "social force" or "instinct" as having definite value for the animal so satisfied. According to this reasoning, the group collected by social instinct would be a "society"; although, since there is no other demonstrable advantage accruing to the members of the group, Deegener at first was inclined to regard such an aggregation of individuals as an "association." Faced with this dilemma, he decided (1919) that associations whose occurrence depends upon a social instinct may be designated as "instinctive associations." They are opposed to aggregations of purely accidental character which are formed not because of instinct but because of limited space or localized food. If the aggregation is formed from obvious mutual attraction but without any recognizable objective benefit to the members, Deegener calls it an "instinctive association," as with young spiders, young ticks, or groups of grasshoppers.

Alverdes (1927) understands by "associations" the chance gatherings produced solely by external factors, such as insects collected around a lamp, while "societies" are genuine communities held together by the force of a social instinct. "In short," Alverdes says, "no social instinct, no society!" According to this point of view, the individuals are collected into an association because of their responses to environmental factors, but they collect into a society primarily because of the presence of other similar animals and only secondarily because of the action of environmental forces. Alverdes would consider the lack of a social instinct all the evidence necessary for calling such a group an "association."

Wheeler (1928), commenting on these two classification schemes, doubts the applicability of Deegener's basic principle of benefit or no benefit, but commends Alverdes' position as being essentially sound.

The ecological use of these and related terms must needs be considered. Modern ecological work has shown that each different kind of a habitat contains a more or less characteristic set of animals which are not mere accidental assemblages but are interrelated communities. When these are geographic in extent, they are usually spoken of as a "formation," within which may be recognized smaller units or "associations," which are composed of groups of habitat strata that are uniform over a considerable area but smaller than a formation. These associations are frequently composed, at least in part, of developmental stages, such that an orderly succession of communities can be recognized. Such a series, forming a unit of succession from initial to climax stages of an association, is sometimes called a "sere"; and the different developmental units of the whole association are called "associes." Thus, we have the animal communities or associes of the open sand, the foredune, the pines, the oaks, and finally the beech and maple forest, forming one developmental sere arranged in the order given, within the beach and maple association near Chicago (Smith, 1928; Shackleford, 1929).

In ecology the term "animal society," according to the most recent usage (Smith, 1928), has been divided into two parts, one of which is called a "pre-society."[1] This is a community of organisms living among the plants of an association and subordinate to the plant dominants. The "plant association" is named for one or more of the dominant plants, while one or more of the predominant animals give the name for the "super-society." The super-society, like its accompanying plant association, generally covers an extensive area with an essentially uniform taxonomic composition. Within such a super-society one finds animal societies which are communities of lesser magnitude and which may be seasonal, or stratal, or confined to a given locality. When these societies, or recognizable subdivisions of them, are composed of animals closely bound together by biotic relationships such as have been described in general terms as

[1] "Super-society" would appear to fit the meaning more exactly.

composing a "web of life," they are frequently called "biocoenoses." Food and shelter relationships, climatic and edaphic factors, are important determining conditions for a given biocoenosis.

With the growing complexity of ecological terminology and the growing precision with which different terms are applied to various recognizable groupings of animals, a need has developed for some term which could be used in a general sense to cover any one of the named units from the largest to the smallest. The word "community" has been reserved for this general purpose, and one may speak with equal propriety of the animal communities of the Amazon rain-forest or of a decaying tree within that forest.

In the border-line field where general sociology meets and overlaps general physiology and ecology, the field which is being considered in the present discussion, it seems desirable to have a term which may be applied loosely, but not incorrectly, to any of the recognized units lying below the groups accepted as definitely social, just as the term "community" is applied by the animal ecologists with equal propriety to strata, super-society, society, association, and what not. It is in this general sense, for this level of social or subsocial life, that I propose to use the term "aggregation." I am not concerned with defining it closely in terms of the association or society of Deegener or Alverdes. It may be used with equal propriety in speaking of a group of frogs collected as a result of sexual attraction during the breeding season; or of a concentration of May flies about a light, where they have been collected by forced movement as a result of their strongly positive phototropism. There is in the term itself a strong suggestion that the groupings involved are not closely integrated, which is in keeping with the facts in the field to be covered.

INSTINCTS

In the course of this discussion we shall have reason to refer to "instinct," a term deservedly in disrepute among careful thinkers because of the slipshod way in which it has been used. Early students of human sociology and recent zoölogical commentators on sociological phenomena have sadly overworked the word by referring any unanalyzed social behavior to the working-out of a social instinct.

That social instinct may be acting in given cases is not to be denied, but there has been an increasing and wholesome tendency to deprecate the use of this term to cover ignorance.

"Instinct" is hard to define. The most satisfactory definition known to the writer is that of Wheeler, who says (1913*a*): "An instinct is a more or less complicated activity of an organism which is acting (1) as a whole rather than as a part; (2) as a representative of a species rather than as an individual; (3) without previous experience;[1] and (4) with an end or purpose of which it has no knowledge."

It is obvious to one who has observed the reactions of animals that there are two types of behavior: the learned and the unlearned. Much of the latter is frequently called "instinctive," with propriety, though in the case of many highly organized animals, including man, there has been an unfortunate tendency to regard, as instinctive or unlearned, behavior that is in reality based on very early training which has been entirely forgotten or overlooked.

In man breathing, swallowing, gland secretion, and muscle contraction are all unlearned; and some of these, for example the secretion of certain glands, cannot be effected by learning. These unlearned reflex actions of parts of organisms seem to be the simplest of a series of unlearned responses whose other categories are those reflexes of an entire organism commonly called "tropisms," and the more complex behavior usually called "instinctive."

It is becoming increasingly difficult to draw hard and fast lines between instincts and tropisms, or between either of these and the general functioning of living cells. It is further impossible to dissociate any of these three categories of behavior from the activities concerned with growth and development. If one considers in this connection the metamorphosis of a larva into an adult, which is usually regarded as the function of growth and development, one finds the processes concerned so inextricably bound with major and minor activities of the animal that the instinctive behavior cannot be clearly separated from the other processes going on at this time. Is the production of the silk cocoon of the moth an instinctive action, while the production of the thickened hypodermis to form the chrys-

[1] Or without modification caused by experience (W. C. A.).

alis of the butterfly is only a growth process? What is the essential difference between the two?

In so far as is possible, we shall avoid dwelling upon the aspects of behavior usually called "instinctive," except in reference to the literature. This is not due to a disbelief in the reality of instinctive social behavior, but rather to a conviction that progress lies in a field where the elements of behavior can be more exactly ascertained.

The drive which leads an animal to exhibit such behavior as is usually classified as being due to the operation of social instinct I prefer to regard, as does Wheeler (1928), as an expression of appetite. Wheeler says in this connection: "It thus takes its place with the other appetites like hunger and sex, though it is feebler and more continuous, i.e., less spasmodic and, therefore, less obvious. It is most strikingly displayed, however, in the restless behavior of the higher social animal when isolated from the continuous, customary stimuli of its kind." From this approach, the strength of the social appetite can become a subject for objective investigation, such as Warner (1928*a*) has recently made for the relative strength of the drives furnished by food or sex hunger; but such an objective investigation of the general social appetite has not yet been conducted.

The scope of the discussion, some concepts, and a part of the terminology having now been considered, we may plunge directly into the mass of material awaiting analysis.

CHAPTER II

CLASSIFICATION OF ANIMAL AGGREGATIONS

It has long been known that animals not naturally bound together in organic union may aggregate into groups or clusters more or less closely associated, in which physical contact may or may not occur. Actual physical contact is normally found as part of the aggregation phenomenon among many Protozoa, as, for example, in *Paramecium;* in flatworms, such as the planarians; in earthworms; in echinoderms, such as starfish; in mollusks; in arthropods; and among many chordates, including ascidians, fish, frogs, reptiles, birds, and mammals.

Among other animals similarly widely distributed through the animal kingdom, collections occur in which physical contact is not the rule. These may be illustrated by the jellyfish, ctenophores, or copepods that may discolor the ocean for miles; by collections of leeches, snails, or ostracods; by the swarms of gnats that dance together like particles in brownian movement; by ants, bees, schools of fish, flocks of birds, herds of ungulates, and groups of various other mammals, including man. The highest development of aggregations not based on physical contact requires the possession of highly developed sense organs.

These two types of animal aggregations are not mutually exclusive, even when reactions associated with copulation are disregarded; for animals may be involved in first one and then the other in different phases of their life-cycle or seasonal history. With many birds the loose flock of the daytime may be replaced by close physical contact during the night roost. At times this may be due to the lack of adequate perching space, and show merely toleration of close proximity; but in other instances, as, for example, the Indian tree swift, there is a positive movement together even in the presence of abundant roosting space. Bats may show the same phenomenon during their daytime sleep.

There are abundant examples of animals that lead wholly or partially solitary lives during part of their seasonal- or life-cycle but at another period come together into flocks or in actual physical contact. This is true of the cowbirds, reared singly from eggs surreptitiously laid singly in the nests of other species of birds. The young cowbirds develop quite out of touch with other members of their own kind and yet collect into definite flocks when adult. Another aspect of the same kind of behavior is shown by the grackles, which nest fairly separately but join in large flocks before the fall migration; by deer, which summer separately or in partial family groups but winter in herds; by frogs, which remain practically solitary during the year except for possible hibernation groups and then aggregate during the breeding season; by solitary bees or wasps, which for the greater part of the year are out of physical contact with their fellows and yet during the summer may form overnight aggregations in closest physical proximity; or, to give one more of many possible examples, by land isopods, which congregate into dense bunches when their habitat becomes dry.

The aggregations of the physical-contact type are, of necessity, transitory in character in motile organisms; but in sessile animals, such as the ascidians, or the marine mussel *Mytilus*, this may well be the normal way of living. The physical-contact type of aggregation finds its most complete expression among the sessile colonial organisms that grow in dense stands of many individuals, which are physically connected with each other throughout life. *Obelia* hydroids represent this growth form.

Collections without physical contact, such as the flock or the herd, may be constant and normal for some species; and the animals in these are usually said to exhibit the social habit. This social habit finds its best development in the insects, such as the ants and termites, among whom division of labor is carried out to its logical end, in that polymorphic forms have evolved of which some do not complete their sexual development while others specialize upon reproduction. These have been well described by Wheeler and Forel.

Animal aggregations may be classified on many other bases beside that of the degree of physical contact. Deegener (1918) has made an exhaustive classification of the different forms of animal groupings (*Vergesellschaftung*) in which he undertakes to arrange logically all such associations, ranging from the relatively simple colonies of the protozoans—*Synura* or *Carchesium*, where all the individuals are similar and all arise from the same parent-cell and are organically connected with each other—to colonies of ants with their complicated social structure, which may include, in addition to the ant castes themselves, their slaves, their commensals, their tolerated guests, parasites, parasites of the parasites, or parasites of other associated forms.

A summary of the classification of animal aggregations as worked out by Deegener is given here at some length, not because I accept it entirely with all its implications, but because it is the most complete classification yet produced and because I am in hearty accord with the principle underlying this scheme of organization: that no hard and fast line can be drawn between well-integrated social organizations and loosely integrated aggregations which are usually regarded as being definitely non-social. Further, experience with presenting this material to seminar students has shown the desirability of wading through a detailed outline, such as that of Deegener's, in order to acquire a comprehensive view at one and the same time of the ramifications of the subject matter and of its inherent unity.

It is the custom at present to ignore this work of Deegener or to fail to appreciate its essential value (Wheeler, 1928) because of obvious defects in its cumbersome terminology, in the criteria used to distinguish between major groupings, and because the categories are not clean cut and mutually exclusive. Many of these faults are inherent in a pioneering classification of subject matter in any field, and others were caused by the lack of definite knowledge in 1918 of the relationships involved. On this latter count we are in a position to make improvements on Deegener's classification at the present time, but we do not appear to be able to refine it sufficiently as yet to repay the trouble involved.

The account given below is not a direct translation of Deegener's 1918 outline; but it follows that outline and gives his point of view, criticisms of which have been suggested and will later be elaborated.

DEEGENER'S CLASSIFICATION OF AGGREGATIONS

Part I. *Accidental unions or associations* are groups of animals without mutual benefit for individual members. "Accidental" is, to Deegener's mind, a better term for these aggregations than "indifferent," because to him it plainly indicates the method of their formation, and also because the members of accidental aggregations are not always indifferent to each other. Accidental aggregations will be seen to be of various kinds, formed in various ways. They may consist of one or of a number of species. One cannot always be sure concerning the proper classification of a given association, which may as yet be merely a matter of opinion. Deegener recognizes that even the major distinctions are not always clean cut and that one of a pair of apparently closely similar groupings may be assigned to the accidental associations while the other is called an "essential society." In the minor categories the methods of formation determine the classification to a considerable degree.

A. *Homotypical associations* consist of members of the same species which have arisen either sexually or asexually, which may have remained together because they are the offspring of the same parent, or which may have become accidentally associated together although of different parentage. The former are called "primary," and the latter "secondary," associations.

Alpha. *Kormogene associations*[1] are confined to invertebrates and do not occur in arthropods, echinoderms, and mollusks. They are those colonial forms in which the different individuals remain morphologically attached to each other. The advantages of the colony are not always clear. In Protozoa, relationships of individuals in the colony are not such as to guarantee nourishment for the entire colony; thus there is no advantage in this respect with this phylum. In the hydroid colonies, nourishment is better assured for the individual by the colonial form. The colony does not appear to be formed necessarily because it is a more favorable adaptation to living conditions but because of the failure of the different elements to separate at fission. The tendency toward colony-building increases

[1] Budded colonial forms, as among the hydroids, cannot be regarded as "accidental" in the usual usage of that word.

as habits become sedentary, and is also more marked in relatively simple animals having strongly developed skeletal parts, as the sponges, hydroids, bryozoans, and tunicates.

I. *Primary colonies* arise as the result of division in which the smaller pieces remain together, or as a result of budding in similar fashion.

1. *Homomorphic colonies* result when the divisions are equal and all members of the colony are similar, as in *Synura*, *Carchesium*, and *Salpa* chains. Such colonies as *Zoöthamnium* may represent true societies, since all individuals may contract if one is stimulated, and so all may escape harm; while *Carchesium* does not, and so is placed in the present category.

2. *Heteromorphic colonies* are formed when the divisions are unequal, as is the case with the strobila of the Scyphozoa, or during the processes of asexual reproduction of certain worms, such as *Autolytus*.

II. *Secondary colonies*, or *concrescence colonies*, arise by the secondary union of individuals which are entirely separate for at least a brief period.

1. *Concrescence colonies having a genetic basis*, in that the individuals composing the colonies originated from the same mother, are shown in *Proteriodendron*, *Dinobryon*, and secondary *Salpa* chains. The fact that identical or related forms have survived and can live as separate individuals indicates that these animals are able to live without the small and perhaps accidental benefit arising from their communal life.

2. *Concrescence colonies without a genetic basis* are those in which the animals that later become attached together in one colony are not descendants of the same mother. These commonly occur in sessile animals, such as the ascidians, sea anemones, sponges, oysters, and *Mytilus*. If no organic union takes place, causing a real fusion between the different animals composing the colony, then the association remains accidental.

Beta. *Associations of free individuals.*

I. *Primary associations* arise through asexual or sexual reproduction when individuals descending from the same parent or parents remain near the place of origin and form an aggregation which varies from a loose to a firm integration. The primary cause of their being together lies in their common origin, but the cause of their remaining together is not of a genetic nature but may depend on the favorable character of the place or on the presence of food. In other cases one must assume the operation of a social instinct which holds the animals together.

1. *Syngenia* are primary associations which arise by means of asexual reproduction. This may be illustrated by *Stentor coeruleus*, which lives on decayed water plants and occurs frequently in such abundance as to give a blue color to the surface of the water. The aggregation is located in space by favorable food conditions. So long as there are only offspring from a single mother present, the aggregation would be called a *monosyngenium;* but when second and third generations appear from the same stem-mother, the group becomes a *polysyngenium*. Other unrelated individuals may wander into this favorable niche, forming a secondary association. Similar relations hold with *Vorticella*, but with both these aggregations there may be some social value accruing to the different individuals, since the combined vortex action of the cilia brings more food to each animal. This does not occur in hydroids, such as the common fresh-water *Hydra*, which reproduces asexually and remains in a purely accidental aggregation in which there is no reciprocal relationship before sexual reproduction begins. Similar relations hold with various other simple coelenterates whose slight powers of locomotion tend to confine them close to the place in which they are budded free, providing it is a generally favorable location.

2. *Primary associations arising from sexual reproduction* may form close unions which may rise to the widely extending reciprocity of the highest types of society found among animals. In the invertebrates these are represented by the conditions obtaining in ant and termite colonies; in the vertebrates, by human societies. This part of Deegener's outline undertakes to consider only the more primitive, purely accidental forms of this family union, in which the parents need not necessarily be concerned. Various combinations of simple families where the young all originate from the stem-mother may be distinguished and divided as follows:

a) *Sympaedium*, in which the offspring of the same mother form the aggregation without the presence of either parent. This condition is seen in some spiders and insects, where the young of the same mother remain together for a longer or shorter period. If the mother remains with the offspring, the group belongs to another category. *Lophyrus* caterpillars, which feed on pine needles, form an aggregation due in the first place to the eggs being laid together. No obvious benefit accrues to the individuals. They are more conspicuous as a result of the grouping and cannot defend themselves better than if alone. The causal factors in such an aggregation are obscure. The fact that the eggs are laid together is not sufficient in itself, since other forms have their eggs laid similarly close together and yet separate immediately on hatching. It may be that the sluggishness of the animals and the lack of disrupting stimuli explain a large part of the behavior; while, on the other hand, there may be a social appetite which holds the groups together. The problem becomes more difficult with those larvae which remain together during the early larval life and separate when partly grown.

Many lepidopterous larvae that remain together during part or all of their larval life spin a common nest. The formation of such a nest may be due to the fact of living together rather than the living together being due to the need or use of a common nest. The ability to spin a common nest does not guarantee the actual building of one, for many spinning animals live alone. These larval colonies are common among animals in which the adults are winged, and hence are readily distributed during that phase of their life-history. Such a sympaedium occurs in solitary bees which lay eggs in cells. The resulting larvae and pupae form an accidental association, living together as offspring of a common mother. When adult, they fly away separately.

b) A *gynopaedium* is composed of a mother and her offspring that remain together for a period. This grouping is not concerned with the relationship between mother and offspring beyond the fact that they remain together without obvious benefits accruing to the group from the association. The aphid stem-mother in the spring gives birth to young parthenogenetically. This gynopaedium, consisting of one female and her immediate offspring, may be designated a *monogynopaedium*. The young also reproduce parthenogenetically, and such a complex group may be called a *polygynopaedium*. These colonies are homomorphic; but as winged forms appear, heteromorphic colonies are formed. In the autumn sexual generations appear and produce a resistant over-wintering egg, which carries the colony over the winter season. In this aggregation there are no benefits immediately apparent. The brood is not cared for by the older members or by each other. The individuals composing a crowd of aphids are more easily cared for by ants of the myrmecocolous species when together, but also are more easily preyed upon by their numerous enemies. The massed aphids also tend to destroy the food plant on which they cluster, to their own disadvantage. Deegener recognizes no social advantage, and therefore regards the aggregation as accidental.

c) *Patrogynopaedia* occur when both parents remain with their offspring in groups. Those with no social benefits for their members belong here, but this type of aggregation often carries with it some social advantage, and so usually belongs in a later category. *Necrophorus* beetles live with their young in decaying animal bodies. This association may confer social benefits under certain conditions, but they are not recognizable in all cases. In these scavenger beetles, the presence of a dead body seems to release a digging reaction whether the individual is solitary or in company with others. Each individual digs without reference to the others. The results may have no significance for the assisting beetles, but only for the pair leaving their eggs with the dead body. Obviously the whole has racial significance, although without significance for many of the participating individuals.

Combination family groups also occur in which the individuals composing the aggregation come from more than one stem-mother.

d) *Synchoropaedia* are formed when eggs laid by different females in a favorable place hatch out and the larvae remain together from the very first, not as separate families, but freely mixed into a common aggregation. Mosquito (*Culex*) larvae in a rain barrel are an example of a synchoropaedium. When larvae of different species are present in the same rain barrel, we have a *heterosynchoropaedium.*

e) Similarly, *symphagopaedia* may result from several groups of the same species laying eggs on the same food material except that here the favorable food rather than the favorable place becomes the integrating factor. This type of aggregation may be illustrated by flesh flies and, according to Deegener, by *Drosophila.*

II. *Secondary associations* may be distinguished from primary associations because they are the result of a coming-together of free individuals rather than their merely remaining together. The classification is based on the integrating factor judged to be most important.

1. *Sysyngenia* arise from the secondary fusion of two or more syngenia.

2. *Sysympaedia* consist of fused "children-families" and arise when one sympaedium meets with another. Deegener observed such in juvenile spiders of *Epeira* (1919*b*). The members of both groups mixed peaceably and gave no sign in their conduct that they were influenced by the foreign spiders; indeed, they did not seem to notice that their membership had been doubled, and new and old alike aggregated into one close mass. Another sympaedium was added to these two with similar results, although it was not ascertained whether or not the individuals of a given sympaedium remained for the most part together.

Two sympaedia of caterpillars of *Malacosoma castrense* L. are not mixable when the larvae of one sympaedium are in the molting period; otherwise they mix without the caterpillars of the two broods appearing to sense the change in their association. Schulz (1926), in studying the reaction of caterpillars of *Vanessa io* L., *V. urticae* L., and *Araschnia levana* L., found, with the methods he used, no recognizable value to rest in the aggregations other than the satisfaction of a social instinct; and this value had lost much of its meaning, since the caterpillars are able to live if isolated, under which conditions they spin small coverings in place of the usual communal nests. They will again take up communal life after an experimental isolation of four days. Marked sympaedia, some of which differed from each other in size of individuals, fused to a single sysympaedium. When this divided late, the resultant groupings usually contained members derived from different original sympaedia.

3. *Sympolyandria* are accidental polyandric associations formed on a synchoric basis, as that of *Alcippe*, a barnacle which dwells on the deserted snail shells occupied by hermit crabs, forming an accidental heterotypical association; but the barnacles, considered alone, form a sympolyandria. Polyandria form a type of essential mating society to be discussed later in this outline.

4. *Synchoria* are locality aggregations formed primarily because of a limited expanse of particularly favorable locations for living. Barnacles gathered together on available rocks are a good example.

5. *Syncheimadia* are hibernating aggregations, such as those of snakes or salamanders.

6. *Synhesia* are swarming aggregations under the influence of the breeding season, as illustrated by palolo worms. Factors concerned here include the simultaneous ripening of the sex cells, a limited favorable area, and the correct external conditions,[1] which are frequently associated with lunar rhythms. Similarly, the swarms of May flies are due at least in part to simultaneous pupation rather than to sex attraction.

7. *Symphagia* are aggregations about a favorable food supply, as flies collect about carrion or sugar. Here there is no obvious benefit from the association.

8. *Symporia* are migration aggregations joined either because they originated in the same place or because they are going in the same direction, and may be illustrated by the migrating masses of fiddler crabs, of butterflies, or of salmon.

9. *Symphotia*[2] occur when the aggregations collect about a source of light. Such a reaction is given by a great many insects, as well as by other animals (Mast, 1911).

[1] Considerations given later, particularly in chapters xvi and xvii, indicate that such swarms have a rather obvious survival value and hence should not be placed among the "accidental" groupings.

[2] If this type of category be included, it is necessary to include similar headings for tropistic collections due to the reaction to other environmental factors, such as heat,

10. *Synaporia* are collections due to unfavorable conditions, as when beetles are collected by the wind and deposited in beetle drifts in the same way that snow is drifted.

Krizenecky (1923) recognizes two different types of synaporia, the passive and the active. The first are formed when the animals are passively carried together, as by wind or wave action. The latter are formed when animals faced with unusual disturbing conditions collect together. Such aggregations may be noted in the worm *Enchytraeis humicolor*, which ordinarily lives singly in the soil but which aggregates into symphagia about decaying food material. If the worms are placed in a dish of water, they aggregate into larger or smaller masses with the worms closely intertwined. Such clumped masses do not remain together; but after the group is closely formed, there comes a disintegrating movement which results in the animals finally coming to rest scattered singly over the bottom of the dish. The animals remain thus scattered as long as the water is undisturbed. When subjected to renewed stimulation by adding chemicals or by mechanically disturbing the water, another aggregation cycle is set up.

B. *Heterotypical associations* consist of collections of unlike species which may occur for the reasons given above, and which may be designated by adding the prefix *hetero-* to the proper term for the homotypical aggregation, as: *heterosymphagopaedium*, *heterosynchorium*, *heterosyncheimadium*, etc. Deegener recognizes also *co-incubatia*, which are breeding aggregations of different species of birds, for example, selecting a common, restricted nesting site. Finally he adds *symphoria*, which are formed when one or more species of animals settle upon another of different species, forming a heterotypical aggregation without obvious mutualism or parasitism, and are well illustrated by the barnacles, hydroids, snails, bryozoans, and others growing on the shell of an old horse-shoe crab (*Limulus polyphemus*).

chemicals, touch, gravity, and the like. Rather, it seems preferable to replace this category by some such term as *syntropia*, meaning those collections which are brought about by tropistic reactions to some environmental factor. Such collections occur, due to a combination of elements, including that of a limited space into which tropistic reactions lead animals to assemble and the incidental presence of numerous individuals in the region at one and the same time. In all these collections there is this time factor working; otherwise we could not recognize them as aggregations.

Some of the overlapping inherent in this type of subject-matter classification appears when one considers a heterosynaporium, a collection of different species due to the action of unfavorable conditions, which Deegener illustrates by the growing concentration of water animals in a drying pond. Obviously, such a collection would be at the same time a heterosymphagium and a heterosynchorium. Apparently, Deegener would classify the animal community of modern ecology as a heterosynchorium, since it is composed of several species occupying the same place, although the individuals of the group are not of obvious advantage to each other. He does not actually say that an ecological community should be so classified; he does use the term *biocoenosis* in connection with his discussion of a coral reef heterosynchorium.

Part II. *Essential aggregations or societies* are communities of species of similar or dissimilar animals which have a real value for the individuals composing them, thereby differing from the "associations" treated in the previous sections.

A. *Homotypical societies* are composed of the same species.

Alpha. *Kormogene societies* have the different individuals composing them organically connected with each other.

I. *Primary colonies* have arisen from the same mother.

1. *Reciprocal colonies* are those in which all the individuals represented stand in reciprocal relationship to each other.

a) *Homomorphic colonies* have all the individuals morphologically similar and may be found among sponges and at certain times among hydroids and bryozoans.

(1) *Colonies formed by division* may be illustrated by the colonies of *Volvox* so long as they remain free from specialized reproductive cells.

(2) *Colonies formed by budding* occur commonly among the Hydrozoa, the Bryozoa, and in many colonial chordates.

b) *Heteromorphic* and *polymorphic colonies* are formed when there is a differentiation between the different members of the colony, as occurs in the hydroid *Hydractinia*, in which feeding, reproductive, and protective zoöids may be recognized. Polymorphism is carried much farther in the Portuguese man-of-war, *Physalia*, and its allies. Here again we may recognize (1) colonies formed by division, as in

Volvox, when reproductive cells appear, and (2) colonies formed by budding, as in the hydroids.

2. *Irreciprocal colonies* must be recognized in which all members do not contribute equally to the welfare of the whole. This is simply illustrated by the case of a budding fresh-water *Hydra*, where the new individual, the developing bud, has a parasitic relationship with the mother.

II. *Secondary colonies* develop by concrescence, as when the young fresh-water sponges developing from different gemmules coalesce, due to their proximity, and form one sponge body originating from several gemmules.

Beta. *Societies of free individuals* may be classified as follows:

I. *Societies based on a sexual or genetic foundation.*

1. *Primary societies:* families in which the young are descended from a common father or a common mother or from common parents, and which remain together from the very first.

a) *Reciprocal families* in which all members benefit from the social connection.

(1) *Sympaedia* are composed of young of the same brood, but without either of the parents present. Such societies may be homomorphic, as in the case of minnows or young birds, or heteromorphic, as in bee colonies after the queen's swarm has departed.

(2) *Gynopaedia* are composed of the mother and her immediate offspring, which may again be divided between homomorphic and heteromorphic groups. The former is represented by the mole crickets (*Gryllotalpa*), the earwigs (*Forficula*), and many birds and mammals; the latter group, by colonies of bees or ants.

(3) *Patrogynopaedia* consist of a male and a female and their offspring, and may be divided into *monomorphic*, *dimorphic*, and *polymorphic* societies. Monogamous monomorphic societies of this sort are common among birds where both parents remain with the young.

Polygamous monomorphic families are similarly common among many large animals, although monogamous families occur there, too, as among foxes. In dimorphic patrogynopaedia the offspring living with the parents are true larvae, as, for example, in the passalid beetles. The best example of a polymorphic colony of this type is given by the termites, where sexually mature males and females of one or more grades occur in the same nest with soldiers and workers.

(4) In a *patropaedium* the male remains with his offspring for some time. Schulz (1926), in his analysis of the situation in the brooding stickleback fish (*Gasterosteus aculeatus* and *G. pungitius*), concludes that the value to the male is in the psychological realm, and quotes Deegener with approval as saying that the nest and young are of lively interest to the male stickleback, their loss is a misfortune, and the nesting and brooding phenomena are a source of inner peace. Obviously, such assertions are not susceptible of demonstration. To the eggs there is the benefit of added certainty of fertilization, of protection from other fishes, of aëration, with resulting protection from fungus growth; while the young find a favorable place for development, passive protection by the nest, and active protection by the guarding male. The relation between eggs and young and the brooding male is essential rather than accidental, and therefore forms a true society. It is reciprocal, and the female is not concerned after the eggs are laid; therefore a patropaedium, which had its origin in a polygamous connubium existing merely as a mating relationship, but this connubium is an association rather than a society. If the male dies, the society becomes a simple sympaedium, which would be accidental in nature, since the association of the young has no value for them. The relation of the young to the nest has a synchorium factor. The existence of the patropaedium is necessary for the well-being of the eggs but not of the young fishes. The relations between the males of the large- and small-mouthed black bass, the bullheads, and the fresh-water dogfish (*Amia calva*) and their nests and young give an opportunity for similar analyses.

b) *Irreciprocal families* are those in which the social values rest only with the young.

(1) *Gynopaedia* of this sort are to be found in the leeches (*Glossiphonia*), according to Deegener; but Schulz detected evidence which led him to conclude that the female leech is somewhat interested in her eggs, and on this account he places these leech gynopaedia among the reciprocal societies. Similarly, careful observation might show

the same sort of value, if such it can correctly be considered, in the other cases cited by Deegener, such as the amphibians, *Hylodes lineatus* and *Pipa pipa*.

(2) *Patropaedia* of this sort are thought by Deegener to be illustrated by the relations in the obstetric toad *Alytes*, in which the male carries the strings of eggs twisted about his legs, and in *Rhinoderma darwini*, a small cricket-like frog of the moist beech forests of Chile. The male of the latter species takes the fertilized eggs and crams them into his singing pouch, which becomes greatly enlarged during the breeding season. Here they develop and transform, hopping forth from their father's mouth as fully developed small frogs (Barbour, 1926).

2. *Secondary societies* are those in which the individuals are not together from the very beginning, or at least those in which the primary social group becomes modified by secondary additions.

a) *Sexual societies of the Protozoa* are such as are shown in ciliate conjugation.

b) *Connubium simplex of the Metazoa* is a grouping in which mating occurs between animals of the same species but of different sexes, or between hermaphroditic animals.

(1) *Polygamy* includes *polygyny*, or the mating of one male with more than one female, as in polygynous birds, such as the domestic fowl, and in many mammals; and *polyandry*, in which several males mate with a single female without the female being free to all males. Among Deegener's examples are the cases of double copulation in insects. In the case of *Alcippe*, a barnacle, the females as a rule live near each other, and from three to twelve dwarf males join each female and remain with her during their lives. Alverdes (1927) states that this sort of relationship is rare, but adds the case of *Bonellia*, a worm of which more will be said in a later section, and with which as many as eighteen males attach themselves to a given female and remain so for extended periods. Polyandry has also been observed among some spiders.

(2) *Monogamy* is fairly widespread, at least in the form of seasonal pairings. It is found among beetles, as for example, the monogamous

Passalidae, which remain with the larvae and the pupae. Alverdes lists also cases of at least seasonal monogamous mating among spiders, fishes, amphibians, reptiles, birds, and mammals.

(3) *Communal connubium*, or *promiscuity*, occurs among many fishes at the spawning grounds, among certain lizards, and among gregarious bats. It is also reported for the American bison, for the American cowbird, and among various other birds (Alverdes). Miller (1928) summarizes evidence that this is a common state among anthropoid apes and certain monkeys; unlike most modern sociologists, he believes this represents the original mating relationship among *Homo sapiens*.

(4) A *conconnubium* is formed when monogamous animals collect during the breeding period, forming small societies that continue during copulation. Deegener gives as examples the viper (probably *Pelias*) and birds, such as gulls, which move at mating time to a restricted location and there form seasonal pairs.

c) *Perversum simplex* applies to those cases where males attempt to mate with each other, as has been observed for drones of the honey bee, after they are driven out of the nest in the autumn, and for various other insects, including certain beetles and house flies.

d) *Preconnubia* occur when individuals of one sex collect at one place before the mating season, or both sexes may be present, but without mating. Such preconnubia occur among many frogs and birds.

e) *Synhesmia* are swarming societies which collect under the influence of reproductive drives. *Androsynhesmia*, male swarms; *gynosynhesmia*, female swarms; and *amphoterosynhesmia*, or mixed swarms, are known to occur.

II. *Societies that are not immediately based on a sexual or genetic basis* are also known, as follows:

1. *Sysympaedia* are combinations of sympaedia, such as occur in minnows.

2. *Syngynopaedia* consist of two gynopaedia which have united as may happen with ants, or seals (*Phoca gruenlandica*), or wild hogs (*Sus scrofa*).

3. *Sympatrogynopaedia* are combinations of at least two patrogynopaedia, and are known in monkeys, marmots, elephants, antelopes, and many other mammals.

4. *Adoption societies* are those in which a female takes offspring from the same species. They are known for birds and mammals, for example among the wild hogs (*Sus scrofa*).

5. *Synandria* are groups of males which herd together. Thus, male birds of several species are known to have this habit; and it is reported to be common also among mammals, as in seals and antelopes.

6. *Syngynia* are similar groups of females, such as are formed by the stickleback fishes.

7. *Symphagia*, again, are feeding societies formed of several individuals, and illustrated by *Necrophorus* beetles during a portion of their life.

8. *Synchoria* are societies united around a common place which has some peculiarly favorable quality or qualities. They are well illustrated by the common bird roosts, as of crows and robins, and, among insects, as wasps and *Mellisodes* bees. (See chap. iv.)

9. *Syncheimadia* are combined over-wintering societies, and may be illustrated by solitary bees and coccinellid beetles.

10. *Symporia*, again, are migration societies, such as swarms of bees or flocks of migrating birds or mammals.

11. *Synepileia* are marauding societies or hunting bands, such as those of jackals and wolves.[1]

12. *Sympaigma* are groups of individuals brought together in order that they may engage in common play. Deegener cites the whirligig beetles (*Gyrinus*) as examples. Schulz (1926) has investigated this aggregation somewhat and concludes that play is not the principal integrating factor; he believes that the greater security furnished is the more important cause. Therefore he places them in the next category. Brown and Hatch (1929) think that the collection of gyrinid beetles is an example of a reaction to a general environ-

[1] The American wolf pack apparently is usually a family affair, but may not always be so (Seton, 1929).

mental pattern which they regard as more important than the biological values involved.

13. *Symphylacia* are societies that furnish protection for the individuals composing them.

B. *Heterotypical societies* are composed of individuals of different species.

Alpha. *Reciprocal societies.*

I. *Integrated by sexual drives.*

1. *Connubium confusa* are societies of both sexes, but of different species, brought together for the breeding season. Thus, male frogs will attempt to mate with females of other species, or with toads, or even with fish. On another taxonomic level, coccinellid beetles of different species have been observed to attempt copulation.

2. *Perversum confusa* are formed when individuals of the same sex congregate during the breeding season, although of different species, as for example, male frogs and toads, *Rhagonycha melanura* Oliv. with *Luciola luistanica* Charp.

II. *Non-sexual combinations.*

1. *Phagophilia* are heterotypical reciprocal societies wherein each species benefits, although at least one of the two receives its food through its association with the other. Thus a passive species is freed of its parasites through the efforts of its active associates, showing one variety of mutualism. This is illustrated by cowbirds following cattle and feeding on the flies which infest the latter.

2. *Synsitia* are also symbiotic societies in which one of the associates lives on the shell or the outer covering of the other, without being parasitic and without the type of relationship found in a phagophilium. Deegener regards the relationship between a hydrozoan and a hermit crab, such as *Hydractinia* growing on the shell occupied by *Eupagurus*, as a synsitium. The former clearly receives transportation and fragments of food, while the latter may be protected by the nematocysts of the dactylozoids, as Deegener suggests.

3. *Phylacobia* occur when two species live together in the same cavities, as *Campanotus punctulatus termitarius* Em., an ant, is said

to live (Wasmann, 1901–2) in the runways made by various termites, receiving shelter and giving increased protection. Wheeler (1913*a*), Emerson, and other students of social insects are agreed that cases of reputed association in compound nests are in need of further careful investigation. Wasmann calls this relationship *phylacobiosis.*

4. *Trophobobia* exist when one species feeds upon the excretions of or the waste of the other, and in turn provides protection for the weaker species. This relationship is found between certain species of ants and aphids.

5. *Symphilia* are formed when one species receives food, protection, and shelter from another, and in turn supplies excretions which are apparently narcotic in nature. This relationship exists between many ants and their myrmecocoles and between termites and termiticoles.

6. *Dulobia* are illustrated by the slave-making ants which raid other colonies and carry off the young, which in time take over the routine work of the colony into which they are carried, receiving in return the advantages of being members of the given society.

7. *Adoption societies* are formed by mutual adoption freely entered into by both species, and without recognizable advantages or noticeable harm for either. The ants, *Formica consocians* and *F. incerta*, are said to form such societies.

8. *Heterosymphylacia*, as in the homotypical symphylacia, furnish increased protection for all individuals as a result of the social union. Thus zebras and ostriches, or giraffes and elephants, are reported to live together, thereby increasing the security of both constituent species.

9. A *heterosynepileium* occurs when more than two species of animals join forces and gain greater hunting efficiency for the group. Different species of storks over-wintering in East Africa have been observed to form common hunting bands and to conduct more or less organized drives for concentrating scattered grasshoppers.

10. *Confoederata* are recognized by Deegener as being societies of unlike species united by mutual friendship or sympathy, and as having no other basis. Crows and jackdaws, alone or with starlings, golden-crowned kinglets and titmice, common creepers and wood-

peckers, are given as examples. Obviously, such a category is without secure foundation, but is perhaps to be expected from a worker who believes that the future belongs, however the present resists, to the psychic and not to the mechanistic (Deegener, 1920*b*).

11. *Heterosymporia* are mixed migration societies, such as occur in birds and mammals, and are especially well marked on the plains of South Africa.

Beta. *Irreciprocal societies* occur when the benefits extend mainly to one species, while the other may be decidedly harmed from the association.

1. *Synclopia*, or thieving societies, are those in which one species feeds upon the stored food supplies of another, as thieving ants are known to prey upon stored termite food, or as thieving species of termites take the food of other termites. Wheeler calls this *cleptobiosis;* Forel designates it as *lestobiosis*.

2. *Syllestia* are societies containing robber guests which prey upon the eggs or the young of the species with which they are associated. Thus staphylinid beetles may prey upon the brood of the ant colonies whose nests they inhabit, as Wheeler's "synechthren." In somewhat similar relations are the hawks that prey upon flocks of migrating birds. The flocks of wandering grasshoppers, springbok, and the like are each set upon by its own particular set of predators which accompanies the food flock on its migrations.

3. *Paraphagia* are societies composed of harmless companions of their host feeding commensually on fragments neglected by the host. *Alcippe*, a boring barnacle, inhabits the snail shells which have been appropriated by hermit crabs, and feeds on fragments escaping from the feeding of the latter. Dermestid beetles occupy nests of other insects, feeding on waste material such as molted skins. The so-called *synoektes* of ants form paraphagia with the ants with which they live.

4. *Synoecium* is the term given by Deegener to the association between certain animals and the nests of other animals. This is known to be a widespread relationship. The crab *Pinnixa* lives in holes occupied by marine mollusks. Birds' nests have many animals regularly living in them; sparrows may build in storks' nests. Fishes build in

the nests of other fishes (Reighard, 1920). Many similar examples could be given for other nests, such as those of ants and termites.

5. *Paroecia*, or neighborly groups, are formed in which the less conspicuous animal species finds protection from the other without occupying a part of its nest. Thus, small fishes are frequently associated with medusae or with the Portuguese man-of-war *Physalia;* while many animals, such as fish, worms, snails, and starfish, have similar relationships with coral colonies.

6. *Metrokoinia* occurs in ants when the fertilized female of one species who has lost the ability to start a new colony joins herself with the fertilized female of another species that has retained this power, and is thus associated with a colony development which she would be unable to secure alone, and to which she contributes little or nothing. This relation has been described for *Strangylognathus testaceus* Sch., which has lost the power of colony formation, living in mixed colonies with *Tetramorium caespitum* L.

7. *Irreciprocal symporia* occur when one animal species attaches itself to the surface of another without becoming parasitic and without contributing aid to the animal on whose back it grows. This relationship may exist between barnacles growing on whales, between hydroids and crabs, and between stalked protozoans, such as peritrichs and suctorians, and the snails, crustaceans, or hydroids supporting them.

8. *Syncollesia* are cemented societies in which one animal cements into its own covering the case or shell of another species of animal without killing off the original owner. Small mussels (*Sphaeridae*) and snails may be worked into the cases of caddis-fly larvae.

9. *Parachorium* is the name given to the relationship that exists when one animal lives within the body of another without being parasitic upon it. Hydroids, sea anemones, polychaete worms, ophiurids, and crustaceans live in the canal systems of sponges; and *Pinnotheres*, a crab, lives in the mantle cavity of *Mytilus*, the sea mussel.

10. *Parasitism* is not easily separated from several of the preceding categories. A parasite, in the restricted sense used here, obtains its nourishment, at least, from the host with whose continued existence the parasite is more or less closely bound. Frequently the nour-

ishment of the parasite comes from the living substance of the host. Many categories of even such restricted parasitism are recognized, and may be found listed in reference works on the subject (Hegner, Root, and Augustine, 1929).

We have given here an outline of Deegener's classification of animal groupings in detail, but it is not our intention to fit the different aggregations to be discussed later into their appropriate niches in this classification. In fact, certain of its more detailed aspects will not be referred to again. But it is upon the idea that there is an essential unity within the phenomena to be discussed that the present summarizing account has been prepared; this concept, although foreshadowed by Espinas, was first fully expressed in Deegener's outline. We shall return to it in the concluding chapters.

CLASSIFICATION OF ALVERDES

When we turn to the analysis of social phenomena by Alverdes, we find, as suggested in the introduction, that the relations composing the first part of Deegener's outline are omitted as without social significance, since in them Alverdes cannot recognize the expression of a social instinct and since the entire discussion of these so-called associations is limited to a definition and slightly more than two pages of text. This omits consideration of much of the material to be presented in the present discussion, and limits markedly the field of general sociology. Even under these sharper limitations, the criterion of social life suggested by Alverdes, that of the possession of a social instinct, must necessarily be vague and easily capable of misinterpretation.

The material which Alverdes believes to form the subject matter of general sociology is organized in the main about sexual relations, in which he recognizes such categories as monogamy, polygyny, father-families, mother-families, and other similar divisions which were also found in Deegener's more inclusive outline. In addition, he recognizes that animal societies may be closed or open. In the former, new members are admitted only under special conditions, if at all; insect states are such. Within a closed community there is frequently an established hierarchy, as has been shown for birds which, to be sure, are only partially closed communities (Schjelderup-Ebbe, 1922, 1923). In the open societies membership is much less exclusive, and chance alone determines whether or not its members shall unite or separate. The open societies may be organized, like those of the saiga antelopes, which have guards and leaders; or unorganized, as in many groupings of what Alverdes regards as, strictly speaking, non-social insects, as when grasshoppers, butterflies, caterpillars, and the like unite in migrating swarms.

CLASSIFICATION OF ESPINAS AND WHEELER

Wheeler, in his discussion of animal societies (1930), gives a summarized scheme of classification of social and subsocial groupings, based upon the work of Espinas, which is reproduced herewith in a somewhat modified form. The principal modifications made have been the placing of all distinctions between homotypic and heterotypic groupings in the third and least important column, the rearranging of the categories under associations, and the substitution of "anthropoid" for "human" in the last category. Wheeler does not believe that the societies arose from associations, although he says that the ancient aggregative or associative proclivities may have been retained by many species and may serve to reinforce their specifically social behavior. This subject will receive more detailed attention in the last two chapters.

CLASSIFICATION ON BASIS OF INTEGRATION

It is illuminating to attempt a classification of social grouping on the basis of the type or the degree of integration of the social group. Some of the available knowledge on this point will be set forth later. From many points of view this seems a most desirable basis of classification, but there is not at present sufficient exact knowledge to justify an elaborate attempt in this direction. When made, such a classification would follow the general outlines suggested by Deegener, at least to the extent that such a scheme would present the social organization of animals from the loosely organized, apparently chance aggregations due to collections around favorable locations or on account of physical limitations which prevent separation, through a series of small quantitatively, rather than qualitatively, different degrees of integration, up to the closely organized societies of ants and termites and the more extensive group societies of man.

Simplified Schematic Arrangement of Types of Associations and Societies

(Modified from Wheeler, 1930)

A. Associations: Loosely integrated, relatively unstable, and temporary systems primarily dependent on the reactions of individuals to environmental stimuli	1. Passively collected aggregations or agglomerations. e.g., wind collected	Homotypic or Heterotypic
	2. Actively collected aggregations or agglomerations, e.g., tropistically collected	
	3. Food chain associations *a*) Predatory *b*) Parasitic	
	4. Commensal associations	Heterotypic
	5. Mimetic associations	
	6. Symbiotic or mutualistic associations	
	7. Communities (biocoenoses)	
B. Societies: More closely integrated, more stable, and permanent systems primarily dependent on reactions of individuals to each other	1. Persons (multicellular)	Homotypic
	2. Organically interconnected colonial organisms forming closed societies chiefly nutritive in function, e.g., sponges, colonial hydroids	
	3. Mainly reproductive societies closed, e.g., subsocial insects and social insects such as bees, ants, and termites	Homotypic or heterotypic; i.e., may be pure or mixed colonies of dominant animals; dominants may be accompanied by social parasites or by various other sorts of associates
	4. Mainly protective societies, closed and open, e.g., flocks, herds, and schools	
	5. Anthropoid societies; group societies	

The outlines of such a scheme of classification can be sketched. In doing so, its limitations in the present state of knowledge become the more evident.

Alpha. Individuals organically connected.

I. Individuals with true organic union, as in the hydroid *Obelia*.

II. Individuals only superficially connected, as in the mollusks

Mytilus or *Ostraea*.

Beta. Individuals not organically connected.

I. Aggregations primarily due to reactions to environment. Animals live at this level of group integration in a common habitat but without marked organization into groups. This category would include the habitat communities of the ecologists. To some extent the plants share with the animals in the organization of this community, usually, in fact, being the conspicuous factors in land communities; hence the modern emphasis by ecologists upon the biota. Further classification would depend on the physical or biotic factors in the environment which dominate the habitat.

II. Aggregations primarily due to reactions to other organisms. These are generally recognized to be more closely integrated than are habitat communities, being bound together by biological relationships as well as by those of habitat. There is no sharply defined line to be drawn between the two.

In addition to the subdivisions based upon the method of integration, three fairly definite subdivisions can be recognized, based on degree of integration.

1. Relatively slightly integrated groups in which the primary (individual) reactions predominate, and whose survival value is apparent only after experimentation. The aggregations of isopods, *Ophioderma*, and *Procerodes*, to be discussed later, are examples. Further classification would depend on method of formation of the group and on the type of integration, as well as on the different sorts of animals of which it is composed. Many of Deegener's groupings could be taken over here and in the next two categories.

2. Moderately well-integrated groups in which the secondary (group) reactions predominate although primary reactions are still strongly in evidence. The survival value of the group is more obvious. Schools of fish, flocks of birds, and the like would frequently come under this category.

3. Highly integrated groups in which the primary reactions are decidedly in the minority and the social value is strongly in evidence. Here would be classified the different insect societies, together with

the societies of man and those of the other vertebrates which approach these standard societies in their social organization.

The difficulties inherent in the further elaboration of this scheme reveal at once the lack of natural divisions between the different levels of organization with which we are dealing. It is apparent that we must recognize that the whole field of interrelationships of organisms must be taken as the content of general sociology; we can only arbitrarily single out some particular level of social appetite, group reaction, community integration, social value, or exhibition of division of labor, as forming the beginning of social life.

CHAPTER III

FORMATION OF ANIMAL AGGREGATIONS

The method of formation of animal aggregations differs with the degree of integration and with the different types of integrating factors. The discussion to be given here is not necessarily exhaustive, but the examples included may serve to illustrate the common methods and some of the problems involved.

One whole group of aggregations of individuals that are ordinarily solitary is caused by tropistic responses to environmental stimuli. Deegener recognizes one phase of this type of aggregation in his grouping called "symphotium" which occurs when individuals collect about a source of light. Aggregations of this general type may be called "syntropia," as suggested earlier. The method of formation of such aggregations attracted much attention in the three decades and a half of J. Loeb's work in this field, from about 1888 to 1923. Loeb and his immediate followers were concerned chiefly with aggregations which result from environmentally forced orientations and movements.

FORCED MOVEMENTS

When exposed to certain stimuli, some animals react as if they were automatons forced by the interaction between their own organization and their environment to move in a certain direction and to aggregate when available space is limited. The term "tropism" was at one time reserved for such reactions. These are well illustrated by the response of the larvae of the annelid worm *Arenicola* to light.

These worms burrow as adults in the sandy tidal flats of the Atlantic Coast south of Cape Cod. The eggs are deposited in large numbers in a jelly-like mass which is attached at the opening of the burrow. The eggs develop into free-swimming ciliated larvae having two eye-spots symmetrically placed near the anterior end. Immediately after hatching, the larvae are strongly positive to light and negative to gravity. Accordingly, they travel to the surface of the water, where they may collect in great aggregations unless scattered by waves or by tidal currents.

These larvae swim in a long spiral path orienting quite accurately to light. The orientation, Mast says (1911), is not entirely accurate but is subject to frequent muscular turnings which result in re-orientations. The general course is toward the light, as shown by the diagram (Fig. 1). The following account of the details of this reaction is taken from Mast's description (1911), since he has been consistently critical of interpreting any animal reaction as approaching automatonism.

"If the direction of the rays of light is changed after the larvae are oriented, they all appear to turn directly toward the source of light in its new position without preliminary trial movements." Ordinarily, these larvae swim so rapidly that the exact details of their path are hard to follow. When caught under a sloping cover slip so that they can no longer swim spirally, if the larvae are caught lying on one side no definite movement is seen except a slight forward motion; in those lying on either dorsal or ventral surface, the anterior end is seen to move constantly from side to side with a slight jerky motion, a movement undoubtedly due to muscular contractions. If light is thrown on such an organism at right angles, the lateral movement toward the illuminated side is at once increased, and the larva turns in that direction. "By using two sources of light so situated that the rays cross at right angles in the region where the specimen is located, and then alternately intercepting the light from each of the two sources, it can be seen clearly that the larva, by muscular movement, turns the anterior end toward the source of light directly. There is no trial reaction in this process. It is an asymmetrical response to an asymmetrical stimulation. The movement of these annelid larvae appear little more voluntary than the precise movement of algal swarm spores."

Galvanotropic reactions frequently produce aggregations in a diagrammatic fashion. Thus *Paramecium*, a protozoan well known to react usually by a reflex type of behavior which suggests Jennings' designation of a "trial and error" reaction, exhibits a forced-movement type of behavior under the influence of a continuous electric current. Jennings (1906), in his discussion of this reaction, says, "When a *Paramecium* is transverse or oblique to the direction of a

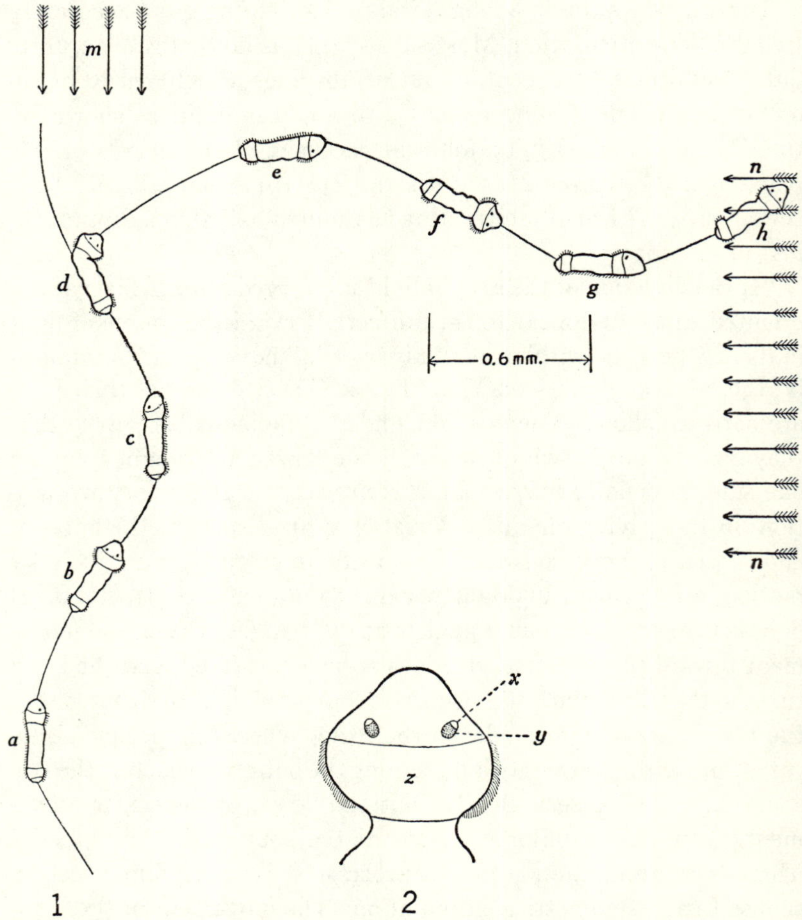

Fig. 1.—(1) *Arenicola* larva in the free-swimming state proceeding on a spiral course. *m*, *n*, Directions of light; *a–h*, positions in the spiral. Larvae react to changes in ray direction in positions *b* or *d*, but not in positions *a* and *e*. (2) Much enlarged sketch of larval head. The eye-spots are composed of a dark-brownish part *y* and a clear part *x*. Note the ciliary bands on (2) which are a part of the locomotor system. (From Mast, *Light and the Behavior of Organisms;* courtesy of Wiley & Sons.)

current at the time when the circuit is closed (Figure 2) certain striking effects are produced. If a current of medium strength is employed, such as causes reversal of about half the cilia, the following results may be observed. On the anode side the cilia strike backward as usual. On the cathode side the cilia strike forward. As a result the animal, when in a transverse position, must turn directly toward the cathode side, the cilia of both sides of the body tending

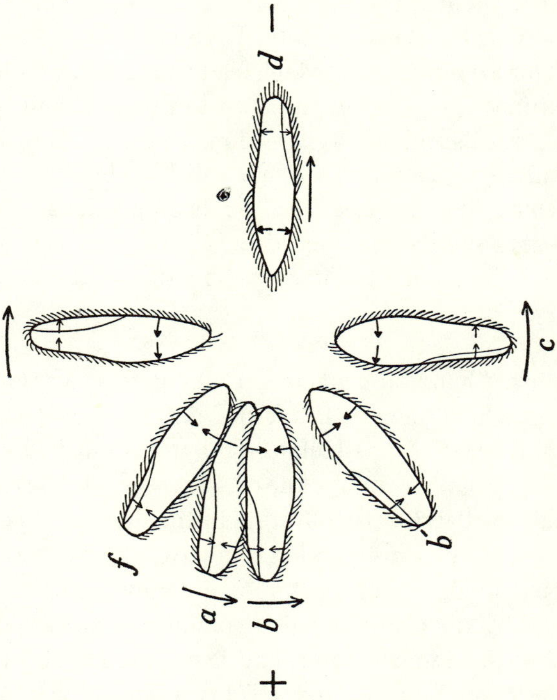

FIG. 2.—Effects of electric current on the cilia of Paramecia and the direction of turning in different positions (large arrows). The small internal arrows show the direction in which the cilia of the corresponding quarter of the animal tend to turn the animal. At *f* the impulse to turn is equal in both directions and there is no result until the revolution on the long axis brings the aboral side to the cathode. (From Jennings, *Behavior of Lower Organisms;* courtesy of the Columbia Press.)

to produce this effect, as indicated by the arrows in Figure 2. This happens even when the oral side is directed toward the cathode (Figure 2*e*). The animal turns toward the oral side—a result never produced by other stimuli, and due to the peculiar cathodic effect of the current."

Once oriented so, the animals swim toward the cathode; if the current is reversed, a reversal is caused in the orientation and locomotion of the animals. Many similar cases of forced orientation and locomotion under the influence of the galvanic current are to be found in the literature; in certain cases the animals move to the anode rather than to the cathode. A general summary of galvanotropic reactions has been given by Loeb (1918).

Similar forced movements which lead to aggregations under favorable conditions are given in response to other stimuli, particularly those of chemicals and of gravity. They are not given by all members of the animal kingdom, and are more likely to be exhibited by those animals which, like the insects, have a disproportionate development of the sensory system in comparison with the central nervous system, so that the animal becomes the creature of its sensation (Kennedy, 1927).

RANDOM MOVEMENTS

On the other hand, animals may congregate as a result of a series of reactions, which suggest the method described by Jennings (1906) as "trial and error" or, as Holmes (1905) has put it, by "the selection of random movements." The classic case is that originally given by Jennings, of Paramecia collecting in the more-acid portion of the water they occupy. This reaction is in part, at least, a trap reaction, in that the animals do not react upon entering the more-acid region, but respond by the characteristic avoiding reflexes when they come in contact with less-acid water, and hence are caught in the region of higher acidity (Johnson, 1929). This reaction by Paramecia is so well known as to have been diagramed in all the current textbooks of zoölogy. It is worth emphasizing that such a method of formation of an aggregation, while less spectacular, is not necessarily less mechanistic than is the type of reaction given by *Arenicola* larvae when they collect under the influence of directive light stimulus. It is also of interest to us that, as the Paramecia aggregate, the carbon dioxide given off as a result of their normal metabolic activities tends to keep the region more acid and thus the aggregation tends to perpetuate itself.

When there is a limited space available, or a limited amount of optimum space, aggregations may form from either of these two reaction methods, the method used depending in part on the nature of the stimulus emanating from the favorable locality, but, in the main, on the reaction system of the animals involved. If the conditions are such that directive stimuli are absent, aggregations, if formed, will result only from the method of "trial." This apparently happens many times in nature and in the laboratory.

Land isopods (Allee, 1926) tend to collect in aggregations in the hot, dry summer and in the cold, and often physiologically dry, winter. These aggregations are frequently such as might result when shelter is limited, provided there is a tolerance for the presence of other similar animals; but at times these animals collect in much closer units than can be entirely explained on this basis. That is to say, the isopods do not occupy all the available and apparently equally desirable space, but clump together in one part of this.

When the method of formation of the aggregations is studied in the laboratory, the grouping is found to be brought about by the "selection of random movement" type of behavior. Usually the isopods wander over the surface of their container, preferably around the margin, and come to rest in the position in which they are apparently less stimulated. Downs (Allee, 1926) made a long series of observations in an attempt to find the method of formation of aggregations when conditions were as nearly uniform in all parts of the container as they could be made. Under these uniform environmental conditions the land isopods usually wandered about until one came to rest for some reason or other. Sometimes inequalities developed in an originally uniform environment; at other times the isopod apparently stopped for internal reasons. After one became quiet, there was a distinct tendency for others to come to rest nearby. These might or might not be in physical contact with the first; frequently they had crawled over it immediately before stopping. In their incipient stages these bunches were frequently quite loose. The isopods would then alternate periods of rest and of motion. During the latter, many, or perhaps all, might start up again; but often a nucleus remained, consisting of the original individual and one or more others. Around such a nucleus the isopods would again gather, and the bunch would at last become consolidated by slight movements on the part of those on the periphery. Partially successful attempts were made to control the place of bunch formation on a uniform field by gluing a recently killed isopod to the substratum.

When a drop of water was introduced on a dry background, the isopods tended to occupy all of that favorable location regardless of whether or not they were in contact. The bunching in close physical contact came later, and might take place as a thigmotropic reaction, perhaps modified by chemical stimuli, or might have been conditioned by the drying of the small moistened region.

Similarly, detailed studies have been made on the bunching behavior of the ophiurid starfish, *Ophioderma brevispina* Say (Allee, 1927), which lives in the eelgrass along the Atlantic Coast of North America from Cape Cod southward. Individuals of this species have not been found in physical contact in nature during the summer and late autumn, but the collectors for the Marine Biological Laboratory report that large numbers may aggregate in late November and December. In the laboratory the tendency to collect in bunches disappears as conditions approach those obtaining in nature. Thus, bunches were absent or rare when eelgrass was present in approximately natural condition. These relations held even under the temperatures of about 10° C. obtaining in laboratory aquaria in late December.

When, however, the *Ophioderma* were placed in bare containers, bunches formed within a short time. The speed of formation was retarded by the slower movement accompanying low temperatures and dim illumination. The effect of changes in illumination are shown by the following example: With a constant temperature near 20° C. one lot formed a compact aggregation in from 1 to 10 minutes in different trials in direct sunlight; in from 14 to 25 minutes in diffuse light, and in from 27 to 56 minutes in complete darkness.

Detailed observations of the method of formation of a large number of these aggregations made under a variety of conditions show that the collections occur in the less illuminated part of the container when there is a difference in light intensity. When conditions are uniform, the starfish cluster about one of the least active individuals of the lot. In both cases the aggregation forms after a large number of apparently random movements in which the individuals react to the others present in much the same way that they do to pieces of glass rods or to eelgrass. Once formed, these aggregations tend to move together and so to form a more compact bunch. This may smack of a social tendency, although similar behavior is shown to occur when isolated individuals are adjusting themselves to the inequalities found in a tuft of eelgrass or a loose pile of glass rods. These bunches of *Ophioderma* are formed in the same general manner already described for land isopods.

Such behavior as that of the land isopods or of these starfish is obviously to a large extent conditioned by the reactions of the animals to their physical surroundings. In the absence of elements usually found in the normal physical environment, animals may so react to each other as partially to substitute for the normal environment; that is, other individuals may take the places usually occupied by non-living environmental items. Two types of explanation have been advanced for this kind of phenomenon, one of which implies some innate social tendency. The other explains such aggregations in more objective terms.

THE FORMATION OF CELL AGGREGATES

Roux (1894), a distinguished experimental embryologist, observed that when cells of the frog's egg are shaken apart during early stages of cleavage and placed in water only a short distance apart, they slowly approach each other until they come in contact. He termed such cell behavior "cytotropism." In normal development this tendency acts to help keep the cells close together in a compact mass. Later Wilson (1910), Galtsoff (1925), and Child (1928), among others, have observed the behavior of dissociated tissue cells of sponges and hydroids. Some of these thoroughly dissociated cells move about and collect in cell aggregations which under proper conditions regulate into new organisms. Galtsoff for sponges and Child for the hydroid *Corymorpha* have concluded that these cells come together as a result of chance movements on the part of certain cells which incidentally collect other cells as they move and by chance come together to form viable aggregations. Galtsoff's statement concerning sponge cells is: "The examination of the behavior of dissociated cells shows that the formation of aggregates is chiefly due to random movement of the archaeocytes which collect all the cells lying in their route." Child is more certain of the absence of definite cytotropism around dissociated *Corymorpha* cells and has observed cells when near together to move apart without aggregating apparently as often as he has found movement in the opposite direction.

The cytotropism observed by Roux can be interpreted as analogous with a very simple social appetite, or at least showing that mutual attraction between living units extends to dissociated embryo cells. From forces similar to those causing such simple mutual attractions of cells, we might expect social appetites to develop. Such a reaction may be regarded as a forerunner of the social instincts of many observers. On the other hand, in the formation of aggregations of dissociated sponge and hydroid cells there is no evidence of such mutual attraction. The method is essentially the same as that just outlined for the formation of aggregations of land isopods and of starfishes. Yet under favorable conditions these cell aggregates formed without evidence of mutual attraction may develop into well integrated animals.

PROTOTAXIS AND INSTINCT

Wallin (1927) has postulated a factor or principle which he regards as of fundamental importance for many interrelations between cells or between whole organisms and which he has called the principle of "prototaxis." This is defined as "the innate tendency of one organism or cell to react in a definite manner to another organism or cell." This reaction may be either positive or negative. The latter results in a mutual repulsion of organisms or cells, for, since organisms may be found separated for a number of reasons, Wallin recognized that negative prototaxis can be demonstrated only if the actual process is observed. On the other hand, positive prototaxis, which is "the affinity of one organism or cell for another organism or cell," may result in such well-known phenomena as those of conjugation, symplasm, cell fusion, parasitism, and symbiosis. Obviously these are real phenomena; but, as detailed information concerning the process by which the cells or organisms come together is lacking, the fact of their being together is no more evidence for the existence of a positive prototaxis than the separateness of other cells or organisms is proof of the existence of a negative prototaxis.

If we waive this objection to accepting the principle of prototaxis as an all-inclusive explanation of all aggregations whether of cells or of organisms, and proceed to examine the nature of prototaxis, we find that, instead of a simple tropism which may be best understood as a reflex action of an entire organism, prototaxis is a compound or complex tropism which Wallin says cannot be analyzed. Certainly we can recognize different elements, such as chemotropism, thigmotropism, stereotropism, as well as reactions due to surface tension, temperature, light, moisture, and electrical potential. In fact, such an analysis indicates that Wallin's conception of prototaxis is merely another name for the type of reactions referred to by many writers as being instinctive, except that no one would ordinarily regard the reaction of tissue cells as belonging in this category. Logically there is no real reason why they should not be so regarded, but the usage has been otherwise.

Wallin's conception of the formation of aggregates, whether of cells or of organisms, as being due to the expression of a fundamental biological tendency or principle, has two merits. In the first place it recognizes rightly that there is no logical line to be drawn between the behavior of tissue cells forming an animal body and that of plants or animals forming a close aggregation like those seen in symbiotic or parasitic relations. This is in line with the conclusions of Espinas and of Deegener, which I believe to be essentially correct, that there is no hard and fast line that can be drawn between the social and the infrasocial. Further, Wallin specifically recognizes that the ideas that have developed about symbiosis and parasitism have usually been based on the utility of the relationships and have also involved the idea of purpose. When such phenomena are considered from the point of view of prototaxis, then parasitism and symbiosis and presumably all their related phenomena are merely different end-responses in the expression of one and the same biological principle, involving therefore only the vague type of utility necessary for the cumbersome working of natural selection and with no more suggestion of purpose than is inherent in scientific conceptions generally.

This analysis of prototaxis shows that it is in the main a renaming of the type of activities usually called "instinctive," with the extension of this sort of action to include the behavior of tissue cells and with a deprecation of the tendency to include a distinct teleological element which is usually present in discussions of instincts. The question immediately arises as to what social instincts or appetites may mean, and whether or not they are capable of analysis.

Szymanski (1913) undertook to investigate this problem by comparing the reactions of isolated caterpillars of *Hyponomeuta* and of *Arye* with those given by groups when placed under the same general conditions. Recognizing the fact that social reactions are not readily analyzed, Szymanski undertook to separate them into two categories: (1) those peculiar to the individual, which, if fortunate, make possible the living-together of individuals as a social group, and which may be called "primary reactions"; and (2) responses which arise as the result of the living-together of many individuals, and which may be called "secondary reactions."

In order to distinguish primary and secondary responses in a social group, Szymanski suggested and used the following procedure. The reactions of the individual are first studied with a view to finding the usual responses given to various stimuli; thereafter one studies the behavior of individuals as members of a group. In the latter study it is frequently possible to recognize elements of behavior which have been observed in the isolated individuals. If all the reactions given by the individuals of a colony can be recognized as primary responses, such as would be given were all the animals isolated, the problem of group behavior is solved without the need for recognition of secondary or essentially social behavior; but if there is a residue of behavior which cannot be recognized as primary, then this is to be regarded as the secondary or true social behavior.

Szymanski so analyzed the responses given by caterpillars of

Hyponomeuta, which inhabit an irregularly rounded web usually placed between several branches of the food plant. The individual larvae exhibit no tropisms except a strong negative stereotropism. When placed singly on the ground, the larvae make a looplike path. They stop at almost every point of the loop and test out their environment with their heads, selecting thus a place to lay their silk thread. Single caterpillars spinning their web behave similarly. The individual reaches out as far as possible from the place of beginning and lays down its thread. This action is repeated and results in a spreading of the web. Such a response to space Szymanski regards as a negative stereotropism.

In one experiment eight caterpillars were observed in nest-building. Six were placed together at one place, and one each at two slightly distant points. All began spinning webs as described above. The six spun a common web, which finally reached and fused with the webs of the isolated caterpillars, so that a joint web resulted. This probably happens in nature. So far, there are no reactions remaining over and above the individual responses, and Szymanski concludes that in the formation of this common web there are no secondary or purely social reactions.

Deegener (1922) disagrees with Szymanski's observations and with his interpretations, particularly the latter. He, too, found that isolated *Hyponomeuta* larvae can spin webs, but concluded that they do not begin spinning as soon as if grouped together, and that the web spun by a group is smaller than that made by the union of webs spun by the same number of isolated larvae. In both respects he would recognize the working-out of a social instinct. Further, he believes that the caterpillars actively seek out the company of others, guided by sensing vibration waves, which may be merely refined touch perception. Back of all this he believes there is a need for association which leads the isolated larvae to seek their own kind. If they do not find their fellows, they build their own individual nests, which later they may abandon, wandering and seeking in order to associate themselves with other larvae. If none are found, they may remain solitary for days without losing their social instinct.

Szymanski (1913) further studied the formation of feeding aggregations of *Arye* caterpillars. Groups of these young caterpillars gather on their species of food plant and arrange themselves on the leaves so that they cling with their thoracic legs on the upper surface while the posterior end hangs down curled around the edge of the leaf. They arrange themselves so, side by side along the margin of the leaf. When the larva at the tip has eaten to the main vein, it may do one of three things: (1) turn around and go to the base of the leaf and begin feeding there; (2) leave the leaf entirely; or (3) cross over to the opposite side and begin feeding there. The older caterpillars tend to lose this regular arrangement and behavior.

By the usual type of analytical experiments the *Arye* caterpillars are shown to be positive to light, negative to gravity, and positive to certain touch stimuli. The method of locomotion consists in the extension of the anterior end and the drawing-up of the posterior. The posterior end shows a definite motor reflex upon stimulation. Thus, if touched at the posterior end, the posterior half of the body is raised. A similar reaction is given if the substratum is gently shaken. If one side is touched, the same response may occur, together with a bending-away of the touched part.

If one tests out the method of colony formation, one finds that when the larvae are placed at the base of the food plant they will crawl up on it, since they are positive to light and negative to gravity. When the first leaf petiole is encountered, they will turn aside onto that because it is narrower than the main stem, and for the same reason they will move along the edge of the leaf. On the leaf they move to the side most strongly illuminated, or to the side farther from the ground, as the case may be.

The larvae crawl here and there over the leaf, passing over each other; or they may touch the larvae ahead and cause them to move forward. Finally one begins to eat, and gradually all settle to eating. The positions taken may be accidental, for wide spaces may occur between larvae, while others are closely crowded. The piece of leaf between two larvae becomes eaten away, so that eventually the head of the second larva touches the posterior end of the first. This causes the latter to raise its posterior end, as in the test experiments described above. The reaction will be repeated whenever the posterior end is stimulated, and only ceases when the abdomen curls over the edge of the leaf. In this way, and as a result of these reactions, the colony takes on its well-organized appearance, which depends on the interaction of the following factors: (1) the crowding of many individuals into a small space; (2) the tropic reactions of the larvae; (3) the character of the anterior and posterior end reflexes; and (4) the manner of locomotion and of feeding.

Here, as in Szymanski's analysis of the group formation in *Hyponomeuta*, primary reactions play the principal rôle in the colony formation; but there are some elements of the behavior of the colony that Szymanski thinks may be due to secondary or social behavior. Thus, when the leaf is shaken, the posterior end of each larva is raised simultaneously. When we remember the great individual differences usual in behavior, the synchrony of this response suggests that there may be a social factor at work. However, it is possible that this, too, is merely an expression of primary or individual reaction, with the synchrony either more apparent than real or due to the proximity of the responding larvae.

These investigations of Szymanski's lead to the same conclusion as my own, formed independently, concerning the method of formation of aggregations of land isopods and of *Ophioderma*. In these cases it is the primary, individual reactions that produce the groupings, not the expression of a community spirit or of a social appetite. The only social trait necessarily present is that of toleration for the presence of numerous other similar animals within the same region. If this analysis be sound, as it appears to be, then one of the early stages of mutual interdependence is the appearance of toleration for the presence of other animals in a limited space, where they have collected as a result of tropistic reactions to environmental stimuli. Once formed, aggregations may persist for a considerable time, merely because of the lack of disruptive stimuli.

The conclusions of Szymanski are supported by Krizenecky (1923) in his work on the transitory aggregations of the enchytraeids already mentioned in the chapter on classification. He thinks that individual reactions are important in the formation of these aggregations, and that thigmotropic reactions are largely concerned. The observations of Essenberg and of Riley on water striders, of Clark on *Notonecta*, of the Severins on *Belostoma*, as well as the tremendous general literature on animal behavior (see Loeb, 1918, for a partial bibliography), show that aggregations do form in many cases without evidence of a positive social instinct or appetite, although this is not to be taken as proof that in other instances aggregations may not form as a result of social appetite.

AGGREGATIONS OF ASELLUS IN NATURE

The analysis of one other case is illuminating. For a number of years I had been seeking a favorable opportunity to apply in the field certain analytical methods worked out in studying aspects of the laboratory ecology of animal aggregations, and accordingly welcomed the information that a great aggregation of the common fresh-water isopod, *Asellus communis* Say, had been found in mid-winter in the Indiana dune country near-by.

At the point where this collection occurred, a low sand ridge had been thrown up to serve as a roadway across an extensive cat-tail swamp, here about a quarter of a mile wide. To the east, the swamp stretched as far as could be seen from the low elevation of the roadway. To the west, there was also a very extensive continuation of the cat-tail swamp for at least a half-mile. The whole formed a major part of the headwaters of a small stream.

The roadway was pierced at several places by culverts, introduced to relieve the water pressure above. These had proved inadequate, and at one place the water had washed away the ridge of sand and flowed over the roadway through an opening about 5 meters wide, with a current there sufficient to prevent complete freezing. When first seen, the ice was about 6 cm. thick and the effective stream was reduced to about 1.5 meters width.

Here on the under side of the ice were tens of thousands of isopods, oriented to face upstream, and showing by their arrangement the definite lines of force of the current below. Thousands of other isopods were resting on the bottom in protected places, and many more were being swept downstream by the rapidly moving current. Sometimes these collected into small balls of from 6 to 20 isopods, which rolled along the bottom until they found a lodging against some obstruction or settled into a deeper pool where the current was less strong. There were many isopods on the sandy bottom of the stream, mostly facing against the current, but making very little progress against it. Chopping through the ice above or below the roadway revealed no comparable collection of isopods, although there was evidence of an increase in numbers as one neared the narrow channels of the washout either from above or below.

After the break-up of the winter ice, the majority of the isopods disappeared, although traces of the aggregation could still be seen, particularly in the sheltered places just below the opening of the stream into the lower swamp. There the isopods were mainly traveling downstream with the current, or were collected in sheltered places in deeper water or about lodged débris. As before, few were found in the open above the roadway.

After the ice was entirely gone, and with the usual rise in water level, the aggregation re-formed. In early April a few were being carried downstream through the washout. Several more were to be seen along the margins, for the most part headed upstream, where some were able to make their way for a considerable distance. At the lower edge of the roadway, great masses of isopods had collected about willow shrubs, old cat-tails, or in deeper pools, wherever they might find a lodging. The largest of these masses was about 75 cm. across the current, 30 cm. up and down stream, and over 10 cm. deep, a solid writhing mass of isopods. This was loosely joined with other similar units, each formed about some basis of support from the force of the current, the whole making an isopod barrier all along the lower margin of the washout, over 5 meters in length and about 1 meter wide. The numbers concerned were unbelievable. They were to be measured by liters rather than by individuals. The mass can be imagined by thinking of the full swarms of some twenty or more beehives settling near each other. Conditions remained much the same for the next 3 weeks, with but a slight variation in the position of the largest mass, depending apparently on the strength of the current.

In late April the water level had again been raised by rain, and in the main current stood about 45 cm. deep, in place of the more usual 15–18 cm. A new, smaller overflow had been formed near-by. The isopods were all gone from their place of aggregation; and although they were still plentiful all around the edge of the fan of sand washed down by the recent rains, they were not collected into the great masses found heretofore. In the slacker current just preceding the rains, isopods were no longer being carried downstream across the roadway; and one could not have collected more than 50 such drifters by watching all day. Now with the higher water level and the swifter current they were again being swept downstream from the upper swamp in numbers.

With the higher water level of early April, smaller aggregations had appeared about the lower ends of the central iron culverts piercing the roadway, but now there, too, were dissipated. With the higher water of late April, the culverts situated at the edges of the swamp showed a marked current for the first time—not nearly so strong as that in the center culverts, but corresponding in strength to the latter when aggregation of isopods occurred near them. Now, for the first time, sizeable aggregations were present at the lower ends of these side culverts. At the upper end of one of these there was a log and much plant débris on and about which isopods borne down from the upper swamp might have lodged; but none were there, while they occurred in large numbers at the lower opening, particularly in eddies out of the main current. The current ceased to flow through the north marginal culvert within a few days, and the aggregations there disintegrated. Those at the opposite margin persisted for about two weeks. The aggregations below the main overflow did not re-form, although many individuals could be seen at any time unsuccessfully attempting to make their way up over the shifting sandy bottom.

The final breaking-up of the aggregations was not observed, though at any time small groups or single individuals might be seen becoming detached and borne away by the current. When the water rose, the increased velocity probably carried the whole lot off in a similar manner. At the end of the season some of the aggregated animals died *in situ*, especially if located at one side where the current became cut off.

An increased flow following heavy rains in late May produced physical conditions similar to those of late April, but no aggregations were formed, although a few large isopods were carried downstream through the main spillway.

The favorable localities were well watched the next winter and spring, but no large aggregations were found. In early April a small aggregation occurred below the culvert at the extreme south side, where the last one had formed the spring before. With the passing of time, the main washout had deepened so that a stronger current was running there than when the isopods had aggregated the preceding year. In general, there appeared to be fewer Aselli in the swamp, and one is led to suspect that there may have been an unusually large production of isopods preceding the formation of the monster aggregation observed in 1927.

SEX RATIO

In early spring one can usually determine the sex of *Asellus* by considering the size, shape of thorax, and presence or absence of the brood pouch. In the laboratory, sexes are easily and accurately determined. Careful observations showed that during the time of the great spring aggregations the ratio of the collected isopods ran as high as 25 males to 1 female, and never ranged below 9:1.

November collections from the scattered isopods, both above and below the culverts, showed a 1:1 ratio. In early April of the next year, random collections from the relatively small aggregation at the lower end of a lateral culvert showed a sex ratio of 12 males to each female. At the same time, similar collections both above and below the aggregation showed a ratio of approximately 1:1.

Five suggestions readily occur to account for the high ratio of males to females in the great bunches:

1. The aggregation may be due to a mating or other social impulse acting more strongly in the males than in the females, which impels them to gather in these large groups.

2. The males may tend to move about more and to come into

contact with the current and be swept off their feet, regaining a foothold only when the current slackens or when they reach a solid footing.

3. The males may possess less clinging power than the females.

4. The females may be carried downstream as well as the males, but may escape from the bunches to the lower swamp.

5. The aggregations may be formed from isopods that start up from the lower swamp and are unable to make progress when the swifter current is encountered. If this is a factor, it would imply that the males are more strongly positive in their rheotropic reaction than are the females.

The last four possibilities would account for the formation of the aggregations through the operation of tropistic reactions of the isopods as individuals, the so-called "primary reactions" of Szymanski (1913); while the first would bring in a secondary or group reaction. The different possibilities may be considered in reverse order.

The rheotropic reactions of both sexes were tested according to methods developed earlier (Allee, 1912). These tests indicated that, in the breeding season at least, the males are somewhat more strongly positive in their rheotropic reaction than are the females, and that they respond positively to stronger currents. In so far as the aggregations form as the trapping of positive isopods moving upstream from the lower swamp, this helps to account for the great discrepancy in the sex ratio. However, this is not the whole story.

There is little evidence for the assumption that the females may be carried down from the upper swamp in the same numbers as the males but escape from the aggregation to the lower swamp. There were very few females found among the many isopods collected while being carried downstream.

The supposition that the males have less clinging power than the females, at least in the breeding season, was subjected to direct experimentation, using the method described by myself in 1914. The results indicated that there is little, if any, difference in the clinging ability of the males and females under the conditions of this test, with whatever advantage that may exist favoring the males. Such results are to be expected from a consideration of the mechanical difficulties of maintenance of position by females carrying a large brood pouch between their anterior thoracic legs.

Of the tropistic non-social suggestions advanced as possible explanations of the greater proportion of males than females in the spring aggregations, one more remains for detailed consideration. This is the suggestion that the males move about more and so come into contact with the current more frequently than the more passive females. Such differential action would result in more males being swept off their feet and carried down from above, and also in more males coming in contact with a current strength which would call forth a positive rheotropic response and so bring them up from the lower swamp. This possibility is supported by the following kinds of evidence. The direction of the current impinging on a large bunch was artificially changed, and the current change resulted in a reorganization of the bunch of isopods in a new position. At a time when the main bunch showed a ratio of males to females of 25:0, 25:3, 25:2, 25:2, with a total of 100:7, the reorganized bunch showed ratios of 50:1 and 45:4, with a total of 95:5, which is nearly twice the number of males per female as found in the bunch of longer standing. Again, I pulled from near a large aggregation a tuft of grass heavily covered with isopods. The sex ratio of those that actively crawled from the grass onto my hand proved to be 4 males to each female. The sex ratio of all the isopods on a similar tuft was found to be 1 male to 3 females. In both cases the males showed a higher degree of activity. It is also true that the vast majority of animals taken while being carried downstream by the current were males, and that the sex ratio of the isopods on the water plants outside the main current, but above the roadway, showed a higher number of females than males.

Regarding the possibility that the males may be responding to a stronger internal sexual stimulant than the females, there is evidence from earlier work that in the breeding season the males do tend to cling to any passing isopod, and apparently have this tendency more strongly developed than do the females. The tendency to collect in bunches is so strong that spring isopods must frequently be tested singly for rheotropism or they will fail to respond to the current at all. I have seen males which were responding definitely to a water current behave as if they perceived another isopod at a distance of some 2–4 cm., discontinue their rheotropic reaction, and move directly to the nearby isopod and cling to it. I have no evidence of such reactions at distances greater than 5 cm., so that their effect would be operative in bunch formation only, after the isopods had been brought close together through the operation of some other factors.

I have no knowledge of such isopod aggregations except in winter and spring, and unfortunately the sex ratios of the winter aggregations were not taken. In this connection it must be remembered that the isopods do not start their breeding season in December in nature. Yet large aggregations were found at that time. The observations show clearly that the ratio of males to females is high in the spring aggregations, and suggest that this is due to the tendency of males to move about during the breeding season, which makes them more likely to be caught in the current and swept down from the upper swamp, and, on the other hand, more likely to come into contact with a current sufficiently strong to cause them to react positively, and so move upstream to the place of aggregation from the lower swamp.

METHOD OF FORMATION OF THE AGGREGATIONS

This subject obviously overlaps consideration of the preponderance of males, and the conclusions reached from much consideration of the problem are the same as those indicated there. As was to be expected, disturbances in the swamp just above the opening of one of the outlets caused a marked increase in the numbers of isopods carried down. These might lodge in slight depressions in the stream bed where the current was less strong; some 50 were observed to collect in a small depression less than 12 cm. in diameter within 5 minutes following a disturbance in the upper swamp.

Others were carried on by the current until they found physical support against rushes or other débris, or against other isopods which were in turn supported by the rushes. Thus, the bunch may be seen to grow on its upstream side, the newcomers using the other isopods as an extension of the support furnished by the lodged débris.

But this is not the only method by which the aggregations are formed. Mention has been made already of the finding of a large aggregation at the lower end of a culvert whose upper opening was well protected by the presence of logs, grass, and other débris, through which the water ran easily, but upon which few isopods collected even at the sides where the current was certainly not of sufficient strength to tear them loose from available support.

In laboratory experiments with artificial streams some isopods, mostly males, traveled against the current and collected in the more quiet water at the upper end of the trough. Similar behavior was repeatedly seen in nature. After the ice left, isopods from the great spring aggregations could be seen laboriously moving against the current over the sandy bed of the stream; while those located below the opening of one of the streams into the lower swamp, if not present in sufficient numbers to form an aggregation of three dimensions, were frequently spread thickly over the bottom, with all individuals headed upstream.

Of all the isopods moving upstream, those near the margin were most successful. Usually, however, all were swept down sooner or later to the main group below. When a board was placed with one end resting in an aggregation so that it furnished a solid substratum on which the isopods might crawl, they immediately started upstream as closely as they could stick on the board. On reaching the upper end, many were immediately washed down by the current, while others would continue over the precarious bottom for a short distance before they, too, lost their footing. If dikes were built so that the current impinging on an aggregation was slackened, the isopods started upstream in numbers, only to be swept down again when a stronger current was encountered.

There is also a fatigue factor which causes the failure of these isopods to continue their journey upstream even in a fairly weak current. The length of time before reversal is roughly correlated with the physical condition of the isopods. In laboratory tests with isopods from these aggregations, reversal in a straight current occurred after an exposure of about an hour.

If the impinging current is cut off completely by the construction of a dam, the aggregated individuals begin a rearrangement which usually results in new aggregations being formed in depressions, or about some quiet individual or a quiet group, just as such aggregations form in the quiet water of a laboratory tank. These groupings are usually less dense than those exposed to the drive of the current. The negative reaction to light is one of the factors conditioning this reaction; positive thigmotropism is another. If grass or other débris is present in abundance, the isopods usually collect in contact with the inanimate matter rather than piling up in great isopod aggregations. There may be some collections due to positive chemotropism, for these aggregations cause measurable differences in their chemical environment.

I was much impressed, in all the observations made upon these groups, by the fact that so large a part of the formation of the aggregations could readily be explained on the basis of individual tropistic reactions to environmental stimuli largely produced independently of the massed isopods themselves. Relatively few of the causes of aggregation were left to be explained by the reactions due to social appetite. In this respect the situation is wholly similar to that found with land isopods, with *Ophioderma*, and with Szymanski's caterpillars. Again the main social trait exhibited appears to be that of tolerance for the presence of many other individuals in a limited space where they have collected, or—one might almost say—where they have been collected. The same idea can be expressed by saying that almost the sole social trait exhibited is immunity to injurious effects resulting from the presence of many others in a limited amount of space. It is interesting to note that there were also leeches, snails, and other animals collected in the same location and, to a large extent, by the same combination of physical forces and tropistic reactions that had brought the isopods together.

GYRINID BEETLES

The reactions concerned in the formation and maintenance of two more complex, more closely integrated types of aggregations have also been made. In the case of the gyrinid or whirligig beetles, giant aggregations may occur on the surface of streams or of still water, where the animals may be resting quietly or where they may exhibit what appears to be a perfect frenzy of erratic activity. As stated above, Deegener regarded these as forming play societies, while Schulz thought of them as having protective values.

From the analysis of Brown and Hatch (1929) it appears that the aggregating behavior of these beetles is largely due to visual stimuli, since the aggregations break up in the dark. Further, the position they occupy in the laboratory tanks, though not necessarily in nature, may be determined by the lighting. These authors believe that the gyrinids are exhibiting a more complex type of behavior than that which is usually called "tropistic," and refer it rather to some sort of configurationist behavior, in which orientation behavior consists of movements so co-ordinated that an invariant relationship is maintained between movements and variations of the visual field.

They find evidence of two sorts of orientation: one in which the body axis is maintained in a relatively fixed position with respect to the base of orientation, and another in which the body is maintained in a relatively fixed region but without body orientation. The former is like the orientation called for by the tropistic theory. The latter bears at least a superficial resemblance to those cases where organisms move along a physical or chemical gradient in one direction without reaction to it but execute negative "avoiding" reactions when moving in the opposite direction, like the trapping of Paramecia in weak acids, as described by Jennings.

They believe that the location of an aggregation in nature is due to habituation to certain visual patterns, possibly of light and shade, to which the animals respond; these patterns are not significant in themselves but are a sign of the location of general environmental conditions which are of vital importance to the beetle and the species. If the patterns are slowly changed, the beetles may remain in a given position; and collections have been observed not to shift their position as much as a meter during a whole day, although the pattern of the field of vision changed radically in that time. If, however, the patterns are rapidly changed by a sudden increase in the complexity of the visual pattern, marked stimulation to activity results, which may cause a breaking-up of the aggregation.

Brown and Hatch, in their report, do not discuss the importance of the presence of other individual gyrinids in the immediate neighborhood in connection with the pattern complex. Rather, they give the impression that each beetle is reacting as an individual to a general environmental pattern, which traps it somewhat as Jennings regards individual Paramecia as trapped by a drop of acid, until a collection is formed.

CATFISH AGGREGATIONS

The young of the silurid fishes, the catfishes and the bullheads, exhibit a striking type of aggregation, which has been analyzed by Bowen (1929). In the species *Ameiurus melas* used in this work the young may be observed in the summer months swimming in close bunches near the surface of ditches or small ponds, packed together in a more or less spherical mass. A single fortunate dip has yielded over 500 of these minnows.

If such a group is scattered, within a few minutes 2 or 3 individuals appear singly and come together somewhere near the original location of the entire group. Gradually they are joined by single fishes or by small groups which come into the same locality, apparently swimming at random. These show no reaction to the larger group until they are within 2 or 3 feet of it, when they swim directly toward the larger aggregation and join it. Within 30 minutes to 1 hour the original aggregation will have re-formed.

Appropriate tests showed that the individual fish were not reacting to a gradient of chemical emanations from the group, for they do not respond to fish-conditioned water (i.e., water in which fish have stood until it exhibits various chemical and perhaps physical changes), even when the conditioning is greater than could be the case with an aggregation in nature. Cutting the olfactory nerves had no effect either on normal or on blinded fish.

With these bullheads, as with the gyrinids, vision is the important factor in the formation of the aggregation and in its subsequent integration. Neither blinded fish nor normal fish in the dark ever aggregate, and normal fish will follow a moving fish model in a way similar to that which results in aggregation when in the company of other normal fish.

Ameiurus melas can sense through the skin the presence of another fish in motion, probably by detecting the vibrations set up by the tail of the nearby fish. Slight positive responses to others of a group are shown by blinded fish; this reaction is not affected by the destruction of the lateral-line organs but is almost eliminated when the skin is anesthetized with magnesium sulphate.

When the bullheads come into actual contact with another object, a positive thigmotropic response is given. The barblets are dragged over the object; and by means of the sensations received, apparently chemical in nature, the fish is able to discriminate between paraffin models and live fishes, but it is apparently unable to distinguish between fish of the same or of different species.

Bowen sums up her observations in practically these words: "In the evening, as soon as it begins to grow dark, the aggregated young catfish separate and swim about, sweeping through the water or along the bottom with the barblets, giving a feeding reaction similar to that given by blinded fish at any time of the day. As soon as it begins to grow light the young fish come together into aggregations in which they remain for the entire day, re-forming in a short time if scattered by a disturbance. Some feeding may occur while the fish are aggregating, but it is doubtful if this occurs to any extent. Usually the fish are in a close bunch actively pushing against each other, or resting at the surface in contact or close proximity. A thigmotactic reaction seems to be at the base of this behavior. Unless disturbed, older fish in the aquarium rest during the day in contact with the substratum, or more often in contact with one another. By means of aggregations the young fish can satisfy their positive thigmotaxis even while in motion. The pushing in a group suggests the importance of this. Catfish will also push against other species of fish, which, however, do not reciprocate. This contact reaction is largely one of pressure, but gustatory response apparently plays some part, as shown by the different responses to paraffin models and to fish. Whether this factor is instinctive or is influenced to any extent by conditioning is yet to be determined. The reaction is apparently not species specific, since there is no evidence that young catfish show different behavior toward members of their own species than toward other forms." The aggregation arises from the tendency of the other catfish to respond by appropriate positive reactions, instead of making-off as fishes of other species do when approached.

With these fairly well-integrated aggregations of young bullheads analysis shows that the social appetite is diffuse rather than specific, and that in the normal fishes, aggregation involves a sight reflex, a touch reflex, and possibly a low-frequency vibration reflex, all of which may be given to other moving objects, non-living as well as living; a chemical reflex, the sign of which is reversed with non-living models; and, finally, reciprocal behavior on the part of the different individual members of the aggregation. The matter of reciprocal responses contributes the distinctly social element in this behavior. The extent to which the combination of these reflexes into a functioning whole depends upon the presence of an inherited social appetite, or upon early conditioned behavior, remains to be investigated.

CHAPTER IV

GENERAL FACTORS CONDITIONING AGGREGATIONS

In many animal species the formation of an aggregation depends on the physiological state of the animal. This may be controlled by internal developments, such as the maturing of the sex products, or by external factors, as when land isopods are made to bunch by controlling the moisture of the substratum; but more commonly the internal and external conditioning factors work together closely. Some of the more outstanding of these are discussed here.

THE BREEDING SEASON

Water isopods.—My own attention was drawn to the general problem of animal aggregations in 1911, while studying the factors controlling the rheotropic reaction in the common water-isopod *Asellus communis*. As spring came on, the stream isopods no longer gave highly regular, positive responses to the water current; but, as stated in the preceding chapter, one might strike across a strong current, guided apparently by sight, and seize another isopod, male or female. From such a beginning one might soon have all the isopods under observation gathered into a compact rounded cluster, rolling over and over in the water.

During the height of the breeding season stream isopods disregard the stimulus of a water current almost completely unless they are relatively isolated. On the other hand, I have repeatedly tried to induce half-grown Aselli to form such a cluster, even placing them in a watch glass with rounded, smooth bottom, where they were continually brought in contact with each other, but no real aggregation resulted. Bunching may be induced in adults out of the breeding season, but many conditions that favor it in April during the height of the breeding season have little or no effect in late May (Allee, 1923).

Mosquitoes and midges.—*Culicidae* and *Chironomidae* form swarms of males which maintain position as groups, although the individuals within the swarm are continually darting from one part of the swarm to another. Such swarms have been known for years, although their significance has not been generally understood. Knab (1906) cites his own observations on the swarming of mosquitoes and reviews the literature to show that the swarms are composed of males which hover over or near prominent objects such as trees, corn shocks, house gables, or people. Enormous numbers of these dipterans may be present in the collections, which occur generally in the early evening of quiet and almost windless days. Straight-flying females dart into these irregularly gyrating swarms of males and emerge *in copula* with one male. One such newly mated pair was observed to emerge from one swarm only to enter accidentally another near-by. The copulating pair appeared to be greatly stimulated and flew into the open as soon as possible. The swarming males which were associated for even so short a time with the mated pair also increased their rate of flying and "danced up and down at a furious pace for some time" before again quieting down to their normal rate of gyration. With growing darkness the activity of the swarms increased, but fewer successful matings took place; the entering female would be set upon by two or three males, and all would fall together to the ground, where they would separate. Later, females ceased entering the swarms, and the males gradually dispersed. Counted sex ratios of *Culex* were 897 males to 4 females (Knab), and of *Chironomus* 4,300 males to 22 females (Taylor, 1900). Mosier and Snyder (1919*a*) interpret the large morning swarms of tabanid flies which they observed in the Florida Everglades as aggregations of males to which females are attracted and into which they dart for the purpose of mating.

Frogs.—With the approach of spring frogs desert their hibernation quarters for breeding places in the shallow ponds (Cummins, 1920). Many hibernate in the mud at the bottom of these same ponds; but others winter elsewhere, perhaps in nearby bodies of water or on land among masses of dead vegetation, or in localities similarly favorable. Cummins suggests that such frogs may migrate to open water caused by the early melting of ice in a pond with proper exposure. Banta (1914), Yerkes (1903, 1905), and Noble (1923) find evidence that frogs may respond to frog calls and splashings, particularly during the spring breeding season. Studies on the breeding migration of toads indicate that with them the voice serves as a sex call (Courtis, 1907; Miller, 1909; Wellman, 1917). Boulenger (1912) concluded that the voices of frogs and toads do not control migrations toward breeding grounds or movements of individuals at the grounds. Cummins later came to the same conclusion as a result of his observations on a partially fenced pond, since he found that heavy migrations followed periods in which there was no croaking in or near the pond, and that, on the other hand, great vocal activity was not accompanied by increased migration. Certainly, vocal activity cannot account for the similar spring migration of the voiceless *Ambystoma*.

The immediate inception of the migratory impulse must be intrinsic and is probably associated with the conditions of the sexual glands. In frogs it is secondarily conditioned by weather, since waves of migration are coincident with high relative humidity and with a temperature of from 41° to 52° F. The migration is independent of daylight. All of Cummins' illuminating observations still give no information as to why the frogs congregate in a given pond or how they learn of its existence. He does record that migration routes are not direct, so that we may assume that we are dealing, at least in part, with random movements, probably controlled largely by temperature. Blanchard (1930) concludes that the external control for the breeding migration is to be found in rainfall rather than in temperature relations.

During the breeding season a gregariousness appears among frogs which does not exist under usual circumstances. This is not entirely accounted for by the tendency which the animals exhibit to seek a similar habitat for breeding, for if there are only a few pairs of frogs in a given place, they force themselves together as closely as possible and the eggs form a continuous mass.

At the height of the breeding season several males will struggle for the possession of a single female (Banta, 1914); the struggles attract other males, and one female may become the center of a struggling mass. One such group which Banta caught had 6 males fastened together about a single female and 5 others nearby but not yet attached. The actual egg-laying and fertilization of the eggs is accompanied by the formation of a close aggregation (Fischer-Sigwart, 1897). In addition to the male that has been *in copulo* for some time, these supernumerary males gather and, despite kicks from the first male, still manage to form a close clump. In *Rana fusca* one may find single pairs, but as a rule fertilization is a community matter. Supernumerary males also crawl over and among the egg masses and effect the fertilization of ova which may not have been reached by spermatozoa at the time of their discharge.

At the close of the breeding season frogs scatter and resume a solitary, non-social existence.

Fish.—Similar breeding clusters of fish have been described by Reeves (1907) with many identical details. With the rainbow darter supernumerary males crowd about the spawning pair and appear also to shed spermatozoa. Reighard (1903) has seen such behavior; but in the main his studies (1903, 1915, 1920) emphasize the orderly spacing of breeding holdings in fish, a phase of the aggregation phenomenon with which the present summary is not greatly concerned. The close contact between males and females of fresh-water animals with external fertilization is made necessary by the extremely short life of the gametes shed into fresh water. Reighard has stated that fish sperm can remain functional for less than a minute under these conditions.

Snakes.—Snakes are reported to form bunches in the breeding season similar to those described for frogs, except that they occur out of the water (Ditmars, 1907; Ellicott, 1880; Ruthven, 1908). Ellicott records: "I first saw such a bunch of snakes on the stony banks of the Patapsco River, heaped together on a rock and between big stones. It was a warm and sunny location where a human being could scarcely disturb them. I reasoned that the warmth and the quiet of that secluded space had brought them together. Some hundreds could be counted, and all in a very lively state of humor, hissing at me with threatening glances and with such persistency that stones thrown at them could not stop them nor alter the position of a single animal. They would make the proper movements and the stone would roll off; all the snakes in this lump were common garter snakes (*Eutaemia sirtalis* L.).

"The second time I noticed a ball of black snakes rolling slowly down a steep hillside on the bank of the same river. Some of the snakes were of considerable length and thickness and as I noticed clearly, kept together by procreative impulses."

Lunar periodicities.—Such breeding aggregations are much more important in fresh-water and land forms, with whom the surroundings are more injurious to shed sperm or eggs, but they do occur among marine animals. With marine organisms the most spectacular expression of breeding aggregations is to be found in the case of the large number of animals whose breeding rhythms coincide to some extent with lunar periodicities. The literature on this subject is extensive (Woodworth, 1907; Fox, 1923; Legendre, 1925; B. H. Grave, 1922, 1927); but while the facts are plain enough, the fundamental causal relations remain unknown. One illustration must suffice, based on the account given by Lillie and Just (1913) for the swarming of the sea worm *Nereis limbata* in waters around Woods Hole.

Nereis limbata has its swarming period only after twilight. Each run begins near the time of the full moon, increases to a maximum during successive nights, and sinks to a low point about the time of the third quarter, again rising and falling to extinction shortly after the new moon. They appear in four periods or cycles during the summer, corresponding to the lunar cycles in the months of June, July, August, and September.

Only fully mature animals swarm. The swarming begins shortly after twilight and lasts for only an hour or so. The swarming animals are attracted by the light of a lantern. Males appear first, darting through the water in curved paths in and out of the circle of the light. Females are fewer in number and swim more slowly. The males outnumber the females hundreds to dozens. In the next few minutes the numbers increase, waning again after about three-quarters of an hour.

New females appear each night, but some males may presumably reappear on several successive nights. A swarming female is soon surrounded by several males. These swim rapidly in narrow circles about her. In a little while they begin to shed sperm, probably in reaction to some secretion from the female, rendering the water milky. Soon the female begins to shed her eggs, shrinking in bulk as she does so, until, a shadow of her former self, she sinks through the water to die. Lillie and Just, following a lead from Hempelmann (1911), assume that the maturing of the animals is dependent on some relation of the life-history to the phases of the moon, involving, probably, through lunar tidal variations, rhythmical alterations of conditions of nutrition.

HIBERNATION

Over-wintering aggregations of animals have long been known. This phenomenon in social bees has been noted in scientific literature for almost two hundred years (Reaumur, 1734–42). Barkow (1846), in his monograph on hibernation written over three-quarters of a century ago, has a short chapter in which he calls attention to the winter aggregations of lepidopterous larvae, adult ants, bees, true bugs, beetles, including the frequently observed case of the coccinellid beetles, carp and the eel-like *Muraena anguilla*, snakes, frogs, and a few mammals, including marmots and bats. Barkow advances no theory to account for the congregation of these animals but does state that there is a suggestion current that the animals come together as a result of response to their sense of smell.

This list of over-wintering aggregations has since been much extended, especially by Holmquist (1926), who has made extensive studies on hibernating arthropods in the Chicago region. He reports that of 329 identified species taken during the winter season, nearly 17 per cent were more or less closely aggregated. Omitting those known to be of a somewhat social habit at other times of the year, about 9 per cent of the species ordinarily solitary in the summer were aggregated in winter.

In the social bees careful experiments have shown that temperature-control results from such clusters (Phillips and Demuth, 1914; Phillips, 1917); and Holmquist (1928) has demonstrated that protection from flooding, and other benefits, may accrue from the cluster formation of hibernating ants.

In many cases these over-wintering groups are essentially shelter aggregations, apparently due to the small amount of serviceable shelter available. Often, however, all the apparently equally desirable space is not occupied, so that the aggregation cannot be entirely explained on the basis of unavoidable crowding. In other cases Holmquist has been unable to find any environmental differences to account for the location of the hibernating aggregation. These groupings are partially under temperature control; but, as with other phenomena connected with hibernation, the temperature control is incomplete, and the problem of the exact nature of the causal factors remains open.

AESTIVATION

Aestivating aggregations have been less studied. Land isopods will form aestivating groups which may be either homotypic or heterotypic. Dr. C. H. Abbott has informed me personally that they collect in large numbers in protected places, and so pass the long, hot, dry summer of southern California.

AGGREGATIONS CONTROLLED BY MOISTURE

The chief controls of the aestivation reaction of these isopods are temperature and moisture. Of the two, laboratory experiments show the latter to be more important (Allee, 1926). When land isopods of various species are placed on air-dry filter paper, they collect in bunches within a few minutes, unless the substratum is too dry, when they will run about actively until at the point of death. If the substratum is moist, the same isopods will remain quietly scattered.

These relations are shown in Figure 3. In the upper picture there are 25 isopods in a crystallization dish photographed 30 minutes after being introduced into the dish, which had the bottom covered with dry filter paper. In the meantime they were in a darkened room and the exposure was by flashlight. The lower photograph shows the effect of adding enough water to make the filter paper thoroughly moist without being sloppily wet. The same animals are shown as in the preceding photograph, but 15 minutes later, and 5 minutes after the background was moistened. The animals not shown in this photograph have crawled up the sides of the dish.

A somewhat similar effect of drought in nature is reported for the California quail (Evermann, 1901). In an unusually dry season these quail do not breed but remain in flocks during the entire summer. The opposite type of moisture control is also observable. Too much moisture may produce well-defined aggregations. Thus *Solenopsis geminata* (von Ihering, 1894), a species of ant which often nests in lowlands, will, if the nest is flooded, aggregate in a ball of some 15–20 cm. in diameter, with the larvae and pupae inside. By constant rotation they avoid too long submergence, and at length may come against some solid object and so escape from the water. Wheeler (1913*a*) cites this case and mentions similar instances in this and other species of ants.

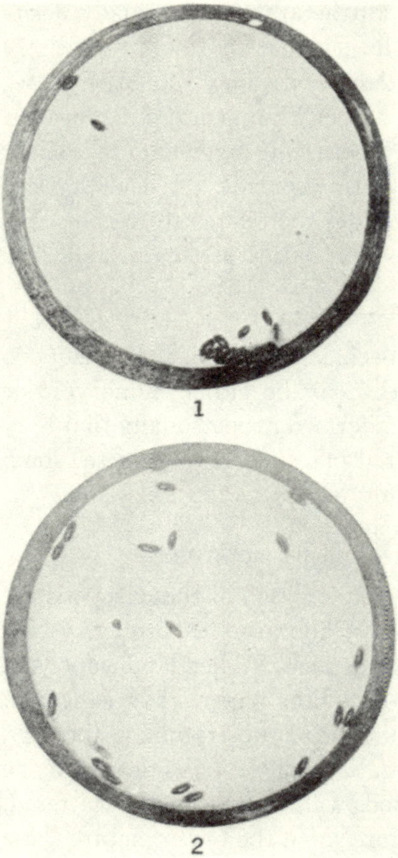

FIG. 3.—(1) Land isopods in darkened room on dry background of filter paper. (2) Same animals and conditions as in (1) except that the filter paper has been moistened.

The formation of the dancing bunches of midges already mentioned, which one frequently sees aggregated in the space of a half-bushel basket, appear to be in part conditioned by the atmospheric humidity, although the absence of wind is another obvious prerequisite.

In both these cases the environmental conditions are uniform; and the animals, in grouping together, react to each other only. There are also the place aggregations controlled by moisture, when animals will collect in a limited area because it provides an oasis of moisture or of dryness in an otherwise overdry or overwet environment. Thus land isopods can be made to collect at will in a given spot by making it moist. Selous (1907) gives a striking picture of the congregating of large ungulates about an African drinking-hole in the dry season. The common fruit fly, *Drosophila*, struggling to escape too great moisture, aggregates in shifting masses at the top of a projection; these masses continually fall apart and re-form as the flies move up again. Under optimal conditions all of these move out of contact with their fellows.

LACK OF NORMAL ENVIRONMENT

The snake starfish, *Ophioderma*, lives in eelgrass in certain locations along our eastern coast. Repeated attempts have failed to find this animal in contact with others of its own kind in nature during the summer (Allee, 1927). They are often found near together but never aggregated.

Ten of these starfish were introduced into a laboratory aquarium made to approach normal living conditions by the introduction of eelgrass. Nineteen hours later 7 of the 10 animals were sighted after a search lasting half an hour. One was found on the bottom at the side away from the strongest light; 6 animals were in the densest part of the vegetation in the same region; and, although not in immediate contact, all of them could probably have been inclosed in a 5-inch cube. The exact location of the other 3 animals could not be observed without disturbing them. These animals in the field may also be close together without actually touching. Only such loose collections were ever seen in this eelgrass aquarium. Extended experience with these animals in the laboratory leads me to conclude that the tendency to bunch is greatly reduced in proportion as favorable natural conditions are approximated, and that the animals so congregated are usually found in regions to which they have been directed by their tropistic reactions.

When, however, *Ophioderma* are placed as they are collected in a glass or similar container, they form dense mats of bunched animals with arms closely interwoven. The aggregations form in the shadiest part of the dish and are to be explained in part by the fact that the lower animals are shaded by the upper ones, and so, having satisfied a negative phototropism and a positive thigmotropism, they remain quiet.

The position of the arms shows the strong thigmotropic reaction of these animals. In the recently formed bunches there are a larger number of free arms than in older aggregations; at first the arms tend to extend out and up into the water. They may be entirely free, or they may touch another arm only where the two cross. Even in the early stages of the bunching some of the arms lie nearly parallel with each other. In bunches of longer duration there are practically no free arms. In one case I saw two starfish with four pairs of arms paralleling each other, and only two free. Larger bunches become ropelike masses composed of parallel arms or of arms intertwined like basketry. In these older aggregations the arms of the animals, at first extending freely, are turned back and interwoven with the others so that the outer edge presents a relatively regular line. When these starfish are isolated and left for a week or more in separate dishes exposed to light, frequently the arms are moved into contact until they present a sort of self-bunching.

Laboratory aggregations occur in a large number of animals. May-fly nymphs, various isopods, earthworms, frogs, and others may readily be observed to form such bunches. The behavior appears due to similar causes to that which results in the collection of foreigners into communities of their own nationality in our large cities; that is, a group of similar animals tend to minimize for each other the disturbing effects of unusual surroundings.

"SLEEP" AGGREGATIONS

The sleep aggregations of insects have been relatively little written about, even in research journals; so it seems important to bring together a more extended summary concerning such slumber aggregations than is needed for the better-known overnight assemblies of birds.

Fabre (1915) found some hundreds of the wasp *Ammophila* (*Sphex*) *hirsuta* assembled under the shelter of a stone on the mountain side, and speculated much concerning this gregarious condition of a solitary wasp. The Raus (1916) found three related species sleeping in such assemblies, from which it would seem probable that Fabre was observing a slumber aggregation. With *Chalybion caerulum* both males and females may be found aggregated at night in about equal proportions. As many as a thousand have been found in one colony. Marked individuals will return to the same sleeping place for at least 2 weeks. No one knows how the male of the species passes the day; the female labors about the nest.

The solitary *Sphex* wasps appear to choose their sleeping quarters independently; but since they select the same sort of place, they tend to form spaced aggregations. *Prionyx* sleeps sometimes singly; sometimes gregariously crowded close together on the top of a weed, with equal numbers of males and females present but without observed copulation. The males and females of the horse fly *Tabanus sulcifrons* are reported also to collect in favorable places to sleep (Hine, 1906). Similar observations are on record for various other insects.

There is no evident protection from enemies in such assemblies. The sleep may be sound, and may extend so late that early birds could pick off the sleeping insects in numbers, as beetles are reported to kill off sleeping butterflies (Floerscheims, 1906).

Schrittky (1922) observed in Paraguay an aggregation of from 20 to 27 butterflies (genus *Heliconus*) that gathered nightly during August and September. The butterflies could be handled in the early mornings without waking them. The temperature then ranged around 5° C. The butterflies were quite restless in the evening long after dark, when the temperature was higher than in the morning. He also observed males of the genus *Tetrapedia* in aggregations at night; females are found in temporary aggregations until the time of fertilization, after which they separate.

Banks (1902) gives observations on males of the solitary digger bees of the genus *Melissodes*. He saw these bees at dusk in his back yard clinging with mandibles and feet to grass blades. He records three or four returning for several nights. He cites the record of Schwarz (1896) to show that *Melissodes pygmeus* clasp twigs with their mandibles. Bradley also records finding *Melissodes agilis* clinging on dried blades of wild oats alongside a newly cut grain field. In this same patch were large aggregations of a number of species of wasps—no two species on the same blade. He accounts for the aggregations of wasps as being caused by the cutting of nearby grain. Boyer and Buchsbaum of this laboratory, from their unpublished observations, think that the *Melissodes* which Bradley found aggregated were present because of the cutting of the flowers on which they ordinarily collect.

Von Frisch (1918) gives observations on 6 solitary male bees, *Halictus*, which returned for 4 days to the same dry stem of a plant. He records that the bees would return to this plant if the weather grew bad or if the temperature became low, even during daylight. For 4 evenings after his original observation he observed 5 bees present on the same stem, although there were other similar stems nearby. One bee had been taken for identification. He cannot be sure that the same five returned each evening.

The Raus have several notes on bees. Concerning *Melissodes obliqua* they say: ". . . . We found the twenty-eight bees clustered near the tops of a small clump of stalks. Since it was now almost dark my presence did not disturb them. They were huddled together in groups of two to five, with only three insects occupying their sites singly.

"The next evening twenty-nine bees, only one more, were asleep on these five stems, all clustered on the apical three inches of the dead plants. At the top of another plant ten feet away, two were at rest. If they had chosen this site for protection alone they would have rested singly on the plants, but since they huddled in groups they must have sought sociability also. They were so close together in some cases as to arouse suspicion about their mating, but a close examination proved the idea false.

"The following night, July 21, twenty-four of these bees were here to spend the night in the same way. On the 22nd, thirty were present. On this evening I marked part of them with white paint. . . . . As fate would have it, the next evening a cow had broken down their chosen stems, so none of the bees were there. However, fifteen were found on similar weeds nearby; seven of these bore the white markings. This gave evidence sufficient to prove that the same bees return to their chosen spot regularly . . . . all were males."

Here Boyer and Buchsbaum took up the problem, using the solitary digger bee, *Melissodes agilis* or *aurigensia*, which Professor Cockerell in a personal communication says are variants of the same species. They found *Melissodes* active in the field only on warm sunny days, with the temperature 16° C. or above, depending on the light. When it is cloudy or cool, the bees remain inactive on the sunflowers *Helianthus annuus* and *petiolaris*, which they frequent. The male bees usually arrange themselves so that two are in contact when two or more become inactive on the same blossom. Several groups of two's have been found on the same flower, isolated from each other.

The bees are invariably inactive between twilight and sunrise. The beginning of activity depends on the amount of light and on the temperature. Controlled laboratory experiments showed that in bright sunlight activity started under stimulation at about 7° C., while in dim light the first activity came at 9° C. Similarly, spontaneous activity began at 18° C. in the sunlight and at 21° in dim light.

Aggregations of males at night were recorded as follows for one particular group of sunflowers:

Year	Singly	Pairs	Groups of 3	Groups of 4	Groups of 5	Groups of 6
1927	100	31	3	1	1	3
1928	39	5	1	0	2	0

Boyer and Buchsbaum marked some of these bees with paints of different colors so that they could follow individual reactions. There were some 500 flowers in this particular group. Each of these was plotted and followed night after night for its bee population. In all, 201 bees were observed in 1927. Thirty-four of these were successfully painted. Of the 20 painted bees which were seen again, 10 bees returned to the flower they occupied when painted. Eight others returned to the same plant but to a different flower. Fourteen returned at least twice to the same plant but to a different flower from that on which they had been painted. In all, 37 returns were noted to the same flower or to a flower within 10 feet of the original one, while 24 returns were noted to some flower more than 10 feet away from the original.

In 1928, 14 bees were successfully painted. Of these only 4 were seen again; and of these, two returned to the same flower where they were painted and the other two returned to a nearby flower. A study of the details of the observations shows that males of *Melissodes* frequently return to the same flower night after night or in cool or cloudy weather. They are generally found in the same vicinity on successive nights, even if not on the same flower. They must necessarily return to a different flower if the one on which they have been staying is destroyed or dries up. No bees were observed on withered flowers. If they are blown to a distant part of the sunflower patch, they tend to remain there in a narrowly circumscribed area for the next several days.

It must be noted that these overnight aggregations in *Melissodes* were composed of males only.[1] They cannot have sexual significance. It seems entirely possible that we are concerned here with an incipient social habit which does not extend to many solitary species and is not found in all individuals of the species in which it occurs.

Swarming locusts of several different species are known to pass the night in dense masses both as nymphs and later when they become adults. Much of this literature is reviewed by Uvarov (1928). Regarding the overnight aggregations of these locusts, Uvarov says: "The night is passed on plants in dense bands, which are extremely conspicuous on the background of the vegetation owing to their blackish general color; during the night the hoppers are in a state of torpor caused by the cold." If the day is cool, the slumber bands do not break up as they do on warm sunshiny days. Even when considerable numbers of the South African locust, *Locustana pardalina*, have become adult, they collect at night near the main nymph swarm, although they may range at considerable distance during the day. The night clusters of the flying adults are not so dense as those of the hoppers.

Faure (1923), in describing the night collecting of nymphs, says they gather slowly together into fairly dense masses, forming clusters

[1] J. F. W. Pearson has taken 3 female *Melissodes* and 90 males from early morning collecting on *Helianthus* in this locality.

that closely resemble close masses of bees. They swarm on the tall grass, or, if this is lacking, they pack together in the low grass, on stones or on the ground. At sunrise the swarm gradually breaks and continues its migration. These night clusters are conspicuous objects showing up as reddish-brown patches on the veldt. Man and presumably other animals take advantage of these aggregations to destroy great numbers of the grasshoppers. The benefits accruing are not known. Nikolsky (1925, *vide* Uvarov, 1928) thinks that they conserve animal heat. Holmquist's observations on mass collection of ants (1928*a*) suggest that they may at least slow down the rate of change of temperature.

The congregation of birds for sleeping has been widely observed (Brewster, 1890; Davis, 1894; Bates, 1895; Widman, 1898, 1922; Allen, 1925), particularly for martins, robins, grackles, and crows. Many other birds are reported to gather in the roosts dominated by martins and robins. Extreme cases of close crowding in these roosts are reported by Baker (*vide* Allen, 1925) for the crested tree swift of India.

"On arriving at their proposed meeting place," Baker says, "they fly round and round, gradually lowering their flight until one bird makes a sweep and settles on some part of the tree near the top. This is the signal for the rest to perch, and in a few minutes they are all dotted about the higher branches. They then begin to close up with the bird which first alighted on the tree, finally collecting in a feathery ball, one on top of the other. Sometimes this happens again and again before they get settled, but at last the twittering stops and they are asleep for the night. It is wonderful how compactly these birds close up; a flock of eleven appeared not to take more than a foot long by half that breadth."

The Indian swallow shrike is said on the same authority to have a similar habit. Sharp records that the colonies of mouse birds of Africa, small birds resembling parrots, roost in small parties that cling together.

It is well known that bats also gather into sleeping aggregations (Goldman, 1920; Howell, 1920; Allen, 1921). They may congregate in clusters comprising only a few individuals, or hundreds may hang with bodies touching. The groups may be homotypic or heterotypic.

To the human senses these bird and bat roosts are easily detected by their odor, and perhaps that is a factor in guiding the bats to the common sleeping place.

Allen (1921) has banded clusters of these bats. He records recovering three of a group of four from the same place where they were banded, after an interval of three years.

These sleeping aggregations appear to be without mating significance. The Raus did not see copulation among the insects they observed; and, in fact, in many cases the sleeping groups were composed of males only. The robin roosts may contain both sexes and all ages of birds above the nestlings. With crows the common roost ends with the beginning of the breeding season, except for the bachelors; and in general these roosts are not occupied by the breeding birds. After the breeding season the birds may return in family groups, a situation to be discussed later at some length. Among bats the sexes are segregated (Howell, 1920) during the time of gestation and of the care of the young, at a time when contact sleeping aggregations were observed.

At the low level of integration of aggregations, with which we are especially concerned, the appearance of social appetite is an intermittent phenomenon. It may be awakened by gonadal activities that precede the breeding season or by the conditions which induce hibernation or aestivation. These varying exhibitions of a stronger social appetite are ordinarily part of an annual rhythm, but in many marine forms the rhythm may be a lunar one during the warmer season of the year. In the slumber aggregations, the periodic strengthening of the social appetite has a diurnal rhythm. Aggregations may be induced or controlled by conditions of moisture or by the lack of normally favorable conditions; this phenomenon may or may not be rhythmical in nature. At this low level of social integration the social appetite is not constant in appearance and in this regard becomes more like the sex and hunger appetites, in which rhythmical or spasmodic appearance is one of the usual characteristics. In more highly integrated social groups the action of the social appetite is steadier, and therefore less spectacular and less easily recognized.

CHAPTER V

INTEGRATION OF AGGREGATIONS

THE COMMUNITY LEVEL OF INTEGRATION

It is instructive to regard an animal as a physiological system of physicochemical processes in dynamic equilibrium. When this is understood, one is prepared for the definition of an "animal society" or an "ecological animal community" as a system of organisms which is in the process of dynamic equilibration.

In the case of the animal considered as an organism, the different parts are integrated more or less perfectly into a unit, which has been receiving considerable attention in the last decade in studies on the organism as a whole as contrasted with the study of different parts of organisms. One can readily see that there are highly integrated organisms under close control of the nervous system or of hormones, the loss of any major part of which will strongly affect the whole system and frequently will cause death; but, on the other hand, there are the lower organisms much more loosely correlated, where the loss of even a major part of the body causes only temporary inconvenience pending the regeneration of replacement tissues. Many of these more loosely organized animals are so poorly integrated that different parts may be in active opposition to each other. Thus, when an ordinary starfish is placed on its back, part of the arms may attempt to turn the animal in one direction, while others work to turn it in the opposite way. With sponges, the pores admitting water to the canal system may be open and the flagella engaged in pumping water into the canals, while the ostia remain closed so that no water can be brought in (Parker, 1919). On account of its loose integration, the sea anemone may move off and leave a portion of its foot clinging tightly to a rock, so that the animal suffers serious rupture.

It is to such relatively slack systems that an ecological animal community is to be compared, rather than to the highly integrated ant or bird or man. In human society we are accustomed to the idea of community integration. Thus a village is composed of a number of families which are connected as a unit not only because they occupy a limited amount of contiguous space but also because they are bound together by social organizations such as church and school, by economic relationships of kinship or of marriage; all of these knit the community into a working unit. The organization is loose. Individuals may come and go. Whole families may depart and others move into the village, and yet the village retains a definite unity, with a more or less marked individuality which may be quite distinct from that of neighboring communities.

In such a community as the village, men are associated not only with each other but with other animals. There are the horses that supply part of the draft power; cattle that give meat and milk; dogs and cats that provide companionship and amusement to man, feed on his surplus food or on other associated animals, add to the dirt of his household and scatter bacteria and parasites; flies that feed on the refuse of man and breed in the excreta of his commensals; mosquitoes that breed in water reservoirs and feed on man and other animals; birds attracted by the nesting sites and food to be found near man; rats and mice similarly attracted, and snakes attracted by the birds and rats and mice; insects that prey on gardens and orchards, and insects that prey upon these; as well as other animals with little direct relationship to the community but occupying the same general space.

If a progressive town board decides to instal a hydroelectric plant, the river is dammed and a breeding place is furnished for thousands of mosquitoes; if some of these are *Anopheles*, the malarial parasite may become prevalent. The breeding range of fish and of pond insects is extended at the same time that the human population is adjusting to the use of cheap water power; the dam is a matter of concern to the whole animal community.

The consequences of an unusually mild winter ramify also through the entire community. One result is that many insects live over which would ordinarily be winter-killed. These attack orchard, garden, and farm, affecting the food of grain- and fruit-eating birds and mammals and of man himself. These are finally checked by the subsequent increase in predaceous insects and birds that live on garden and orchard pests, and the rough biotic balance characteristic of animal and plant communities in nature thus tends to be restored.

These instances are enough to illustrate the interdependence and general type of organization of an animal community of which man is a dominant element. The removal or introduction of animals, whether by accident or by purposive action by man, may upset the whole equilibrium, as has happened with the introduction of rabbits into Australia.

A similar organization has long been recognized to exist in animal communities of which man is only a minor part, or perhaps no part at all. This type of organization has been called by J. Arthur Thompson the "web of life." Frequently the food relationships are the most easily demonstrated in a group of this kind. A partial idea of the complexity of such organization is given by a consideration of the food relations of the black bass as summarized from Forbes' excellent essay on "The Lake as a Microcosm."

INTEGRATION OF THE BLACK-BASS COMMUNITY

The organization of animal communities is more marked in the case of inhabitants of small bodies of water than of equal bodies of land, since conditions tend to isolate such aquatic animals and since, through long evolution, they have become closely integrated and highly independent of the newer societies of the land. The life in such a body of water represents an islet of older, lower life in the midst of the higher, more recent life of the surrounding region. It forms a microcosm, a little world in itself. The play of life is full, but on a smaller scale and less confusing to observe.

In such a community one can see fully illustrated the degree of sensitivity characteristic of an organic complex, which has just been demonstrated for a man-dominated community. Whatever affects one species must have its influence on the whole assemblage. It thus becomes apparent that it is impossible to study any one animal completely if it be out of its relations to other animals and to plants, even though the animal selected for study belongs to what is usually regarded as a non-social species.

It is relatively easy, though sufficiently exciting to be called sport, given the right body of water and the proper season and bait, to lift a large-mouthed black bass from the water; but if one should undertake to trace out all the interrelations from which the black bass has suddenly been removed, he will have seen the whole complicated mechanism of the aquatic life of the locality, both plant and animal, of which the black bass forms a part.

In the food of the black bass are to be found fishes of different species at different ages of the individual, representing all the important orders of the fishes; insects in considerable number, especially the various water bugs and larvae of the May flies; fresh-water crayfishes, shrimps, and a multitude of the small crustaceans called Entomostraca, of many genera and species.

Looking at the food of the fishes upon which the black bass feeds, one finds that one of these eats mud, algae and Entomostraca, and another takes nearly every animal substance in the water, including mollusks and decomposing organic matter. The crayfishes are nearly omnivorous; of the other Crustacea, some eat Entomostraca and some algae and Protozoa. The insects eaten by the bass eat each other, other insects, and Entomostraca. At only the second step, therefore, do we find the black bass directly related to every class of animals, many plants, and the decaying vegetal matter of the water.

Turning now to competitors, which are extremely numerous, we find that all the young fishes, except the suckers, feed at first almost wholly on Entomostraca, so that the young black bass finds himself at the very beginning engaged in a scramble with almost all the other fishes in the lake for food and, in fact, not only with the fishes but with the insects and mollusks and larger crustaceans that also live on these small entomostracans. The Mollusca are not in such direct competition; but they do compete, since they feed upon the microplankton which the Entomostraca themselves take as food.

But the competitors of the bass are not limited to those which take the same food, for predaceous fishes, turtles, water snakes, wading and diving birds, and the large beetles, dragon-fly nymphs, and giant water bugs feed on the young bass at every opportunity.

An illustration of remote and unsuspected rivalries is found in the relation of the black bass to the bladderwort (*Utricularia*), which fills many acres of the northern Illinois lakes. Upon the leaves of this plant are small bladders, several hundred to the plant, which are tiny traps for the capture of entomostracans and other minute animals. The plant usually has no roots and lives largely on the animals taken through these bladders. Ten of these sacs, taken at random, upon examination gave 93 animals of 26 different species, of the Entomostraca and insect larvae. Hence, the bladderwort competes with the fishes for food and, by destroying large amounts, helps keep down the number of black bass in an otherwise favorable lake; and they have an especial advantage since, when the Entomostraca become scarce, they may grow roots and live as other plants.

These simple instances suffice to illustrate the intimate way in which the living forms of a lake are united.

A different phase of the story is shown by the study of fluviatile prairie lakes which are appendages of river systems and form in oxbow cut-offs or bayous, or in other regions where the usual deposition of materials has been retarded. Normally they are connected with each other during the rainy period and for a longer or shorter time during the summer. The amount and variation of animal life in them is dependent chiefly upon the frequency, extent, and duration of the overflows. In them we may see illustrated the method by which the flexible system of the animal community adjusts itself to widely and rapidly fluctuating conditions.

Whenever the waters of a river remain for a long time outside its banks, the breeding grounds of the fishes and other animals are correspondingly extended. The slow and stagnant waters of such an overflow, frequently enriched by sewage to a limited extent, form the best possible place for the growth of myriads of algae and Protozoa. This development allows a similarly great development of Entomostraca. These animals increase with tremendous rapidity due to the pace at which their life-circle is run and to their high rate of reproduction. The sudden development of food resources allows a corresponding increase in the rapidly breeding, non-predaceous fishes; and at last the game fishes which derive their principal food from the non-predaceous fishes also increase in numbers. Evidently the multiplication of each of these classes acts as a check on the one preceding it. The development of Protozoa and algae is arrested and sent below normal by the swarm of entomostracans; the latter are met and checked by the vast swarm of minnows, which are in turn checked by the increase in predaceous fishes. In this way a gradual readjustment of the conditions will occur; but usually, long before this new equilibrium is reached, a new disturbance of the water level results in the recession of the water. As the lakes grow smaller and the teeming life they inclose is daily restricted within narrower and narrower bounds, a fearful slaughter ensues. The predaceous fishes thrive for a time, since their food is more easily caught; but finally they too are thinned out by the lack of food and of space.

Year after year in such lakes and in other animal communities there is a fairly steady balance of organic life. The community remains in dynamic equilibrium. The rate of reproduction about equals the death-rate. Every species must fight its way from hatching to maturity. Adults are as rare as human centenarians; yet no species is exterminated, and each is maintained at the average number, for which we have reason to think there is sufficient food year after year. Two ideas explain the order that is evolved in such communities. First, there is the background of common interests among all elements of the community. New evidence concerning the nature of some of these common interests will be presented shortly. Second, there is the struggle for existence and the elimination not only of the less fortunate but, at times, of the less fit animals.

Upon such a foundation as this, modern comparative sociology is built in part, and must be built in entirety if it is to be solidly grounded. With this conception of the type of integration existing in ecological animal communities, and with the realization that even such loosely knit communities can be regarded as constituting a unit, we are better prepared to search for integrations in animal aggregations and to evaluate those found.

AGGREGATION INTEGRATION

As has been said before, a decided advance toward social life is made by the appearance of tolerance for other animals in a limited space, where they have collected as a result of random movements or of tropistic reactions to their environment. This may occur in connection with some phase of breeding activity, but it may also be exhibited without sexual significance. Some of the less complex of these aggregations may exist because there is an absence of dissociating factors among a group of animals that have been hatched out in a restricted locality or that have been brought together by any other process. Thus, some of the aggregations resulting from tropistic responses may well owe whatever permanency they possess to the absence of disruptive factors rather than to any inherent gregarious tendency or apparent advantage.

Another advance in social life is made when these groupings confer especial survival values upon at least some of the individuals composing them. Such an advantage is illustrated by the slower rate of moisture change in an aggregation of land isopods out of water equilibrium with the surroundings. Under conditions of drought this results in a definite prolongation of life for the members of a group. Other examples will be discussed later.

The land isopods and *Ophioderma* have gone little beyond such a stage in their social development. There is some slight evidence of mutual attraction, but the experiments to date do not indicate how much of this would also be exhibited toward similar inanimate objects. There is also slight evidence of integrated group behavior, in that the bunch shows occasional periods of activity apparently originating in one individual and passed mechanically through the group. Such activity may be the beginning of disintegration of the group; but it frequently results in a closer aggregation, because the animals may move closer together during their brief period of activity.

The state of development of integration by means of which the group, once it appears, acts as a unit is a very important criterion of the degree of social development it has attained. When there is no integrative action, one is dealing with a crowd, a mere collection of individuals within a limited area. Apparently it was this aspect that Szymanski (1913) had in mind in distinguishing between primary reactions, the reactions of the individual, and secondary reactions, the reactions of the individuals as members of a group.

TACTILE INTEGRATION

The simplest form of group organization is found when animals in physical contact respond as a group to touch stimuli passed from one to another. Such organization may be sufficiently refined for the whole group to show definitely synchronous behavior. Collections of *Liobunum*, the harvestman, have been observed by Newman (1917), and later by myself, to give such reaction. One group was found by Newman resting on the under side of an overhanging shelf of rock on a steep hillside. The harvestmen were closely packed together within an area of about 5 feet in diameter. When first seen, they were hanging from the rock roof perfectly motionless. When the observer came nearer, they began a rhythmic stationary dance practically in unison. This died down shortly but could be started again by appropriate mechanical stimulation.

When the colony was first seen, the long legs of neighboring individuals were interlocked, which would sufficiently account for the transmission of stimuli through the group. It should be noted, since we are interested in the state of integration of the aggregation, that the rhythm was not perfectly synchronous at the beginning but became practically so after a few seconds.

Such integration, due to tactile transmission, must be present to some degree in all cases of aggregation in close physical contact. It is highly developed in the sleeping groups of bats (Allen, 1921), which may hang in compact clusters, as already mentioned. If one is touched, the whole cluster may drop. Allen caught eighteen by holding an insect net under the group and touching only one of the outer bats.

The effect of physical contact in establishing synchrony in two reacting systems is illustrated by the observation of Fischer (1924), who found that two pieces of embryonic heart planted out in tissue-culture media beat at different rhythms even when taken from the same individual and kept as far as possible under identical cultural conditions. When two such pieces succeeded, by outgrowths from each, in establishing close organic union, the two beat in unison. Such a modification of behavior may involve factors of transmission distinct from those we usually regard as tactile.

CONTACT AND ODOR INTEGRATION

Sex recognition frequently causes animals to give characteristic group reactions; often there are only two animals forming a diminutive group. Sex recognition is frequently accomplished by contact relations alone. Such behavior is recorded for crayfishes (Pearse, 1909; Andrews, 1910), spiders (Montgomery, 1910), frogs (Banta, 1914), amphipods (Holmes, 1903), as well as others.

Among other methods of sex recognition, that due to chemical sense deserves prominent mention. This is well illustrated by the long distances certain male moths will fly to cluster about a female ready to copulate (Kennedy, 1927). Animals may aggregate at other times than the breeding season, due to the same sort of stimulus; and this stimulus is also frequently effective in maintaining the aggregations once formed. In fact, it is common for the principal stimulus causing animals to congregate to be the effective one in integrating their aggregation.

These two senses, odor and contact, are sufficient means of group integration to form the basis of well-unified societies. Much of the social organization of the ants and the termites appears to be based on them. The ants apparently live in a world of contact-odor shapes, as we live primarily in a world of color-shapes (Wheeler, 1913*a*).

VISUAL INTEGRATION

Sight plays an important rôle in the organization of many animal groups. When one vulture, soaring aloft, sees another swoop miles away, he moves over and also swoops; his action is seen by others, and thus these scavenger groups congregate rapidly, although they are practically lacking in a sense of smell.

Aggregations of male frogs in the breeding season will follow and frequently tightly clasp any moving object, whether salamander, fish, or other males; and this reaction is based at least in part on sight. Aggregations of young catfishes are primarily integrated by sight and secondarily by water vibrations and chemical-touch sensations (Bowen, 1930).

Other instances might be multiplied; but one spectacular one, that of the synchronous flashing of fireflies, must suffice. A considerable controversy has been waged over this subject, but the observation experiments of Hess (1920) seem to have established the fact of its occurrence. He found a valley of fireflies flashing in unison, with the flash apparently initiated on a hill at one side, from which it spread almost instantaneously over the valley. The next night in the same place the observer was able to obtain at least partial control of the flash and to alter to some extent the intervals between flashes. With a pocket flashlight he gave the initiating signal just before it would normally have occurred, and the insects followed the artificial lead until the interval was reduced to three-quarters its original duration, and then one-half. At the second trial at one-half the original period fewer insects followed the flashlight, and after that the flashing in unison was broken.

Such synchronous flashings of fireflies are apparently more common in the orient. Morrison (1929) has published a recent note upon their occurrence in Siam, based upon three years' experience there. His account follows:

"During the months of July, August, September, and until the heavy rains set in, on any dark night it is possible to see whole stretches of the river or canal banks lit up by the flashing of myriads of insects. These areas of synchronism may extend for several hundred yards at a stretch or may be confined to single trees, glowing and being extinguished with surprising regularity. Actual timing of this intermittence showed that luminescence occurs at the rate of approximately 120 times a minute. During the period between the flashes the light of the fireflies reached almost complete extinction, the intensity being so low that at a few feet from a tree of actively luminescing insects it is quite invisible.

"Perhaps one of the first things which is called to the attention of the observer is the fact that this synchronism is confined to localities bordering on streams, or to low, water-saturated ground. . . . . Around Bangkok it is commonly known that the synchronal flashing of fireflies is confined to one particular tree, the 'ton lampoo' of the Siamese—*Sonneratia acida*. In all of the observations which the writer has made, no exceptions to this have been found, but whether this particular tree is the gathering-place of the insects in cases of synchronism reported from other parts of the East is a question.

"The fact that *Sonneratia acida* is the tree on which the insects congregate around Bangkok leads one to question the statement that has been frequently made to the effect that the synchronal flashing of the fireflies is a mating adaptation. *S. acida* is found both in mangrove associations, and also as a solitary tree growing along the banks of streams. In these latter cases the roots of the tree are often immersed in water, the tree at times standing several feet from the bank. If the females of the species are wingless, as is the case with the majority of the North American *Lampyridae*, there would be no opportunity for them to approach the tree. Furthermore, at no time have females been found on a tree of actively synchronizing insects, or within its vicinity. Observations on this point have been repeatedly made and have been corroborated by local entomologists who have become interested in the problem.

"Perhaps one of the most popular theories as to the cause of synchronism is that of 'sympathy.' According to this idea there is some particular insect which acts as a pace-maker for the rest, and they follow him, regulating their flashes by his. However, due to the fact that the insects are scattered quite generally over a tree and are not within sight of any one particular animal, this appears to be quite impossible. Furthermore, any follow-the-leader action on the part of the insects would result in a wave of light passing over the tree and originating from a definite point, a fact which is not the case once the synchronism has begun.

"It is possible to inhibit the synchronism of a tree of insects by exposing them to a bright light for about a minute. When the light is turned off, the synchronism returns having its origin, apparently, in some individual or group generally located in the central part of the tree. From this group, then, the synchronism extends over the entire tree in an irregular wave until all of the insects are flashing in unison.

"Synchronism usually begins shortly after darkness has set in, the fireflies emerging from the nearby thickets and flying in an indirect course to the *Sonneratia* trees. During this flight to the trees there is no sign of a concerted flashing, the actions of the insects being similar to those found in our local forms during flight."

It seems probable that with these fireflies we are dealing with a phenomenon of two distinct aspects (Blair, 1915). One is a recovery response similar to recovery from fatigue. Such flashing would rarely be synchronous or near-synchronous. On the other hand, there appears to be a releasing stimulus which, in the cases observed by Hess, might come either from the pace-setting flash of a firefly or of an electric torch. This brings up the problem of the leader in group integration, for which we have not space here. It is discussed at some length by Child (1924).

INTEGRATION BY SOUNDS

Among many animals group organizations occur as the result of sound production. To be sure of this, one must have evidence that behavior is altered as a result of sounds. The fact that collections of animals, such as frogs or insects, are producing sounds which are loud to the human ear is not good evidence that they have group significance (Lutz, 1924). There is evidence that among some animals sounds may be used in sex recognition. Perhaps they are more often of sexual significance in general sex stimulation which, while of advantage to the group, may yield no advantage to the producer of the sound; and may even result disastrously in the case of the young deserted by a nesting bird which had been stimulated to renewed sexual activity by an outburst of song. Such cases have been reported by creditable ornithologists (Sherman, 1924).

Ohaus (1899–1900) and Wheeler (1923) report that the *Passalus* beetles, which have the habit of boring in logs, are kept together by auditory signals; and Professor Wheeler has more than once spoken of his observations, indicating that aërial sounds may play a part in the organization of ant colonies. But on this point there are other observations to the contrary (Fielde and Parker, 1905).

Beebe (1916) thinks that there is a close correlation between habitat and habits of tropical birds and the development of their voices, which are popularly supposed to be one of the most striking attributes of tropical birds. He reports that solitary birds, living in the open country where the view is more or less uninterrupted, have a tendency to possess negligible voices. Inhabitants of dense jungle, if relatively solitary, have remarkable vocal powers, with loud staccato calls or with insistent rhythm, by means of which they communicate with their unseen fellows. Such birds may be nocturnal in habit. Birds living in pairs or in families have, for the most part, vocal organs which they use to good effect; but they lack the superlative voice development of solitary birds. Birds living in flocks have voices that are still less in evidence, though there are notable exceptions to this rule, as, for example, the parakeets.

In the matter of vocal performance, as with tactile and visual integrations, group unisons have been reported. The group singing of the western meadowlark is an example among birds. One of the most interesting cases is that of the snowy tree cricket, which has been much studied and which Fulton (1925) reported to effect changes of chirping rate in order to chirp in unison.

Shull (1907), a careful and critical observer, concluded early in his studies that real synchrony does exist in the chirping of the tree cricket; but later he somewhat modified this opinion, saying that while he still believed that the singing insects do influence one another, he believed that cases of exact synchronism were usually accidental. Lutz (1924) was skeptical both concerning the fact of synchronism and concerning its importance with tree crickets. Fulton (1928, 1928*a*) in recent studies appears to have furnished conclusive evidence that the *Oecanthus* song is both rhythmical and synchronous. After the usual listening tests, revealing almost perfect synchronism, a number of the singing insects were placed in another cage at some distance, and the front tibiae containing the auditory organs were removed. This effectively broke up the synchrony except at those times when the individual rhythms appeared to coincide for a brief period. Fulton records that "when three or more mutilated males were singing at once an utter confusion of notes resulted, so that the rhythmical quality of their songs was entirely obscured." The removal of the tibiae did not seem to affect the general health of the insects. The loss of one or more legs appears to be a matter of relatively small importance among these insects; they lived as long as did those with the ordinary quota of legs. Similar observations were made by Fulton on a katydid and on a grasshopper known as the "Nebraska conehead."

Synchronic behavior may, of course, merely mean that the group, while reacting as individuals, receive the stimulus at the same time and so react simultaneously. This is illustrated by the responses Minnich (1925) obtained when he exposed aggregations of caterpillars to various sounds. Such synchronism has no bearing upon the problem of group integration; but synchronism, such as described by Fulton, of responses by members of the group to each other may well have group significance.

Buxton (1923) records an observation made some years before upon the production of rhythmical sounds by termites. "I noticed," he says, "small numbers of winged termites emerging at one p.m. from a subterranean nest under stones in a shady place by the roadside. The ground round the mouth of the nest over a radius of three feet was covered by thousands of small soldiers and a small number of large soldiers. All of these were making a rhythmical sound which resembled the noise made by sand falling on brown paper and which was caused by tapping their heads on the dead leaves on which they were standing. The sound was produced in perfect time at a rate of about 48 beats per minute, and in the intervals between the beats there was complete silence. This remarkable performance was not disturbed by my collecting a considerable series of the performers, but an hour later when I passed the spot, the emergence of winged adults had ceased and not a soldier was to be seen above ground."

The termites were determined by Silvestri to be *Acanthotermes militaris* Hag. Buxton does not believe that the rhythmical nature of the sound production could be explained by substratal vibrations, since the termites were standing on many different dead leaves scattered over a considerable radius. Gounelle (1900) had previously described the sound produced by termites by tapping their heads on plants as being like the sound produced by a pinch of sand hitting paper, but he did not record synchrony. Emerson (1928) found that, despite the possession of the so-called "auditory organs" on the tibiae, *Nasutitermes guayanae* did not respond to a wide range of aërial sounds but did react to substratum vibrations.

Much emphasis has been placed on the rôle played by the human voice in the integration of human society; some social psychologists prefer to define man as a language animal. In this, man does not appear to be unique except in the degree to which language has been developed in his species. Craig (1908), in discussing voices of pigeons as a means of social control, finds that in animals with so highly developed instincts as birds there is still much of the social life that cannot be explained on an instinctive basis. The reaction of the individual pigeon must be adjusted to meet the activities of other birds, its parents, its mate, its young, its neighbors, and chance strangers. The adjustment is very delicate and requires that each individual must be susceptible to the influence of others, an adjustment which is largely accomplished by vocal means.

Perhaps more time has been spent on the vocal-auditory method of group integration than is justified by the conditions obtaining at the aggregation level with which this study is immediately concerned. Its interest by reason of its importance with the higher animals must be the excuse.

INTEGRATION BY LOW-FREQUENCY VIBRATIONS

Much experimentation shows that animals that give little or no indication of perceiving sound vibrations coming through the atmosphere respond definitely to vibrations of similar or lower frequency coming to them through water or through the substratum. With catfish, Bowen (1930) finds that blinded animals give definite reactions to the passing of another fish or of a model with a posterior part vibrating somewhat as does the tail of a fish. Such reactions are dependent on the presence of the sense organs of the skin. When these are anesthetized, the blinded fish respond very little, if at all, to the passing of others.

Various insects and other animals give no responses to aerial vibrations easily detected by the human ear, but readily respond to the same sounds when their receptacle is placed upon the piano producing the vibrations. Emerson (1929) has demonstrated that in the social termites mechanisms exist for producing substratal vibrations which can be detected at times by the unaided human ear, and easily when a microphone is used. He suggests that this may be one means of communication between these insects. Rabbits have long been known to signal by ground thumpings. The extent to which this kind of vibration is used in the aggregations with which we are specifically concerned awaits investigation.

Buxton (1923) records an instance of co-ordinated movement among arctiid moth larvae which illustrates some of the possibilities of this type of integration. These caterpillars live in webs on herbage in groups numbering several scores. If the web is disturbed, the larvae jerk the anterior ends of their bodies sidewise with a sharp flicking movement. All jerk together and maintain the reaction at a rate of about twice a second for as much as 20 to 30 seconds. Then they cease this movement and resume feeding. If they wander even an inch or so from the web, they do not take part in this movement. If an elongated web is chosen for the experiment (for example, a web 4×12 inches), the movement of the larvae is not simultaneous, but waves of movement may be seen to pass through the mass of larvae from the point of disturbance so that the movement is organized but not synchronous. Obviously, the stimulus is conducted along the web. More mature larvae that have left the web do not generally give this movement when disturbed, although they may so respond when another crawls over them.

POSSIBILITY OF BIOPHYSICAL INTEGRATION

It is probably too early as yet to speculate with profit concerning the possibility of other, more subtle methods of group integration, such as the observations of Gurwitsch (1926), Borodin (1930), and others suggest may result from exploration of the field of biophysics. These workers believe they have demonstrated that rapidly growing plant and animal tissues give off radiations which are able to stimulate other tissues completely separated from the so-called "senders" by being inclosed in quartz tubes so that these "detectors" show a decided increase in mitoses on the radiated half as compared with the non-radiated part of the same stem.

A favorite experiment consists in placing a moist onion root, attached to part or all of the bulb from which it grew, into a small tube made of quartz. One or more onion roots from different bulbs are introduced into open-ended glass tubes and are also kept moist. The former is to be the detector; the latter, the sender. The sender is carefully centered so that its growing root tip points directly toward and at right angles to the detector root, and is allowed to remain so for from 1 to 2½ hours. The detector is then marked with India ink on the side away from the sender and is killed in an appropriate fixing-solution and sectioned. In fixed and stained sections the number of mitoses on the exposed and non-exposed sides are compared. Most of the work reported to date shows uniformly a greater number of mitoses on the exposed side; the work of Rossman (1928) is an exception. These are believed to have been induced by the action of mitogenetic rays. Definite but conflicting wave-lengths have been announced for these rays.

This field needs further clarification before we can begin building, with a sense of security, upon the suggestions opened by this work. If the presence and importance of mitogenetic rays are finally established, we shall then have to inquire carefully whether or not we have similar subtle means of group integration in the field of biophysics which may help us resolve the problems in social and subsocial behavior that are epitomized by Maeterlinck's phrase "the spirit of the hive."

HARMFUL EFFECTS OF AGGREGATIONS

CHAPTER VI

HARMFUL EFFECTS OF CROWDING UPON GROWTH

Our knowledge concerning the methods of aggregating and the factors conditioning the formation of aggregations has grown steadily and gradually, as has our information concerning their integration. On the other hand, marked advance has been made since 1920 in the investigations of the physiological effects which such aggregations produce upon the individuals of which they are composed. The type and extent of such effects make one of the crucial tests of the importance of the phenomena. If these aggregations are merely forced reactions resulting from limited space or from blind tropistic behavior, or if they result only as an expression of a social appetite or instinct, their significance is more remote and the problem of their origin is more difficult of solution than if they can be shown to have group value even in their poorly integrated stages. Failure to observe such values for many aggregations led Deegener to conclude that their formation must be due to some inexplicable instinct.

In the investigation of this problem we must first inquire whether or not the aggregations with which we are dealing have positive or negative survival value which can be recognized. Even if positive survival value is found in a number of cases, the problem is by no means solved; but the methods to be used in its solution will be more clearly indicated than if we are forced to rely upon the postulate of a former survival value, of which the only remaining evidence is a weak social appetite persisting frequently in the face of present negative survival values.

Even with the recently devised methods of analysis, the harmful effects of such aggregations are frequently more easily apparent than are the benefits. To the eye of the naturalist depending on field observation for his data, benefits do not become obvious until the aggregation is sufficiently well integrated so that members may be warned of the approach of danger by some group attribute, such as the multiplicity of eyes in the group, or can attack or defend themselves more effectively by the multiplicity of claws or of teeth. Most of the experimental approaches to this subject have been similarly limited.

The impressive array of facts, accumulated by observation and experiment, which indicates that loosely integrated aggregations have harmful effects will be summarized in the present chapter so far as the rate of growth is concerned. The next following chapters will give other facts concerning harmful effects upon the rate of reproduction and upon longevity.

Dermestes beetles feeding on a limited amount of carrion exhaust their food sooner when more than one is present. This is also true of leaf-eating caterpillars, sap-sucking aphids, or tissue-filling parasites. It is only with well integrated groups of predators catching lively creatures as food that the feeding aggregation becomes of value. A school of young minnows is much more likely to catch a given *Daphnia* than is a single individual, and each member of the group is more likely to feed upon the *Daphnia* stirred up than if he swam alone. This type of group advantage increases with group organization, as shown by the grasshopper drives of African storks.

The same number of relatively defenseless individuals are more easily gobbled down by an enemy when aggregated than when scattered. One of the insect sleeping-clubs described by the Raus would provide a substantial breakfast for the proverbial early bird, and a hungry centipede would have easy picking in a group of aestivating land isopods. In locust control measures, men take advantage of the tendency of locusts to collect in dense overnight aggregations.

There is a general ecological assumption that the accumulation of the waste products of a given species in their habitat tends, with most animals, to limit the time of their occupancy, at the same time preparing the way for another species to come in. This is sometimes considered one of the major biological factors causing ecological succession, a process well illustrated by the sequence of fauna in a protozoan infusion.

PLANT TOXINS

It has long been thought that one of the major causes of such succession among plants is the accumulation of more or less specifically toxic root secretions. Almost a century ago De Candolle, the French botanist, suggested that the reason for the decrease in yield following the continued growth of the same crop on the same soil is due to the accumulation in the soil of harmful material given off by the growing plants. Liebig apparently adopted this view for a time but abandoned it later, thinking that the observed benefits of crop rotation were due to the different nutrient requirements of the crops rather than to the accumulation of poisons in the soil. Pickering (1917) gives conclusive evidence that root excretions may have a toxic effect upon growing plants. In his work he used mustard plants growing in earth, on the surface of which rested a tray with a porous bottom, with a large central walled opening through which the plants grew. This tray held 5 inches of earth. The presence of such a tray made practically no difference in the growth of the plants in the pot below, even when the tray itself contained a growth of mustard plants, providing their roots were kept out of contact with the soil of the lower pot and that water from around the roots of the upper plants was not allowed to reach the lower soil. When washings from the upper growth were allowed to drain into the lower pot, carrying leachings from the plants grown in the upper tray, growth of the experimental seedlings was reduced to 0.01 of that given in control pots. Pickering found such results common and widespread, and especially well shown by the effect of grasses on the growth of apple trees. In summarizing all the evidence on the subject, Russell (1927) concludes that, while a toxin can be shown to be present, the toxin concerned is not stable and is non-specific.

CESSATION OF GROWTH IN BACTERIAL CULTURES

The long-recognized failure of cultures of micro-organisms, such as bacteria and molds, to continue growing indefinitely has been attributed to three main causes: the exhaustion of foodstuffs, the accumulation of metabolic wastes or specific "autotoxins," or the limitations imposed by actual physical crowding. Henrici (1928) gives a good summary of the present state of knowledge in this field.

The possible effect of physical crowding in limiting growth must be excluded because much more dense growths can be obtained with organisms on filter paper or on agar than when they are grown free in liquid broth; and when the organisms are repeatedly filtered off so that the physical effects of crowding are periodically eliminated, the growth period is not thereby prolonged. There can be no doubt but that the exhaustion of food materials does play an important rôle in the limitation of cultures, but the question as to how important this is in comparison with the accumulation of waste products or "autotoxins" has not been decided.

Henrici says, "The idea that growth is limited by the accumulation of some toxic substance is the one that seems to be most generally accepted, though the evidence for it is far from being convincing." The evidence supporting the idea that the toxic substances are important is as follows: Eijkman (1904, 1906, 1907), working with a number of species of bacteria, grew them in gelatine until the culture was densely crowded. He found then that if he took a part of this, heated it to boiling and, after cooling, reinoculated it, it would then support growth; but that another part, heated only slightly and then allowed to resolidify, would not produce growth in a new surface inoculation. Since heating to boiling-point would add no new food material, Eijkman concluded that he was dealing with some thermolabile product of metabolism or a more specific growth-inhibiting substance.

Further experiments showed that the toxic material would not pass through a porcelain filter, that heating which killed the living organisms destroyed the toxicity of the medium, and that treatment with such volatile agents as ether and ammonium sulphide not only killed the bacteria but rendered the medium again capable of supporting growth after the volatile material was driven off. They also showed that if gelatine in which *Bacillus coli* had grown was resolidified into a plate and reinoculated with more of the same organisms and covered with a layer of fresh gelatine, there would be no growth; or, if fresh gelatine was inoculated with *B. coli* and then had one part covered with fresh gelatine and another with the so-called "coli-gelatine," growth would take place in the former only. Inoculation of paper dipped in agar and then placed over the coli-gelatine did not yield a growth unless the coli-gelatine had first been heated.

Rahn (1906), with *B. fluorescens liquefaciens* and three other species of bacteria, obtained essentially similar results except that in his experiments treatment with ether killed the bacteria without removing the unstable toxic material. The ether could be evaporated off but no growth would occur on reinoculation. Appropriate controls made by treating sterile broth with ether, which was then evaporated, showed growth on inoculation. Heated cultures gave growth when similar cultures treated with ether gave none, showing apparently that the food value of the medium was not exhausted.

Chesney (1916), working with pneumococci, found that if the organisms are removed by centrifuging a rapidly growing culture, those remaining will continue to grow at the same rate; but that if the culture be similarly treated after the period of maximum growth is past, growth is delayed and some of the cells may die off. Filtrates from 24-hour cultures inhibit growth of new inoculations of similar organisms, but lose this property if the filtrates are allowed to stand for a time in the incubator. Chesney concludes that the cells do produce an unstable, toxic, growth-inhibiting material. Some investigators have believed this substance inhibiting growth in pneumococci is fairly specific in its action; but Henrici cites later work showing that the limitation of growth in pneumococcus cultures is due to three factors: the accumulation of acid, the production of peroxide, and the exhaustion of the nutrients. The first two come under the general heading of "toxic products of metabolism," which are here shown to limit growth of this organism. Hajos (1922) also demonstrated growth-inhibition due to the accumulation of products of metabolism among the colon-typhoid type of bacteria.

Curran (1925), again using *B. coli*, found a thermolabile growth-arresting material readily adsorbed by bacterial filters. The first 50 cc. of filtrate from a 200 cc. solution which had supported bacterial growth for 3 days was found to support growth on reinoculation much better than did the last 50 cc. of the same solution. It would appear that the filter became loaded with the growth-inhibiting material and so allowed material that was stopped at the beginning to pass at the end of the filtering. Using a different sort of organism, Kuester (1908) finds that molds grown in a nutrient solution produce conditions which check the growth of further inoculations before the nutrient supply is exhausted.

In the face of this evidence, Henrici is still unconvinced of the general applicability of the theory of the production of a toxic material serving to limit growth, and holds, rather, to the idea that the exhaustion of the nutrient material is the crucial point. He suggests rightly that the Eijkman type of experiment may only show that media may contain sufficient material to support a heavy population without growth and still be able to support growth in a smaller population; that heating the medium to kill the bacteria may cause a release of nutrient material, making it available for the reinoculated organism; and finally cites the work of Graham-Smith (1920) in which he was able to revive staphylococci by adding concentrated meat extract and thus inducing new growth after the period of maximum growth had passed, and could postpone the death phase indefinitely by small daily additions of meat extract. Henrici is probably correct in concluding that different factors may limit growth in different cases and that there is no sound basis for believing in the production of *specific* autotoxins.

EVIDENCE FROM TISSUE CULTURE

In tissue-culture work Carrel and Ebeling (1923) and Mottram (1925) report a substance which inhibits growth of explants present in extracts of all adult tissues, in serum and even in extracts of embryos. The latter have usually been found to favor growth in such cultures. Heaton (1926) has found such a substance in yeast extracts and in a number of adult-animal tissues, especially the liver. He thinks that the failure of adult tissues to grow easily *in vitro*, and the stoppage of growth of connective tissue *in vivo*, as contrasted with the continued growth of epithelia, is to be attributed to this growth-inhibiting substance. Heaton finds it to be thermostabile, though destroyed by heating up to 125° C. It is soluble in water and alcohol up to 75 per cent strength, but is insoluble in 97 per cent alcohol. It seems to be destroyed by autolysis. Its action is greater on older than on younger embryonic tissues.

GROWTH INHIBITION IN ANIMAL CULTURES

The work on animal cultures most closely connected with these investigations on bacteria and on tissue culture is that dealing with the growth in a protozoan infusion. In two studies (1911 and 1914) Woodruff demonstrated that Paramecia excrete substances that are toxic to themselves when present in their environment and that probably play an appreciable rôle in determining the time of maximum number, rate of decline, and other characters. Similar conclusions were reached as a result of work with the hypotrich as regards their own excreta, but they are immune to the effects of *Paramecium*

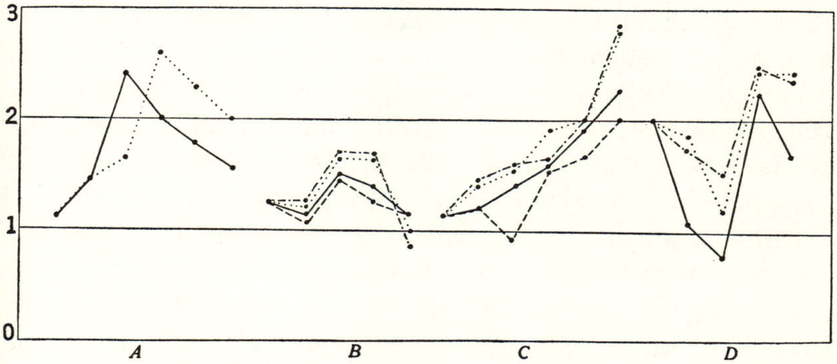

FIG. 4.—A record of the rate of division of *Paramecium aurelia* in a series of four experiments (*A*, *B*, *C*, *D*) to determine the effect of different volumes of culture medium, changed every 24 hours, on the rate of reproduction. The ordinates represent the average daily rate of division of the four lines of organisms in the respective volumes of medium, averaged for 4-day periods. Rate of division in 2 drops, - - - - - -; 5 drops, ———; 20 drops, ...; 40 drops, -·-·-·-· (From Woodruff, 1911.)

excreta. In a protozoan infusion the appearance of dominant Protozoa at the surface runs in this order: Monad, *Colpoda*, Hypotrichida, *Paramecium*, *Vorticella*, and *Amoeba*. This ecological sequence is due in part to accumulation of toxic material and in part to the supply of available food.

This problem is closely related to the consideration of the effect of the size of the effective environment, whether lake, pool, or laboratory container, upon the contained organisms, which in turn is closely related to the whole problem of crowding.

In 1854 Jabez Hogg with some right apologized to the London Microscopical Society for taking their time with observations on the subject of the pond snail *Limnaeus stagnalis*, which had already been well studied; but in the midst of his tedious record he states that a snail kept in a "small narrow cell will grow only to such a size as will enable it to move freely." This is the first recorded observation that has come to my attention of the limiting effect of volume on growth.

Semper (1874, 1881) took up the problem twenty years after Hogg's observations, using the fresh-water isopod *Asellus* and the pond snail *Lymnaea stagnalis*. With the former he found that when animals living in a balanced aquarium were sealed into glass dishes, they might be left for nearly 2 years, with an adequate food supply and, he believed, an adequate oxygen supply for the three or four generations that would be produced; but under these conditions the last generation was abnormally small. With the snails Semper divided the same mass of eggs into different lots, which were placed in variously sized containers ranging from 100 to 5,000 cc. Food was kept at an optimum, but the snails placed in the containers of smaller volume grew more slowly than did their fellows placed in the larger vessels. Similar results were obtained regardless of whether the snails were isolated into a given volume or were put in groups, so long as the volume per snail was the same in both cases. Semper found that optimum growth lay between 400 and 500 cc. per snail and that increases beyond this point gave no further increase in growth. The effect of increase in volume was much more marked in the smaller volumes. Later workers are agreed that relatively large volumes of water per snail are necessary for optimum growth.

Semper recognized the complex nature of the problem and attempted, by chemical analyses made by a trained chemist, to find a chemical cause. Failing in this attempt, he advanced the hypothesis that some substance unknown to him was present in the water, probably in a very minute quantity, "which, by its relations to the water which holds it in solution, and by its osmotic affinity to the skin of the animal, can be absorbed only in a determined and extremely small quantity. . . . . Since, according to this hypothesis, the amount of the substance absorbable in a given time depends on the volume of the water . . . . the attainment of full size within a definite period would only be possible if the volume of water were so great that the *Lymnaea* could at all times absorb this unknown stimulant from the water." This hypothesis, in some form or other, has been proposed, apparently independently, by a number of workers since Semper's time.

Semper seems to have been certain of the evolutionary significance of the limitation of growth by volume. He found it impossible to obtain full-sized individuals from snails stunted during the first year of their lives; and if the causes checking growth were repeated regularly through the succeeding generations, he felt that a dwarfed race must arise. Whitefield (1882) came to the same conclusion, using *Lymnaea megasoma* from Vermont. Whitefield continued the crowding for four successive generations, during which time the snails became successively smaller and more slender, so that an experienced conchologist did not recognize their relation to the shells of the parent stock.

Yung (1878, 1885) concluded from his experience in raising tadpoles in containers of various sizes and shapes that dwarfing is due to a lack of aëration. De Varigny (1894) took up the problem with *Lymnaea* again and in general obtained the same sort of results reported by both Semper and Yung. A snail kept in a liter of water with a surface of 257 sq. cm. for 5 months was nearly twice the length of one kept in the same volume of water but with a surface area of 3.14 sq. cm. In order to facilitate the analysis, De Varigny suspended a glass tube 2–3 cm. in diameter in containers of various sizes. The glass tubes were closed over the bottom with muslin, and each contained a single snail. Each day these tubes were lifted from the water and replaced two or three times in order to secure complete mixing of water. Even so, the contained snails grew approximately the same regardless of the volume of water with which they were in contact through the muslin screen. In one instance the growth was the same in such a muslin-bottomed tube as compared with that of a snail in a corked tube which prevented all exchange between the inner and the surrounding water. From these experiences he concluded that Semper's explanation would not hold and that the size to which snails grow depends in some way on the actual volume to which they are exposed and on the surface area of such water. His explanation was that in the small tubes the snails needed to move about less to obtain their food and that, with this decrease in exercising, there came a decreased rate of growth. According to De Varigny, dwarfing from crowding is not so much due to the actual numbers in the vessel as to the "psychological" influence of numbers, which inhibit exercise, just as a man is less likely to walk a considerable distance on a crowded street than on a deserted one. He also believed dwarfing to be affected by the accumulation of faeces.

Willem (1896) bubbled air through his snail cultures and found growth of the contained snails greatly increased. He concluded that aëration is important because, even in lung-breathing pond snails, he believed cutaneous respiration to be more important than lung-breathing and alone sufficient for the animal.

Davenport (1899) reviewed much of the evidence on the relation between crowding and rate of growth, and concluded with Hogg that in respect to the size attained, as in other qualities, the snail has the power of adapting itself to the necessities of its existence.

Vernon (1895, 1899, 1903), working with echinoderm larvae, concluded that dwarfing is due to a concentration of the excretory products in the media. He found that if eggs of echinoderms were allowed to develop in water which had previously contained other eggs for a considerable period of time, the larvae of the second batch were diminished in size as compared with the control. The growth of the larvae appeared to be reduced by their own excretory products, or especially by those of adult echinoderms, the more so if these belonged to the same species. On the other hand, he found that the excretory products of two species not closely related were favorable to growth.

Warren (1900), working with the common entomostracan, *Daphnia*, found that continued breeding in small aquaria with the medium unchanged caused dwarfing. This result he attributed to the action of the excretory products, which he found to be somewhat specific, since ostracods and copepods flourished in cultures of *Daphnia* in which the latter were dying out. Such results are similar to those reported by Woodruff and others for protozoan infusions.

Legendre (1907) returned to the problem of the effect of crowding on the growth of snails, using *Lymnaea stagnalis* and *Planorbis corneus*, raised in one series of experiments in stagnant water and, in the other, in water changed periodically. As in the case of previous workers, he found the smaller shells in the stagnant water, and attributed the cause to the accumulation of excretions. In further work reported the following year, using another species of *Lymnaea*, Legendre changed the water every 2 hours in order to avoid the accumulation of excreta, and varied the factors of volume of water, surface area, and number of individuals. After 51 days he obtained the same shell size in all such experiments. He recognized that a number of factors might bring about retardation in crowded animals, but laid particular emphasis upon the retarding effect of the excretions.

Colton (1908) continued work on the effect of crowding on growth in *Lymnaea*. Food was recognized as an important element, but just how important Colton's work does not reveal. He did find that snails need a certain amount of sediment to aid in grinding their food, and that certain salts, for example calcium sulphate, aid growth. Colton found that washed and filtered snail faeces placed in aquaria hastened the growth of the snail, probably due to the increase in algae caused thereby. His aëration experiments support the conclusions of Willem that these pulmonate snails have a large proportion of cuticular respiration. Concentrated excretory products caused dwarfing; accordingly decreases in volume of water per individual present, whether in isolations or in crowded cultures, caused a decrease in growth rate. Popovici-Baznosanu (1921) minimized the effect of excrement, thinking the amount of food more important.

Crabb (1929) has recently reinvestigated this entire problem with the pond snail *Lymnaea stagnalis appressa*, taking care that his snails were free from trematode parasites, and supplying them with food known to be favorable for growth in laboratory conditions. He used eggs from the same egg mass for experiments run simultaneously; since this snail reproduces by self-fertilization, individuals obtained from the same egg mass would be expected to have similar genetic constitution. He concludes that food insufficiency and foul media are the most common growth-inhibiting factors in snails reared in otherwise favorable media. Extreme crowding markedly retards growth, but the individuals rapidly reach normal size after transfer to standard conditions, unless they are too old. The volume of medium has little effect on the growth of isolated snails providing foulness is not permitted. Aëration promoted growth through reducing foulness rather than by increasing the respiration of the snails. *Daphnia* introduced into the culture are beneficial to snail growth, since they retard fouling of the medium. He found no evidence that environmentally induced dwarfing is transmitted, though on this the experiments were not continued through enough generations to be conclusive.

Crabb, in his work, continued the general methods of study of this problem which have been used since the time of Hogg, adding refinements which make his conclusions the more trustworthy. Unfortunately, he did not take advantage of the method originated by De Varigny (1894) and used extensively by Goetsch (1924), which allows a separation of the factor of available space from that of available volume. In this procedure Goetsch placed animals in the experimental aquaria in separate tubes thrust through corks to keep them afloat and covered at the lower end with gauze, which allowed diffusion connection with the entire aquarium while limiting the amount of available space.

Goetsch was led to this method by the experience of Bilski (1921), who found that the relatively active tadpoles of *Bufo* and of *Rana esculenta* grew less rapidly when subjected to frequent changes of water than they did when metabolic wastes were allowed to accumulate. Bilski also found that an increase in numbers slowed down the rate of growth more than would be expected by the change in volume relations involved, when the rate of growth was compared with that given by an equal number of animals placed in different aquaria.

Goetsch experimented upon sessile *Hydra*, upon the relatively slow-moving flatworms, and upon amphibian larvae which are capable of rapid locomotion. As might be expected, he finds different factors important for different animals. Thus, with *Hydra*, volume per animal is the controlling factor because of the restriction of food which it conditions. There is no stimulation or depression caused by the crowding of *Hydra* into a narrow space; and, within reasonable limits, concentration of excretory products are not effective. With *Planaria* food is again the most important factor, but growth is inhibited by the concentration of excretion products or of stale food. With the active amphibian larvae, if food is controlled, the major limiting factor is furnished by the more frequent collisions in a dense population or in a restricted area, and the concentration of excretory products plays a wholly secondary rôle.

Church (1927) extended these experiments to include the rate of growth of the tropical fish *Platypoecilus maculatus rubra* in connection with other experiments upon the effect of crowding upon the rate of growth of fishes. Eight liters of water were used in glass aquaria, each of which contained 2, 8, or 16 fish. In each series of experiments, one set of aquaria contained small fish 8–10 mm. long, another set held fish 12–14 mm. long, and the third set was supplied with fish 20–23 mm. Adult *Platypoecilus* range from 30 to 35 mm. The amount of oxygen and the pH of the different aquaria did not differ significantly. The water was left unchanged during the entire experiment, which ran in some cases as long as 70 days, except that there were slight additions to replace the small amount lost by evaporation. The fish were fed the same number of *Daphnia* per fish per day.

Under these conditions the large fish always grew less rapidly the more fish there were present in a given container. With the small and medium fish there was some indication of more rapid growth early in the experimental periods among the fish grouped 8 to the aquaria; but as the experiment progressed, the rate of growth was always greatest when the fewest fish were present. Shaw (1929) has repeated these experiments, with similar results. The experience of these two workers demonstrates that when there is sufficient concentration of waste products the rate of growth is retarded.

In following out the Goetsch type of experiment, Church placed transparent celluloid containers in the center of each aquarium.

These were 4.5 cm. in diameter and were covered with coarse scrim at the bottom. They were suspended by wires so that each extended 1.5 cm. below the surface of the water, thus giving to the contained fish a volume of 24 cc. in which to move about, as contrasted with the 4,000, 1,000, or 500 cc. volume per fish to which the 2, 8, or 16 fish were exposed in the surrounding aquaria. A single medium-sized fish was transferred to each of these tubes, regardless of whether those in the surrounding aquaria were large-, small-, or medium-sized.

Under these conditions the fish within the small tubes grew less than did those in the aquaria. At the end of the first 10 days the average length of the 34 fish in the tubes showed 1.35 per cent increase, while the medium-sized fish in the surrounding aquaria grew 6.51 per cent. At the end of 20 days the difference was still more striking. The 25 fish in the tubes had grown in this period on the average 2.78 per cent, while those of the same original size in the larger volume of water had grown 12.83 per cent. So far as known, the size of the container was the only variable. The meshes of the scrim cloth were open throughout the experiment; but to guard against the possibility of lack of adequate diffusion, the tubes were raised once daily to insure a complete change of water. Such results are similar to those Goetsch secured for the relatively swiftly moving tadpoles, and are probably due to the effect of overstimulation caused by frequent contact with the walls of the small tube.

As stated above, Bilski (1921) points out that when limitation of growth rate is caused primarily by stimulation from repeated contacts, and when the number of individuals present is proportional to the different sizes of the vessels, the rate of growth is not the same. If we take two vessels of different sizes, a and b, and populate them with a and b number of animals respectively, so that each animal has the same amount of space available, in the simplest case the stimulation will come from the contact, or near approach, of two animals. The relation of the size of the two containers will be $a:b$, which in a simple case might be 2:3. The stimulation possibilities from group interference would be $a(a-1):b(b-1)$. Substituting the values suggested above, we get a stimulation possibility of 2:6. Under such conditions one would expect to find growth retardation with increase in numbers to be much greater than if volume relations alone were the responsible factor.

Inspection of his experimental results in comparison with a simple formula built on the assumption that the growth would be inversely proportional to the group stimulation,

$$y=k\left(\frac{1}{x(x-1)}\right),$$

in which y represents size, x stands for the number of animals in a given space, and k is a constant, shows that the influence of the stimulation is not on this order but is approximated by taking an exponential value of x, namely $x^{3/2}$. The equation then becomes

$$y=k\left(\frac{x^{3/2}}{x(x-1)}\right)$$

or

$$y=k\left(\frac{\sqrt{x}}{x-1}\right).$$

Values calculated from this formula fit fairly well with Bilski's observations on the effect upon the growth of differing numbers of tadpoles in jars of equal size; the observations of Semper on the growth of snails in relatively small vessels; and, according to Bilski, with the observations of Hoffbauer on growth in carp. Another formula derived by a continuation of the same reasoning better fits Semper's results with snails in larger volumes.

Bilski recognizes the general significance of his results and believes that such diverse phenomena as the reported dependence of size of mammals upon available land, and other similar relationships, including even a correlation between the size of children and available space, may depend upon an application of this principle. Farr (1843, 1875) worked out an equation essentially similar to that of Bilski to describe the relation between death-rate and the density of human populations. Brownlee (1915, 1920) finds that Farr's law fits a wide range of biological and biochemical relationships, including even the relation worked out by Kennealy (1906) between the racing record for a particular distance and the length of the race. Pearl (1925) finds that essentially the same equation describes the effect of crowding upon the rate of reproduction in *Drosophila*.

The problem is obviously complicated by many factors, but it is interesting and probably significant that the relationship can be expressed mathematically in a similar way for such a wide range of phenomena. It is almost an anticlimax to have to record that physical disturbance due to numbers is not the only factor controlling growth in rapidly moving animals, such as fish, under crowded conditions. The careful work of Church and of Shaw, already summarized, demonstrates that the accumulation of waste products is also effective with fish, just as a long line of evidence culminating in that given by Goetsch proves that it is effective in the slower-moving planarian worms.

More recently Willer and Schnigenberg (1927) and Kawajiri (1928) have independently tested the effect of crowding on the rate of growth of young trout in running water. Both report essentially similar results; the work of the former will be reviewed here, since it is the more comprehensive. These workers used young of the brook trout during their prehatching, yolk-sac, and early feeding stages. In their experiment they tested a wide range of conditions. They used the same number of eggs or of young in different volumes and with different surface areas, and in other tests used different numbers in the same volumes. All experiments were carried on with water running at a rate of from 3.3 to 65 cc. per second.

Their results show that moderate crowding after hatching has no adverse effect upon fish whose prehatching development has been in equally crowded conditions. In fact, under these conditions, one set of experiments show an apparently beneficial effect. On the other hand, crowding the eggs produces definite retardation in length and perhaps also in weight at hatching time. Such retardation is correlated with the volume of water rather than with the area of the screen on which the eggs rest.

Exposure of uncrowded eggs to water that has flowed over a mass of developing eggs is found to produce about the same degree of retardation as is furnished by crowding. Under these conditions the dwarfing effect must be a result of toxic materials accumulated in the water. The general importance of these results is enhanced because of the fact that they have been obtained from animals grown in running rather than in stagnant water. There was an indication of a condition of optimum crowding[1] in the early experiments which was not sustained by later work, although specific experiments designed to test this point were not attempted.

Peebles (1929) has taken up the problem of effect of numbers present upon the rate of cleavage of echinoderm eggs, and upon the rate of growth of arms of plutei, in the light of developments in tissue-culture work. She finds, as did Vernon (1895) and Springer (1922), that embryo-water contains substances which check growth, but adds the observation that some of the inhibiting effect is counteracted when living larvae are present. She produces evidence that the growth-inhibiting substances are associated with the lipoids and that, after their removal, growth-promoting substances can be demonstrated to be present. These latter will be discussed in chapter ix.

The relation between the size of the effective environment and that attained by the animals living therein has more than laboratory interest. The belief is widespread that fish grow larger in large lakes than in small ones. Pearse and Achtenberg (1920) report such a correlation between size of lake and size of contained yellow perch. This correlation is not uniform, for numerous exceptions could be cited; for example, Jewell and Brown (1929) find no such relationship holding between size of fish and the size of the small Michigan lakes in which the fish live.

Hesse (1924) states that the same relation holds for mammals with regard to the size of available range: those living on small islands attain a smaller adult size than related forms on larger bodies of land. In many cases the reduced amount of available food in the smaller habitats has been recognized as being sufficient to explain the observed phenomena. Semper (1879) critically discussed this

[1] Kawajiri reports that the survival-rate increases as the number of fry in a box increases.

general situation long ago and left the impression that the suggested relationship was either not proved or only indirectly related to the suggested space factor. The idea that there is a direct connection between available space, and size attained in land animals, still has, however, considerable vitality, as is shown by Bilski's suggestion (1921), following his careful statistical analysis of the relations between available space and growth in tadpoles, that the smaller size of children reared in slums, as compared with that of the children of more fortunate parents, is to be accounted for by the smaller space available per child for the former and the resulting greater degree of stimulation by repeated contacts, such as have been shown to result in decreased growth in tadpoles, fish, and other rapidly moving animals.

There can be no doubt that crowding decreases the rate of growth in many instances, and any interpretation of the facts to be presented later concerning beneficial effects of crowding up to an optimum population must take this fact into consideration. When one attempts to summarize the evidence concerning the factors causing the retarded growth in crowded conditions, he finds a decided lack of unanimity among the different investigators, indicating that in all probability there are many factors which may produce the same result.

It is instructive to review the retarding factors suggested to date. They are of two kinds: the vague and the definite. In the former category one must put the suggestion of Hogg, working with snails in 1854, that they adapt themselves to the necessities of their existence, which Davenport, 45 years later, said still summarized the state of knowledge on the subject at that time. There is also Semper's postulated *X*-substance necessary for growth in snails and water isopods (1874, 1881); the autotoxins of the bacteriologists; and the growth-inhibiting substances of the tissue culturists (Heaton, 1926) and of Peebles (1929) for echinoderm larvae; as well as a "space factor" seriously discussed by many observers (cf. Willer and Schnigenberg, 1927). As commonly used, this space factor is about equivalent to Hogg's conception.

Regarding this group of suggested retarding factors, the best we can say at present is that they are unproved. We shall find the suggestion of an *X*-substance made in many different connections before we have finished this discussion. It is useful as a hypothesis but is not to be confused with concrete fact. However, the recent developments concerning the importance of small traces of vitamins, and the work upon "bios" and upon tissue-culture inhibitions, will keep us from dismissing this hypothesis too hastily.

Of the definite factors suggested, we have lack of sufficient aëration, in addition to undernutrition, reported as operating in crowded tadpoles (Yung) and among snails (Willem, Colton, Crabb). There can be little doubt but that insufficient aëration is an effective factor under many conditions. The suggested harmful effects of lack of exercise in snails (De Varigny) now appear groundless. The accumulation of excretory products reported as an effective agent by many workers appears to have undoubted and marked influence, whether in echinoderm larvae (Vernon, Peebles), in *Daphnia* (Warren), in snails (Legendre, Colton, Crabb), in planarians (Goetsch), or in fish (Church, Willer and Schnigenberg, Shaw). Evidence in favor of this conclusion will accumulate as we proceed.

The reduction of available food correlated with crowding, whether caused by increase in numbers or decrease in volume, is another undoubted factor in the situation, as shown for snails by Colton and Popovici-Baznasanu and for *Hydra* by Goetsch. With some animals, such as *Hydra*, it may be that this is the only factor operating. With rapidly moving animals, the effect of frequent contacts resulting in overstimulation of some sort also contributes to the retardation of growth in crowded animals, as in tadpoles (Bilski, Goetsch) and in fish (Church).

CHAPTER VII

RETARDING INFLUENCE OF CROWDING ON THE RATE OF REPRODUCTION

In the preceding chapter we have assembled evidence to demonstrate that among many animals overcrowding tends to produce dwarfed individuals, and have discussed the factors that have been suggested as operating to produce this effect. As might be expected, there is frequently a retardation of the rate of reproduction as well as of the growth-rate of the individual. In many respects the two phenomena overlap. The evidence for the slowing-down of reproductive rate under crowded conditions will be examined in part in the present chapter. At another place consideration will be given to the data brought forth by Robertson and others which indicate that under certain conditions the rate of reproduction is increased in early stages of protozoan or other cultures when more than one animal is present in a limited amount of medium.

REDUCED DIVISION RATE IN INFUSORIA

Balbiani (1860) reported from a single experiment on *Paramecium* that this infusorian must be in not less than 2–3 cc. of medium for the greatest productivity to be realized. Kulagin (1899) suggested that this is due to the accumulation within the medium of excretions analogous to toxins, which gradually accumulate until the nucleus is affected.

Woodruff took up this problem in 1911 in an effort to find the effect of excretion products of *Paramecium* on its rate of reproduction. Since the experiments of Woodruff usually form the starting-point for present-day citations on this subject, they deserve to be given in some detail.

The reproduction of *P. aurelia* was followed for from 16 to 20 days in four volumes of hay infusion: 2, 5, 20, and 40 drops, which were changed at 24- and 48-hour intervals in different series of experiments. The results are given graphically in Figure 4. For the experiments in which the medium was changed every 24 hours the Paramecia in 5, 20, and 40 drops are shown to have divided 2.4, 6.4, and 7.4 per cent more rapidly, respectively, than did those in 2 drops. When the medium was changed every 48 hours, the percentages for the same volumes were 5.3, 9.3, and 9.25. The results are given throughout as averages for 4-day periods.

From these experiments Woodruff concludes, "The rate of reproduction of specimens from pure lines of Paramecia when bred under identical conditions of temperature and culture medium is influenced by the volume of the culture medium (within the limits tested in the experiments) and the greater the volume, the more rapid is the rate of division." The slight discrepancy with the 40-drop cultures changed every 48 hours is unexplained, but the suggestion is offered that the bacteria always found in such cultures, and which are used as food by the Paramecia, develop so rapidly under these conditions that they may exhaust their own food or produce sufficient excretion products to be injurious to the associated Paramecia. Otherwise, Woodruff believes that by his culture methods, which included cross-inoculations between the different cultures, he has eliminated the bacteria as agents causing the observed difference in rate of *Paramecium* division.

The conclusion that the recorded effects are due to the accumulation of *Paramecium* waste products rests on three lines of evidence. In the first place, as we have just seen, the rate of division is higher, for the periods and amounts tested, the larger the amount of available medium. Second, the rate averaged 8 per cent greater in the 2-drop cultures changed daily than in similar cultures changed every 48 hours. The other cultures similarly showed a 6 per cent increase if changed daily. Finally, culture media in which Paramecia had flourished for 10 days before removal were shown to have a depressing effect upon the reproductive rate of Paramecia replaced in it, as compared with the effect of an infusion which had contained no Paramecia but which otherwise was as nearly comparable as the two could be made.

Woodruff (1913), as a result of further experience, concluded that the substances which Paramecia excrete into their medium are essentially species-specific, at least to the extent that they do not uniformly influence the rate of reproduction of the hypotrich *Stylonychia*. This hypotrich produces conditions within its own culture medium which are definitely depressing for hypotrich reproduction and without necessarily affecting the rate of reproduction of *Paramecium*.

The question of species specificity has not attracted the work it deserves; but, stimulated by the researches of Robertson, to be reported in a later chapter, several workers have retested the effect of crowding upon the rate of reproduction of Paramecia and of other protozoans. Without exception, all the workers reporting so far, Robertson included, have confirmed the conclusions reached by Woodruff for cultures running the length of time for which his averages were taken. A detailed discussion of this later work is postponed for the present.

From general considerations it appears highly probable that the relationships outlined above and in the preceding chapter, if properly adjusted, could so affect an animal (for example, *Paramecium*) that, while it might be able to continue to live, its powers of reproduction would be lost. Crampton (1912) tried this experiment. He found that a single *Paramecium* confined in a capillary tube could be kept from fission for as long as 32 days, while controls relatively unrestricted as to space were dividing at a rate that would produce 4,300,000,000 animals in the same time. He recognized three factors as working to produce this effect: lack of sufficient nutrition, accumulation of waste products, and stimulation from the narrow limits. That lack of sufficient or proper food is not the sole cause is shown by his experience that the confined animals could be released to swim about in a culture of *Bacterium termo* daily for as long as 12 hours out of the 24, without division, if the remainder of the day were spent in the confinement of the tube; and that they could be held so without division for a week, while controls were dividing on an average of once a day. Such Paramecia remained plump and well nourished in appearance; those left in the tubes for long periods without changing became transparent and emaciated. *Stylonychia* gave similar results. It is significant that Crampton centrifuged his animals, which must have brought them into violent contact with the walls of the capillary tubes.

Crampton's work was in many ways an extension of Conklin's earlier observations concerning the size attained by the gasteropod *Crepidula plana* with relation to the amount of available space. The dwarfing of these snails when crowded, Conklin thought, should be interpreted as due to space inhibition of cell division.

These facts were reported by Conklin in 1898. *Crepidula plana* lives within the shells harboring hermit crabs. If the shells are small, the contained *Crepidula* are few in number and are dwarfed; if large, the *Crepidula* may be present in numbers and be large. Since there may be but 1 small individual in the small shells, while there may be 4 "giants" in a large one, Conklin believed that the difference in size is not due to differences in available food; nor is it due to the presence of accumulations of excreta, since both shells are equally open to the surrounding ocean. Neither is the result due to the lack of room to move about in, since both large and small *Crepidula* are relatively firmly attached to their substratum. Rather, there is a space retardation of cell division, since the cell sizes of the one are no larger than the other. If the small *Crepidula* are transferred to a larger space, they will increase in size. The stimulus acting to retard cell division in these dwarfed *Crepidula* is more obscure than in the case of the rapidly darting *Paramecium* confined in a capillary tube.

Kalmus (1929) has added two other factors to Crampton's three, in reporting his own studies on the effect of inclosing *Paramecium caudatum*, *Stylonychia*, and *Spirostomum* in capillary tubes. He finds that the age of the culture and the solubility of glass in the culture medium have decided effects.

Capillary tubes made of two kinds of glass were used: "Schotschem, nr. 20" and the Bohemian glass made by Cavalier. The tubes measured 100–200 μ and the length of the contained column of liquid was from 8–30 μ. Some of the observations are summarized in Table I. These results and others show that the type of glass in which small amounts of culture medium are held may affect the condition of the contained animals.

Kalmus concludes that his observations show that the retardation of division in small volumes is approximately proportional to the ratio of total surface to the volume of medium. Animals from young cultures are more sensitive to limited volumes than are those from old cultures. A fully bacterized medium retards the poisonous effect of small volumes by furnishing more food and by tending to keep the Paramecia out of the most toxic region next the glass, and by binding the toxins present, thereby rendering them relatively harmless. These toxins may be of two sorts: there are the poisons which may leach out of the glass into the limited amount of medium in

TABLE I

	Schotschem Glass	Cavalier Glass
24 hours:		
Division of 25 pairs	11 animals	1 animal
Conjugating of 25 pairs	2 pairs	4 pairs
Dead of 25 pairs	4 animals	9 animals
48 hours:		
Division of 25 pairs	17 animals	9 animals
Conjugating of 25 pairs	1 pair	3 pairs
Dead of 25 pairs	5 animals	15 animals
72 hours:		
Division of 25 pairs	19 animals	16 animals
Conjugating of 25 pairs	....	....
Dead of 25 pairs	14 animals	29 animals

sufficient quantity to have decided effects, and there are the metabolic products of the Protozoa themselves. The question of the fixing of toxins takes us somewhat afield from our present considerations and will be left to be taken up later in detail.

Unlike Crampton, Kalmus found that divisions of protozoans are possible even when they are contained in small capillary tubes. Apparently he did not subject his animals to the action of the centrifuge, which may partly account for the difference in results. However, when there are so many different factors operating, such as composition of glass, age of culture, and bacterial flora, one cannot be sure of the precise factor or factors causing the differences in observed results. From his observations, Kalmus challenges the entire conception that a small amount of available space, acting directly, may limit the rate of cell division and thereby the size of metazoans. In this he overlooks the important results obtained by

Goetsch[1] on the effect of stimulation by contact with the walls of a small container in limiting the growth of active animals, even though the contained liquid be effectively connected with that of a much larger vessel.

Undoubtedly there may be a limiting toxic effect of materials leached out of glass, particularly from soft glass. The dangers resulting from the use of such glassware have been known for years. In addition there may be a physical as well as a chemical effect from the glass walls of an inclosing vessel. Such effects are shown in the recent work of Drzewina and Bohn (1927). They base their experiments on the report of Norrish (1924, *vide* Taylor), who found that bromine combines with ethylene about twice as fast in contact with a surface of stearic acid as with one of glass, and that, on the other hand, the reaction within a paraffin-lined dish is about one-thirtieth of that given when the exposed surface is one of stearic acid. Using the marine flatworm *Convoluta*, Drzewina and Bohn found that these small worms survive only about half an hour when placed in sea water in a glass dish coated with stearic acid. In this instance the worms are not affected by dissolved chemicals, since stearic acid is insoluble in water. There is no change in the pH of the water, and water which has stood in such dishes is non-toxic when removed. A glass dish coated with paraffin becomes less toxic than is a plain glass dish. If the *Convoluta* in a glass dish on a white background are exposed to sunlight, they do not maintain their normal activity so long as when they are in a paraffined dish. There is also greater protection in the latter against the toxic action of metallic silver. Paramecia behave similarly. They die more rapidly in a dish covered with stearic acid, whether in light or in shade. Paraffin protects them against the action of metallic silver and of neutral red, even though they take up as much neutral red in a paraffined dish as in a plain glass dish. Drzewina and Bohn conclude that stearic acid catalyzes reactions of living animals but that paraffin inhibits them. They suggest that the action is similar to the action of paraffined glass in preventing the coagulation of blood, and advance the theory that

[1] It is not yet definitely proven by chemical tests that the tubes with one end covered by cloth do allow free diffusion of excretory products.

both effects may be attributed to the electrical charge carried by the paraffin.

Warren (1900), in his work on the effect of crowding on *Daphnia*, had previously found that media in which excretory products are allowed to accumulate cause a decrease in the number of generations and the number of offspring in a brood, and that reproduction ceases long before the animals die. Such water is injurious, though not usually fatal to fresh *Daphnia;* and the reproductive power of the newly introduced *Daphnia* is soon reduced. The injurious nature of the water seems to pass off after a sufficiently long period.

Our experience in growing *Daphnia* in quantity for fish food in a considerable volume of water, of perhaps 10–100 liters, accords with the experimental results of Warren. Events run as follows: A month or 6 weeks after having stocked such an aquarium with a few *Daphnia*, conditions being favorable, several hundreds of animals may be living in good condition and reproducing. Then suddenly a change begins. The greater number die, young and old alike. Perhaps from 1 to 3 per liter survive, and these will live for months without producing eggs. After a very considerable time eggs are formed and *Daphnia* may become fairly plentiful again, but the second swarm is never as numerous as the first. During the time when the *Daphnia* have ceased to reproduce and have, for the most part, died off, the water may be teeming with other entomostracans, ostracods or copepods. This indicates a certain specificity in the effect of the *Daphnia* metabolic wastes. The duration of the period of depression of reproduction is greatly shortened by keeping the food value of the medium at a high level.

EFFECT OF CROWDING ON RATE OF EGG-LAYING OF HENS

The effect of density of population upon rate of reproduction in a different medium and with animals far removed in habits and in the evolutionary scale from Protozoa or Entomostraca was reported by Pearl and Surface (1909) from the experiments of Professor Gowell of the Maine Agriculture Station. These men report the result of investigations concerning egg production extending over several years. The chickens studied were kept in pens containing 50, 100, and 150 hens each. The pens with the smaller flocks provided 4.8 sq. ft. of floor space per hen. In the largest flock this was reduced to 3.2 sq. ft. per individual. The number of pens is shown in Table II.

In all there were 700 pullets placed in the 50-bird pens, 500 in the 100-bird pens, and 750 in the 150-bird pens. Conditions varied somewhat from year to year, so that Pearl and Surface warn that "wherever comparisons between years are instituted, great caution must be exercised in drawing conclusions."

Due care was taken to select the members of the different pens with hereditary constitutions equally disposed to egg-laying, so far as this factor could be regulated. All were from the same breed, and

TABLE II

Year	50-Bird Pens	100-Bird Pens	150-Bird Pens
1904–5	6	1	1
1905–6	4	2	2
1906–7	4	2	2

individuals were distributed among the different pens so that the percentage from ancestors of different productivity were the same throughout. The experiment with which we are concerned ran three seasons. Results are graphically given in Figure 5, which shows the mean annual egg production per hen. An inspection of this figure shows that each year there is a trend toward reduced egg production in the pens with the greatest number of birds. During two of the three years the decrease in rate of laying is practically the same between the 50- and the 100-chicken pen as it is between the 100 and 150. The results obtained the first season, 1904–5, are different and affect the mean differences, as is seen from the fact that the pens with 50 birds produced on the average 129.69 eggs per season; those with 100 produced 123.21, while those with 150 gave 111.68. The mean difference between the pens with 50 birds and those with 150 amounted to 18.01 eggs per year.

The difference between the 100-bird pens and the 150-bird pens, where there were two factors acting—increase of numbers and decrease of floor space—is approximately twice as great during these three seasons as is the difference between the 50- and the 100-bird pens, where numbers only were varied. The experiments on the whole indicate, as Pearl and Surface conclude, that the mean annual egg production is much influenced by the differences in environmental factors present in the experiment.

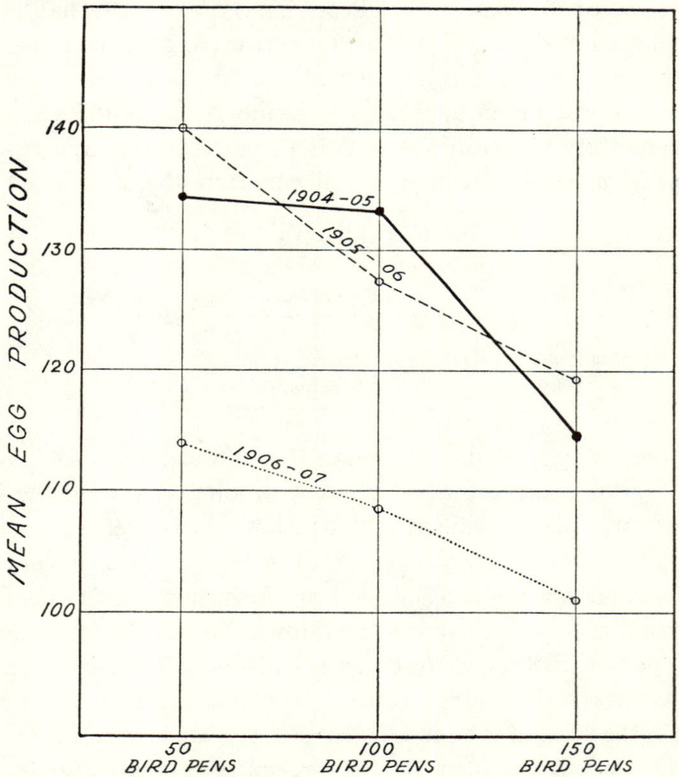

FIG. 5.—A graphic summary of the relation between size of flock and mean annual egg production in the domestic fowl. (From Pearl and Surface, 1909.)

In an attempt to get at the underlying factors they suggest that there is another element involved besides the physical density of the population, which they are inclined to place on the psychological level, and which works even when the amount of floor space per individual is equal. The conditioning of the surrounding medium is of a different type from that of crowded aquatic animals, where the excreta and glandular secretions are dissolved in the surrounding liquid medium and come of necessity into intimate contact with each of the contained animals. Presumably, with chickens we are free from inequalities in food, although in the larger pens some may have fared better and others more poorly, especially in view of the flock organization which Schjelderup-Ebbe (1922) has described. Availability of equal floor space does not insure equality of use, and crowding was probably greater the greater the numbers present. Even so, the significance of these observations is not lessened, and the conclusion of Pearl and Surface may be justified that we are here dealing with physiological effects on the reproductive system produced by physiological effects on the nervous system of the order usually spoken of as "psychological."

It becomes important to follow the differences in egg production during the course of the year with these pullets housed with different degrees of crowding. The results of such analyses are published by Pearl and Surface (1911) and are summarized in Figure 6. The months from November to July are based on the averages of records for 4 years; from July through September on the records of 3 years. October is not included because records for only 2 years were available.

The data summarized in these graphs show that there is no harmful effect from keeping pullets in large and crowded flocks during early winter egg production near the beginning of the laying period. In fact there appears to be a significant advantage accruing from the crowding in the first really cold winter month, December. On the other hand, the 50-bird pens show a distinctly better production than do the other lots in late winter and early spring, about the time of heaviest egg production. This difference does not obtain between the birds kept in lots of 100 and those in lots of 150. The harmful effects of summer crowding on egg production shows plainly when the most crowded pullets are compared with less crowded lots. Overcrowding affects summer egg production in a distinctly adverse manner.

There would thus seem to be three distinct aspects of the effects of crowding on egg production in the domestic fowl. First, in early winter at a time of relatively low egg production, when the nights are becoming increasingly cold, the large crowded flocks apparently

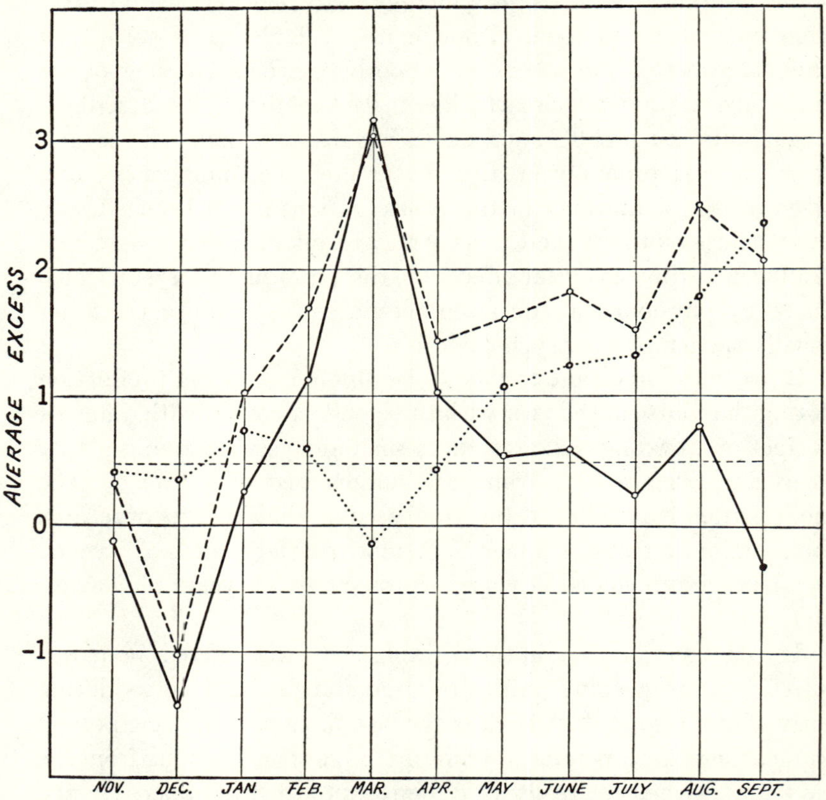

FIG. 6.—Diagram showing the average excess in mean egg production of different sized flocks of barred Plymouth Rock pullets in the first 11 months of their first year of laying.

———, 50-bird pens compared with 100-bird pens.
- - - - - -, 50-bird pens compared with 150-bird pens.
.........., 100-bird pens compared with 150-bird pens.

The broken lines running parallel with the zero line approximate the mean probable error. Points below the zero line indicate that the larger number per pen gave a higher average egg production. Points above the zero line show that the pen with the smaller numbers gave the higher average. (From Pearl and Surface, 1911.)

conserve animal heat, so that greater egg-laying occurs in the more crowded pens. Second, as the period of maximum egg production is reached, crowding has the opposite effect, for reasons not yet clear. The nights are still cold, frequently colder than in December, when the opposite results were obtained. Probably the differential effect of crowding is associated with acclimatization to the cold; the general physiological condition of the hens must be different at the height of the laying season from that at its beginning, and this shift in physiological state may account for the reversed effect of crowding; still, perhaps the psychological factor invoked by Pearl and Surface to cover admitted ignorance may be the only feasible suggestion as yet. Third, following the approach of warm weather and the coming of the hot summer months, the birds of the crowded pens probably have difficulty in maintaining comfortable temperatures, particularly while roosting.

In concluding this discussion of the effect of crowding on the rate of egg production in chickens, it is of interest to note that the variability in the rate of egg production increases with crowding when the annual egg production is taken as the unit. When this is broken into monthly periods, it is seen that the greatest effect of crowding is to be found at the beginning and the end of the laying-year, at a time of low production. From February to July, at the time of heaviest laying, the environmental differences implied by flock size as used in these experiments do not affect the relative variability of production. Unfortunately there are no data concerning egg production of chickens isolated in pens with the floor areas per individual used in these experiments.

EFFECT OF CROWDING ON RATE OF REPRODUCTION IN DROSOPHILA

Pearl and Parker (1922) have contributed another bit of significant evidence to our problem by their work upon the influence of the density of population upon the rate of reproduction in *Drosophila*. In this work mass matings were made from a given line. The offspring from this mass mating were used in making up the matings in the experiments to be described. Half-pint milk bottles were used as containers. The procedure was definitely standardized throughout. Sets of four bottles were started, each containing 1, 2, 3, ... 9, mated pairs of flies. Sets of three bottles contained, respectively, 10,

12, 15, 20, and 30 mated pairs; two bottles held 50 mated pairs each; and one bottle had 25 mated pairs. At the end of 8 days at 25° C. the surviving parent flies were transferred to fresh bottles for a second breeding period of 8 days. The only variable known to be significant throughout this series was the density of the population. All the offspring from the two breeding periods were counted and sexed. The results tabulated as the rate of reproduction per female per day during the first 16 days of life are shown in Figure 7.

In this figure the circles give the observations, and the curve is the graph of the following equation fitted by the method of least squares:

$$y = 34.53\, e^{-0.018} x^{-0.658} ,$$

which in logarithmic form becomes:

$$\log y = 1.54 - 0.008x - 0.658 \log x ,$$

when y signifies the flies per mated female per day and x is the number of mated flies per bottle, taken over the whole 16-day period the experiment ran.

The observations include a total of 23,922 progeny flies, which is a large enough number to cause the results to be treated with respect. Further, it is apparent that the curve fits the observed facts closely. In the preceding chapter, I have called attention to the fact that this formula is related to that which Bilski developed to describe the effect of crowding on the rate of growth in tadpoles, to that of Kennealy for the relation between length of race and the record established for that distance, and to that of Farr for the relation between density of human population and the death-rate. These phenomena must be based on a common fundamental biological relationship.

When these results are analyzed further, we find that the greatest drop in rate of reproduction of adult flies per female per day comes as the number of original mated pairs per bottle increases from 1 to 2, and the next greatest drop comes between the bottles having an initial population of 2 and 3 mated pairs. This result cannot be due to larval crowding, since the bottles containing 9 mated pairs of flies

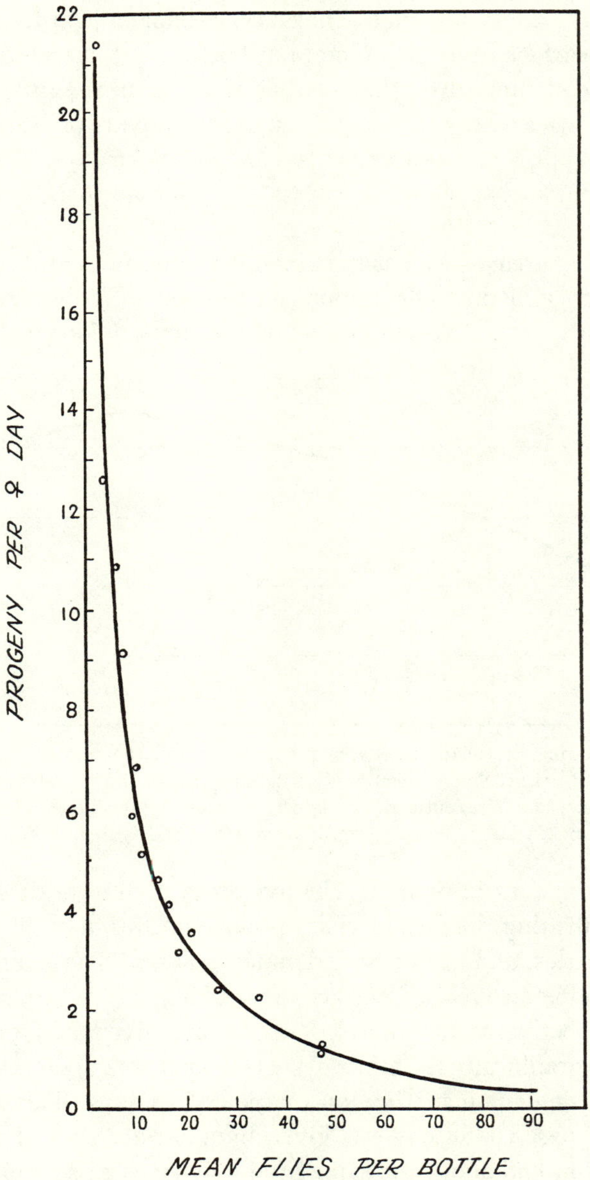

FIG. 7.—Pearl and Parker's (1922) curve showing the decrease in rate of reproduction in *Drosophila* as cultures become more crowded.

produced 2,117 adult offspring. The 80 cc. of banana-agar food with an exposed surface of 23.76 sq. cm. per bottle must therefore have been capable of supporting this number of larvae in the time available. The bottles with 1, 2, and 3 original mated pairs produced, respectively, 1,348, 1,124, and 1,877 total imagoes in 16 days. The food available would have allowed at least 2,117 larvae to pupate and produce adults.

The exposed area, as well as the amount of food present, has been shown to have a distinct effect upon the numbers of *Drosophila* produced. Harnly (1929) varied the area of standard food with the

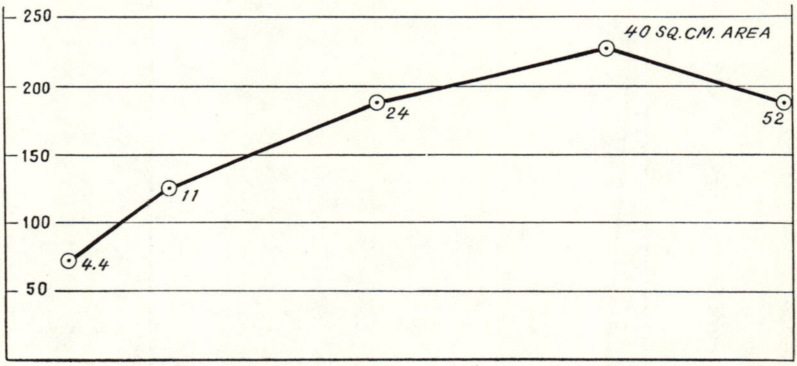

FIG. 8.—Showing the relation between productivity and the area of food with *Drosophila.* The vertical column of figures gives density; figures on the graph show the area of food surface in square centimeters. The corresponding total volume capacities of the containers are: ______, 1181, 2365, 473, and 250. (Data and figure from Harnly.)

depth kept constant at 25 mm. The five areas tried were those furnished by culturing the flies in vials, 4-ounce bottles, half-pint, and pint milk bottles, and in 250 cc. Erlenmeyer flasks. These different culture-containers gave food surface areas of 4.4, 11, 24, 40, and 52 sq. cm., respectively. Summarized results are given in Figure 8, which shows graphically the effect of surface food area upon the total yield from a single pair mating for a period of 10 days. The largest yield under these conditions was given by a surface area of 40 sq. cm.; 52 sq. cm. had about the same productivity as 24 sq. cm. The viability was greatest in the flies reared with the largest amount of space.

The explanation of Harnly's results is not necessarily obvious or easy. It may be that there is actually a surface-population optimum which stands below the largest surface and volume of food available. A possible factor may be that with greater area and volume wild yeasts or molds grow too rapidly for the *Drosophila* to control. Before coming to this conclusion, it is well to note the sizes of the different containers, which were: vials, size not stated, 118, 236.5, 473, and 250 ml. The population curve may be a result of the available. space rather than of the food surface acting alone. Such an interpretation would be in line with the data of Pearl and Parker shown in Figure 7. More work is needed, however, before one can draw assured conclusions.

The tendency toward universality of the effect of crowding upon the rate of reproduction is shown by the fact that Hill (1926) and Sarles (1929), working with hookworms, have reported counts on population density of these parasites in relation to egg production which show that as the number of worms in a given host increases, the egg output per worm decreases.

Pearl and Parker conclude the account of their work upon crowding and the rate of reproduction in *Drosophila* with the following statement, which Pearl repeats in a later book: "In general there can be no question that this whole matter of influence of density of population in all senses, upon biological phenomena, deserves a great deal more attention than it has had. The indications all are that it is the most important and significant element in the biological, as distinguished from the physical, environment of organisms." With this position I am in complete accord.

CHAPTER VIII

CROWDING AND INCREASED DEATH-RATE

Measurements of growth, reproduction and length of life, sum up many of the physiological processes that may be affected by crowding; the first two of these have already been considered in some detail. These three functions are closely connected, and it has been impossible to keep their treatment entirely separate. Thus, in the preceding chapter, the discussion of the effect upon the rate of reproduction of confining Paramecia within a small capillary tube was extended to include a partial discussion of such treatment upon the longevity of the animals in order to get sufficient control of the available evidence to be able properly to evaluate factors affecting the decrease in rate of reproduction brought about by crowding.

Inspection of the material previously presented demonstrates that the ability of adult organisms to live is not necessarily the same as their ability to reproduce. Kuczynski (1928), in studying the balance of birth and deaths among the human population of Western Europe, describes the differential effect of changing conditions upon fertility and upon the death-rate, and concludes that human fertility has become a problem in itself largely divorced from the problem of mortality.

The experience of Warren that *Daphnia* lose their reproductive capacity long before they die, and that in such a condition they may be able to live through adverse conditions produced by overcrowding and again take up reproduction when conditions become more favorable, is a case in point. Kalmus adds observations upon Paramecia along the same line. The usual interaction between fertility and mortality is such that in a given amount of liquid medium the population increases to a maximum whose size depends on the volume of the medium and the concentration of the food material, and then gradually falls to complete or nearly complete extinction. This course of events is shown in Figure 9, which is taken from the work of Myers with Paramecia.

If the initial volume is relatively large (16 drops, or about 0.8 ml.), Myers finds that fission begins at about the same rate, so far as 12-hour periods of observation show, regardless of whether the seeding be with 1, 2, or 4 individuals. The population of such cultures rises rapidly to a peak which is essentially the same for all the seedings just mentioned; at its peak it ranges from 123 to 126 individuals and

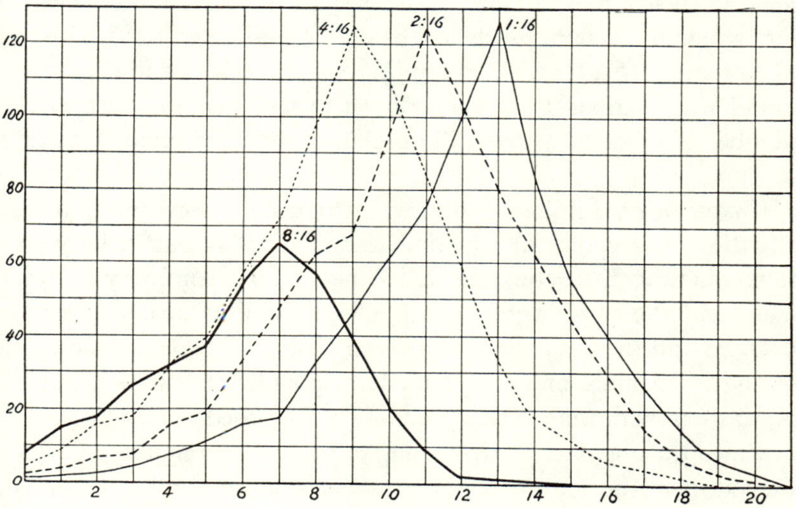

Fig. 9.—Showing the rise and decline of populations of Paramecia in 0.8 cc. of culture fluid by, respectively, 1, 2, 4, and 8 individuals. The horizontal axis shows successive periods of 12 hours each; the vertical axis gives numbers of individuals in the populations. (From Myers, 1927.)

then falls off at about the same rate. The cultures seeded with a single individual take longer to reach the maximum than do those seeded with 2 or 4, but otherwise the course of the population history is similar.

When 8 individuals are introduced in place of 1, 2, or 4, the maximum in Myers' cultures showed a population of only half that given with the smaller seedings. The reason for this difference is not clear. On the surface it appears that the larger initial seeding either exhausts the food supply more speedily or poisons the culture medium more rapidly than do the other seedings or acts in both ways at the same time.

The point of especial interest to us in these observations is the fact that with certain initial densities of populations, even in an unchanged medium, the maximum reached is practically identical and independent of the numbers originally introduced.

The effect of the exhaustion of food supply has been eliminated by Chapman (1928) in his work with the confused flour beetle, *Tribolium confusum*. Chapman introduced varying numbers of these beetles into a definite amount of whole-wheat flour and found for this insect, as Robertson, Cutler and Crump, Myers, and others had previously found with various Protozoa, that there is a definite limit to the number of organisms that will develop in a unit volume of culture medium.

Chapman's work does, however, introduce one new fact into the situation. His choice of experimental material is particularly fortunate in that the beetles can be screened out of their floury environment and the eggs, larvae, and pupae, as well as imagoes, can be counted and the flour renewed at each observation. Hence, in place of the usual more or less symmetrical population curve found by other workers dealing with a population composed of all age groups, in which the population, after rising to the maximum determined by the nature of the culture medium and the amount of space available, falls away to approximate or total extinction on account of the exhaustion of food or the addition of excretory products, Chapman is able, by periodically renewing the environment, to carry his beetle population along for extended periods, perhaps indefinitely, with approximately the same number of individuals present per gram of flour. In his terminology, "a condition of equilibrium is attained in which the biotic potential is equalled by the environmental resistance and the population remains relatively constant."

Appropriate tests showed that the stationary character of the population when in equilibrium was not due to absence of eggs or to their lack of fertility. Rather, the lack of increase in population beyond a certain point was due to the eggs, pupae, and, to some extent, the larvae, being eaten by the adult beetles. When eggs were placed in flour cultures containing only male beetles, the percentage of eggs eaten varied directly with the population of adults per gram of flour.

Chapman's experience with *Tribolium* can be shown in a number of ways; perhaps as significant as any is the result of carrying to equilibrium a series of beetle environments of different sizes, of 4–128 grams of whole-wheat flour, seeded with 1, 2, 4, 8, 16, and 32 pairs of beetles each, making one pair of introduced beetles per 4 grams of flour. This experiment may be followed in Table III, which

TABLE III

BEETLES (*Tribolium confusum*) PER GRAM OF WHOLE-WHEAT FLOUR
(Data from Chapman, 1928)

Days	4 G.	8 G.	16 G.	32 G.	64 G.	128 G.
0	0.5	0.5	0.5	0.5	0.5	0.5
15	15	17	20	17	21	19
30	30	25	26	22	24	23
50	35	33	32	35	32	34
64	39	39	34	39	40	37
78	35	41	39	36	37	39
101	40	46	38	44	49	39
114	48	45	36	43	40	40
134	37	50	41	41	48	45
156	38	49	46	44	45	47
171	46	49	46	43	42	40

gives the total number of the beetles present per gram of flour at different times in the history of the cultures, regardless of the developmental stage of the beetles.

In all the conditions tested by Chapman, the mean number of individuals per gram of flour after equilibrium was established was 43.97, with a standard deviation of 4.27, and a probable error of 2.88. Chapman found, as will be shown in detail later (chap. x), that the time taken to reach this equilibrium differed with the initial seeding per gram of available flour, while the equilibrium population remained constant per gram of flour; and that, with the same initial seeding throughout, the equilibrium population per gram of flour remained constant but the time taken to reach equilibrium varied directly with the quantity of flour available. This equilibrium is primarily a food relation, or a food and space relation, since the metabolic products are removed by the periodic changes of the flour. The equilibrium is apparently based upon a competition between adults and larvae, as is shown by the fact that in one 16-gram environment the number of adults was accidentally reduced on the seventy-eighth day of the experiment and was never returned to its place in the geometric series, while the number in its total population—eggs, larvae, pupae, and adults—did so return.

Following current tendencies, Chapman interprets his findings in terms of a mathematical formula, $C=Bp(R)$, when C is the concentration of insects, R is the environmental resistance, and Bp is the biotic potential, which he defines as the mean maximum rate of reproduction in a given period under given conditions. Substituting and solving, we find:

$$R=\frac{(43.97\times 8.4).25}{43.97}=2.1\,.$$

The concentration of insects per gram of flour is 43.97. The average number of eggs laid per day in these experiments is 8.4, and one-fourth of the population are egg-producing females. The formula so given represents the state of equilibrium only.

The work we have been discussing summarizes the effect of the environmental factors associated with crowding upon the total population. The work which deals most directly with the harmful effects of crowding on length of life is that of Drzewina and Bohn. In connection with their studies on the relation existing between mass of toxic liquids and the contained mass of animals, Drzewina and Bohn have found that many cases of protection are furnished by increasing the numbers of animals present in the same solution. These will be reported later. In some instances they record the opposite results (1921*d*, 1922).

When KCl was used as a toxic agent with cultures of *Convoluta*, a small marine planarian, other things being equal, those in the solution containing the larger number died first. Similar relations hold when the same number of individuals are placed in differing amounts of the same strength of KCl solution: those in the smaller amount of liquid die more rapidly. The fresh-water planarian *Polycelis nigra* reacts similarly. These investigators believe that the planarians give off a substance which causes autodestruction, and that, if this be true, such destruction is hastened by increasing the mass of individuals in proportion to the amount of liquid.

Their interpretation is supported by the observation that if fresh Planaria are introduced into a solution of KCl which has already contained others, their death is hastened. If, after some time in such a solution, a part are removed to a new solution of similar strength of KCl, these die more slowly than do their fellows left in the original solution, which contained not only KCl but also some substance or substances given off by the worms themselves.

Later (1928) they observed that around cytolyzing flatworms the H-ion concentration (acidity) of the solution is greatly increased, and they considered this to be the factor which causes the increased mortality of the groups. The larger the number of cytolyzing individuals, the more rapid and the greater the increase of the H-ion concentration, and consequently the more rapid and pronounced are the lethal effects involved. Such a process accelerates itself in the presence of many individuals, or, on the other hand, is not effective in the case of a few scattered animals. If the latter die, they do so because of the lethal effect of the KCl alone.

Fowler (1927, 1931) undertook to test the effect of a large number of electrolytes upon the rate of survival in certain crustaceans, using mainly a species of *Daphnia*. His results show that there is a distinct correlation between the survival value of the group and the degree of toxicity of the salt solutions employed. Tests upon the rate of oxygen consumption have shown that crowded *Daphnia*, in the concentrations tested, use less oxygen per individual than do isolated *Daphnia* under similar conditions. When the toxicity is sufficiently great so that death occurs within a relatively short time, this group depression tends to favor group survival. On the other hand, when the concentration is low and the effect of the chemicals is deferred, the isolated individuals live longer than the group. Such results are in accord with Child's (1915) differential susceptibility findings when planarians are subjected to relatively strong or relatively weak concentrations of various toxic agents, particularly KCN.

Under conditions of high toxicity the animals, or, in the case of Child's worms, certain regions, which have the lowest rate of general metabolism are least affected by the toxic agent and survive longest. With weaker solutions, on the other hand, the most vigorous individuals, or, in the case of the worms, the most vigorous regions, can acclimate most readily and hence survive longer. Fowler's results show that depression due to crowding may have definite survival value under certain conditions but that with weaker concentrations of even the same salts crowding decreases the chance of survival. In the latter aspect his results support the facts reported by Drzewina and Bohn in their experiments with KCl. Fowler's results fail to support the hypothesis that a specific autodestructive material is produced. They extend the later explanation of Drzewina and Bohn by indicating that the unknown autodestructive substance is the carbon dioxide produced by the animals, which does raise the H-ion concentration as Drzewina and Bohn determined.

For further consideration of the relation between density of population and the death-rate it seems best to take up the case with respect to man, since with human populations this relationship has attracted particular attention for a considerable period of time. We have already referred to the generalization known as Farr's law; this states that if the death-rate be represented by R and the density of population per unit area by D, then $R = cD^m$, where c and m are constants.

Brownlee (1915) rehabilitated this law by showing that the statistics used by Farr, which came from the decade 1861–70, compared favorably, so far as the relation between population density and death-rate was concerned, with those of the decade 1891–1900, as given by Tatham. Brownlee's republication of these tables and his calculations are given herewith; see Table IV.

Brownlee calls attention to the fact that the values of m correspond roughly for each type of analysis in the two periods but that in the case of the life-table death-rates they correspond to the third decimal place, which is as much as could be statistically expected. He concludes that Farr's law is thus shown to be a definite law operating independently of the changes due to sanitary progress. Regardless of improvements in sanitation and in medicine, the exponent does not vary, but only the multiplying constant. Therefore *m* represents the law, while *c* represents rather the coefficient of

TABLE IV¶

No. of Districts	Persons per Sq. Mi.	Corrected Death-Rate	Same Fitted by Least Squares	Crude Death-Rate	Same Fitted by Farr	Life-Table Death-Rate	Same Fitted by Least Squares
A. Showing Figures Relating to Density and Death-Rate, 1861–70							
		*		†		‡	
53......	166	15.50	16.70	16.75	18.90	19.90	20.73
345......	186	17.02	17.00	19.16	19.16	21.07	20.96
137......	379	20.52	18.99	20.87	20.87	23.97	22.51
47......	1,718	24.35	24.03	25.02	25.02	26.09	26.19
9......	4,499	27.94	27.92	28.08	28.08	28.54	28.84
1......	12,357	33.98	32.67	32.70	32.70	32.67	31.92
1......	65,823	40.55	42.39	38.74	38.74	37.17	37.74
			E%=3.79§		E%=2.70		E%=2.01
			△=1.17		△= .90		△= .61
B. The Same for 1891–1900							
			**		††		‡‡
27......	136	11.63	13.06	14.20	14.16	17.38	17.18
112......	161	12.54	13.43	15.05	14.51	18.01	18.12
121......	181	13.44	13.70	15.44	14.68	18.62	18.33
92......	261	14.52	14.56	15.46	15.38	19.36	19.02
53......	407	15.53	15.68	16.08	16.28	20.05	19.90
56......	457	16.53	15.99	16.67	16.52	20.24	20.13
31......	734	17.58	17.32	17.64	17.56	21.45	21.12
40......	1,303	18.53	19.05	18.04	18.88	22.10	22.31
31......	1,705	19.42	19.93	18.61	19.54	22.71	22.99
21......	2,339	20.37	21.00	19.50	20.35	23.36	23.72
18......	4,424	21.56	23.37	20.21	22.08	24.18	25.31
13......	4,884	22.36	23.76	20.69	22.35	24.72	25.56
6......	4,194	23.48	23.16	22.05	21.93	25.49	25.10
5......	2,925	24.33	21.80	23.29	20.94	26.07	24.21
5......	7,480	26.54	25.51	24.74	23.60	27.58	26.68
4......	55,563	34.82	32.67	23.67	30.49	33.25	32.58
			E%=4.3		E%=3.8		E%=2.03
			△=1.05		△=1.14		△= .63

¶ From Brownlee, *Journal of Hygiene*, XV, 16.

* $R=7.534D^{.15571}$ † $R=10.234D^{.11998}$ ‡ $R=12.419D^{.10018}$

§ E%=mean experimental error; △ = $\sqrt[2]{}$ of the mean of the squares of the errors.

** $R=12.40D^{.16715}$ †† $R=13.57D^{.12755}$ ‡‡ $R=10.83D^{.10078}$

sanitary and general conditions of health. The coefficient *c* decreased from 12.42 in the earlier period to 10.83 for the later. In other words, in the same general area, conditions of living have so improved in the thirty years' interval shown in Table IV that density of population had only 0.875 the effect during the 1890's that it had in the 1860's. Here we have evidence that relatively mild crowding affects longevity in men.

This is to be expected when we consider the relatively greater ease of transmission of contagious diseases in the more crowded areas. Such dangers from the crowd are illustrated in a simplified form by one of the *Ophioderma* experiments to be reported in full in another connection. In these experiments the survival of 8 isolated brittle starfish, each in a 1-liter Erlenmeyer flask, was compared with a group of 8 similar starfish in an 8-liter bottle. Usually the group outlived the isolated individuals; but on one occasion one of the members of the group died soon after the daily inspection and change of water, and so polluted the whole 8 liters that all the remainder were dead on the following morning. When an isolated animal died similarly, the effects of its death could not extend beyond the limits of its single flask.

When we pass in review the materials presented in the chapters of which this is the third, we find much evidence supporting the generally accepted dictum that crowding is harmful for poorly integrated groups of animals, breeding and hibernation seasons excepted. We have seen that crowding may slow down the rate of growth and may result in dwarfed individuals, that the rate of reproduction may be decreased, and that the death-rate may be greater. These effects have been reported for so many different animals from such a wide range of the animal kingdom that there can be no doubt of their general significance. But this is not the whole story. In many of the experiments to be reported in our next section, we shall find that crowding does not always produce harmful results; and that under many conditions there are distinctly beneficial results, providing the crowding be not too great. When considering these beneficial results, we must, however, always keep in mind the harmful effects of overcrowding.

BENEFICIAL EFFECTS OF AGGREGATIONS

CHAPTER IX

STIMULATION OF GROWTH BY CROWDING

Having sketched in some detail the harmful effects resulting from the crowding of many animals into a relatively small space, it is now possible, with a better perspective, to look into the more recently accumulated evidence that harmful results do not necessarily follow the formation of such aggregations and that they are often useful and even necessary to the welfare of the individual.

The extent to which the phenomenon of aggregation affects the rate of growth in a positive manner has been relatively little investigated. In the work of Colton (1908) upon *Lymnaea* it will be recalled that he found crowding generally decreased the rate of growth in snails. He found, however, that the snail faeces, if washed free of easily soluble material and placed in weak solutions with snails, tended to increase the rate of growth. With concentrated solutions of faeces the results were reversed. Similarly, weak solutions of urea favored snail growth, though stronger solutions retarded it.

Popovici-Baznosanu (1921) also found that under certain conditions snails grew more rapidly in stagnant water, conditioned by the snails, than in fresh water. In short experiments (1914) 10 young *Lymnaea* attained a length of 9.5 mm. in fresh water while those living in stagnant water grew to 10 mm. Later he tested this effect for a longer period. The young *Lymnaea* from three egg masses were placed in three culture dishes of identical dimensions as regards volume and surface of water; after a long sojourn, when the water was thoroughly snail-conditioned, Popovici-Baznosanu took half of the individuals and placed them in better conditions of existence, in culture jars with a large volume and a relatively large surface, and containing fresh pure water. *Elodea* was used as food both for those in the stale and those in the fresh water, and in the same quantity for both. After 106 days the results were as given in Table V.

In only one of the three cases was there a clearly significant difference; yet Popovici-Baznosanu interpreted these results as meaning that in the stagnant, snail-conditioned water, the higher plants present, as well as the walls of the jar, are covered by growths of microflora, which he regards as forming the chief food of the snails; and that the snails therefore grew more rapidly in cultures containing a rich microflora than in those with only a scanty supply. Colton had interpreted his results similarly.

The observations of Eigenbrodt (1925) that *Drosophila* grow larger in small culture vials when present in numbers of from 8 to 16 than

TABLE V*

Brood	Surface of Water (Sq. Cm.)	Volume of Water (Cc.)	Condition of Water	Length of Largest Shell (Mm.)
I	113	1,300	Conditioned	19
	133	1,900	Raw	9.5
II	113	1,300	Conditioned	19
	208	4,510	Raw	18
III	113	1,300	Conditioned	15
	116	1,640	Raw	15

* Data from Popovici-Baznosanu.

at other population densities may be explained on the assumption that too few *Drosophila* larvae per culture fail to control the growth of harmful elements of the yeast or bacterial flora as well as optimal numbers do, while overcrowding overcontrols the growth of the food plant. This would result in a growth optimum occurring, as suggested, at a relatively low population density but distinctly above the minimum populations studied. These results should be compared with the relation between numbers present and *Drosophila* survival given in chapter xiv.

Bilski (1926) tested the effect of crowding upon the rate of regeneration of the tails of *Rana esculenta* tadpoles. Five of these tadpoles were kept isolated, and five similar ones were placed together in the same sort of dish and with the same amount of water which was given to each of the singles. Although in most cases there was a decrease in length of body from tip of head to the root of the tail, there was growth both of the tail stump and of regenerated material.

The proportions of decrease and of growth or regeneration differ between isolated and grouped animals. The results for the 7 days the animals were observed are given in Table VI. The results indicate, as much as a single experiment is likely to, that there is a greater regeneration with the decreased volume per animal, which is compensated by the greater growth of the tail stump when the animals are isolated. Bilski states that this experiment is supported by his general experience in many similar experiments in other phases of the work, but cites no direct support of these results.

TABLE VI

SHOWING THE EFFECT OF CROWDING ON THE RATE OF REGENERATION OF TAILS OF FROG TADPOLES IN 7 DAYS. (IN MILLIMETERS)

(Data from Bilski)

Conditions	Difference		Percentage of Difference*		
	Body Length (*B*)	Tail Stump (*S*)	Body Length (*B*)	Tail Stump (*S*)	Regenerated (*R*)
Isolated..........	−1.4	2.5	−11.8	15.1	15.1
Grouped..........	−0.7	2.0	− 6.2	12.6	18.3

B, length from tip of head to root of tail.

S, length of tail stump after cutting. *R*, length of regenerated material.

* Percentage *B* is calculated on the basis of the original body length; percentages *S* and *R* are in terms of the original tail length before operation.

CROWDING IN TISSUE CULTURES

Work with tissue cultures has yielded pertinent evidence concerning the beneficial effects of crowding on growth. The literature in this field is enormous; and no attempt will be made to cover the different ramifications of the subject, with most of which we are not immediately concerned. It has been known for some years (Carrel, 1924) that tissues to be grown *in vitro* must have a proper background on which to creep. One of the most used backgrounds is of fibrin network. In many recent studies this is placed as blood-plasma in a thin layer over the bottom of a special culture flask. The tissue to be cultured is introduced aseptically into this sterile medium, which is then covered with a sterile fluid that has Tyrode solution as its main ingredient but which contains other materials such as serum or a saline extract of embryonic tissues. The latter, or some fraction thereof, appears to be necessary for real growth of such cells as fibroblasts or epithelial cells. Extracts of sarcomas are superficially similar to extracts of embryos in their growth-producing qualities. Leucocytes (macrophages) behave in reverse fashion, growing permanently in pure serum and being inhibited by the presence of embryonic extracts.

Inorganic substances, oxygen excepted, apparently do not affect growth-rate of cells *in vitro* when present in approximately the same concentrations as in the blood of the animal furnishing the tissues under cultivation. Any departure from such concentrations yields adverse results. Only approximately isotonic media allow indefinite survival. The exact nature of the growth-promoting substance found in embryonic extracts is still unknown. It appears to be associated with the protein fraction and is particularly associated with proteoses which result from a brief digestion of the protein with peptone (Carrel and Baker, 1926). Prolonged digestion destroys the effectiveness of this material. Willmer (1928) has been unable to confirm this work, but concludes from the evidence furnished by Carrel, Baker, and others that tissues can get some energy from amino-acids but that their nitrogen supply is chiefly obtained from proteoses (embryo extract contains both elements). These are heat-stable substances; but most workers find that there is present in embryonic juice a thermolabile growth-promoting substance which is easily destroyed by heat or is adsorbed when heated, which does not pass through a Chamberlain filter, and which is destroyed by prolonged shakings. Carrel (1924) has called such substances "trephones"; Fischer (1925*a*) calls supposedly similar substances "desmones"; and Burrows and Johnson (1925) named them the "archusia."

Tissue-culture workers appear to be agreed upon the necessity of keeping the cells from normal tissues in numbers, for successful cultivation *in vitro*. Harrison (1928) says in this connection: "It is a very interesting and at present inexplicable fact that single somatic cells isolated in culture media do not proliferate. Experiments to this end made in my own laboratory some years ago but not published did not succeed and other workers have reported similar experience. As Fischer puts it, a colony of fibroblasts cannot arise from a single cell even when the nutrient conditions are most favorable. Likewise small groups of cells if isolated do not undergo division and their growth remains at a standstill. On the other hand certain tumor cells (Rous chicken sarcoma) are capable of multiplying and producing colonies when isolated singly." Similarly, we know that in nature single egg cells will grow. The germinal area of the hen's egg is an excellent example of an isolated bit of protoplasm which, under favorable conditions, will grow. It is of interest to us to note that Wright (1926) has found by dialysis a growth-stimulant in the incubated yolk of hen eggs which is not shown when such yolk is added directly to tissue-culture medium without dialysis.

Haberlandt (*vide* Fischer), in his work with plant cells, could secure increase in size from certain isolated cells but did not find cell division in such cultures. He (Haberlandt, 1919–22) reports a direct relation between the size of the piece of plant tissue transplanted, or the number of cells within it, and the number of cell divisions. From these studies this investigator has concluded that the inciting to cell division comes from substance given off by injured cells, which he terms "wound hormones" or "division hormones."

A dramatic instance of the effect of heterotypic crowding upon growth of tissue cells *in vitro* is furnished by Carrel and Ebeling (1923). Cultures of leucocytes and of fibroblasts were made together in the same flask of plasma. As usual under these conditions, the fibroblasts did not grow, while the leucocytes grew well. In time they spread until they came in contact with the languishing fibroblasts, when a marked revival and initiation of growth took place in the latter cells. This agrees with the generally known fact that in the tissues of early embryos, when growth is taking place most rapidly, there is a mass of growing tissue tightly packed together which is supplied with a relatively small amount of blood. In tissue cultures growth takes place best when the cells are present in relatively large numbers in a small amount of medium which is stagnant but properly supplied with oxygen. Both kinds of observations suggest that the cells forming metazoan tissues are dependent greatly upon one another for their growth.

Fischer (1925) has suggested that this dependence is due to the slow diffusion of products of metabolism or secretions from one cell to another. He thinks these travel by protoplasmic bridges and are independent of Carrel's "trephones," since fibroblasts that cease to grow in the presence of an abundance of these trephones may be restored to rapid growth by the presence of active healthy cells.

Burrows and his co-workers (1925, 1926) have put forward an interesting and ingenious suggestion which explains many aspects of the interrelations between cells and the fact that they must be present in numbers before growth will occur and at the same time explains other characteristic activities of cells in tissue cultures. These workers suggest that in the presence of a sufficient amount of oxygen, about one-third of an atmosphere, the cells secrete a hypothetical substance or group of related substances which as stated above are called "archusia," which are supposed to function somewhat like the desmones of Fischer except that the function of archusia is profoundly modified by their concentration. If present in high concentration, they display an enzyme-like action which causes self-digestion of the tissues; if the concentration is somewhat lower, the presence of archusia allows the cells to digest fats and proteins and to grow, providing the medium is otherwise suitable. In more dilute solutions, tissue growth ceases; but the cells display their characteristic migrating ability, which is frequently shown in cultures, or parts of cultures, in which no growth is going forward. In yet more dilute concentrations, the cells lose their power of carrying on their ordinary activities, round up, and become dormant.

Archusia are water soluble, are secreted by cells, and can diffuse through cell membranes to the outside medium. They tend to collect in quantity when many cells are together in a minimum volume under stagnant conditions, which are known to favor growth of tissue cultures. When too great a volume of medium is present in proportion to the number of cells, or if such cells as fibroblasts are isolated, archusia escape into the surrounding medium and growth ceases. Cells isolated into sufficiently small volume should grow, according to the implications of this hypothesis; but the needed volume may be so small that other complicating factors arise. Such a substance would also be carried away by repeated washings, which are known to be harmful to cells whether grown *in vitro* or *in vivo*. Archusia have properties resembling bios and vitamin B and have been thought to be identical with the latter.

The whole concept of archusia is in the hypothetical stage at present, and more evidence is needed before coming to a definite conclusion concerning its validity.

Heaton (1926) has worked upon the effect of vitamin B upon the growth of cells *in vitro*. He finds two elements present in extracts of yeast and of liver—one which stimulates growth and another which depresses it. The two can be separated by their different solubility in alcohol. Burrows and Jorstad (1926) think that vitamin A is necessary for the functioning of cells and is produced when cells are digesting fats and growing under the stimulus of relatively high concentrations of archusia (vitamin B?). They regard vitamins A and B as antagonistic, and balanced in cells that are functioning. In fact, most observers are agreed that fats and lipoids are associated with substances which inhibit the growth of cells, while some portion of the protein molecule is associated with the promotion of growth.

EFFECT OF CROWDING ON GROWTH OF SEA-URCHIN PLUTEI

Certain of these results of the tissue culturists have been applied to the problem of the effect of crowding upon the rate of cleavage and of the growth of the arms of sea-urchin plutei by Peebles (1929). By treating extracts of sea-urchin eggs and larvae with alcohol or with acetone, a growth-inhibiting substance was obtained which definitely retarded the rate of growth of eggs or of plutei. When this fraction containing lipoids was partially removed, growth acceleration was observed, as shown in Figure 10. Further experiments showed that there was a decided difference in growth, depending on whether the alcohol-soluble or alcohol-insoluble fractions of extracts of echinoderm plutei were used. These results are shown graphically in Fig. 11. Peebles was also able to remove growth-inhibiting substances from such extracts by adsorption, as shown in Fig. 12, but has not been able as yet to isolate either the growth-inhibiting or the growth-promoting principle.

Peebles, in summarizing her work, says: "The eggs and larvae of

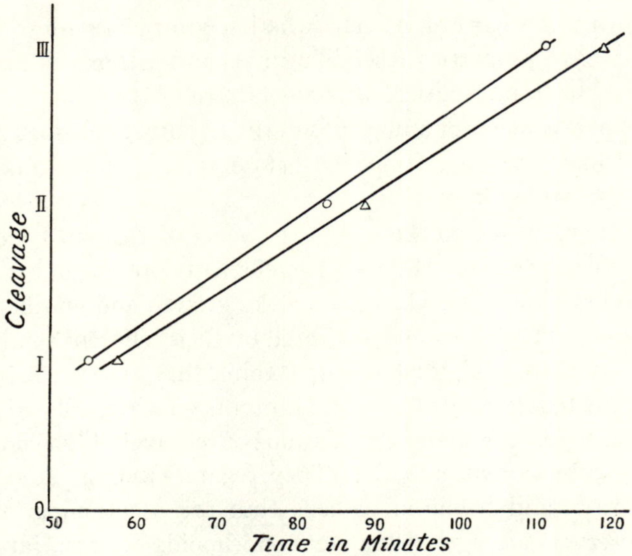

FIG. 10.—Showing the rate of cleavage in eggs of the sea urchin treated with acetone extract from which the fat has been partially removed. Acetone (extract) minus fats, ○. Control, Δ. (From Peebles, 1929.)

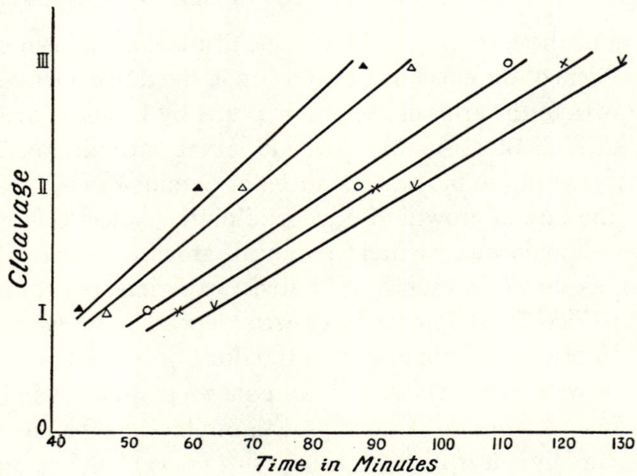

FIG. 11.—A comparison of the results obtained by Peebles (1929) by using the filtered alcoholic extracts of plutei (▲Δ) with that of the precipitate (x, v). Ninety-seven per cent ethyl alcohol (filtrate), ▲; 75 per cent ethyl alcohol (filtrate), Δ; control, ○; 75 per cent ethyl alcohol (precipitate), x; 97 per cent ethyl alcohol (precipitate), v.

the sea urchin and starfish contain growth promoting substances. These substances pass out into the surrounding water during segmentation, and later stages of larval development. . . . . There is

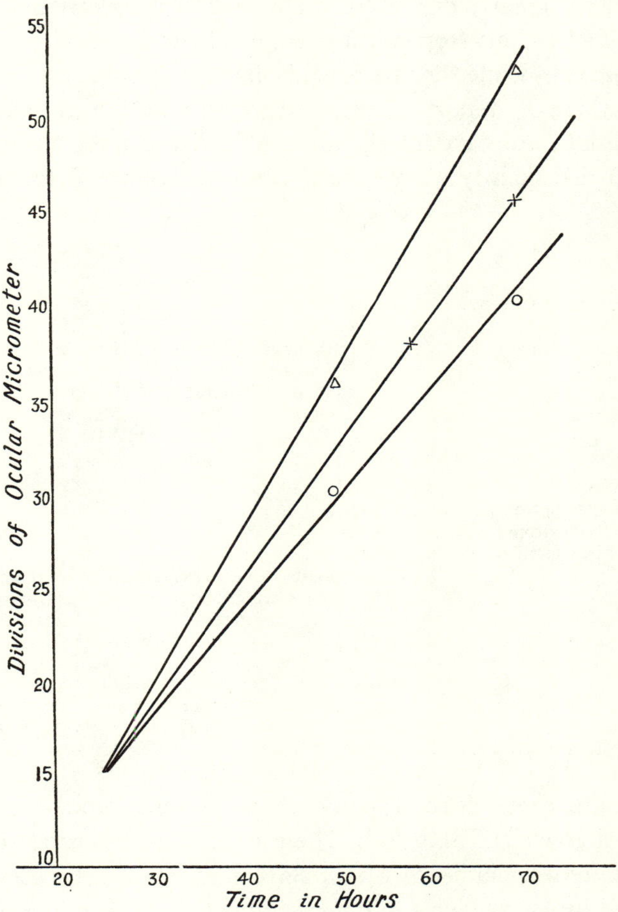

FIG. 12.—Figure showing growth in length of plutei in the presence of animal charcoal (Δ) and fuller's earth (x) compared with those growing in sea-water (○). (From Peebles, 1929.)

some experimental evidence in favor of the conclusion that the inhibiting substances are associated with the lipoid constituents and the accelerating factor is contained in the protein molecule. . . .

The retarding effects of secretions of growing embryos are removed in the presence of animal charcoal and fuller's earth. The percentage of normal larvae resulting from eggs grown in the presence of these adsorbents is greatly increased, while mortality is decreased."

By using hanging-drop cultures, a part of which contained isolated sea-urchin eggs while the others held small groups of eggs, Frank and Kurepina (1930) report accelerated growth in the grouped eggs. These results are particularly noticeable if the temperature is allowed to rise slightly above the normal. A résumé of two of their

TABLE VII

A

Number of eggs per drop	1–2	16–20
Number of such drops	10	3
$8\frac{3}{4}$ hours after fertilization	75% have 8 blastomeres 25% past 8 blastomeres	0 % at 8 blastomeres 12% past 8 blastomeres 88% at 16 blastomeres

B

Number of eggs per drop	1–2	3–6	10–20
Number of such drops	10	7	5
$42\frac{1}{4}$ hours after fertilization	75% no movement 25% slight movement 10% plainly moving 0% gastrulae*	0% no movement 26% slight movement 40% plainly moving 34% vigorously moving 0% gastrulae	0% no movement 4% slight movement 12% plainly moving 60% vigorously moving 24% gastrulae

* Percentages as reported in original work.

experiments, showing the type of results obtained under these conditions, is given in Table VII. These results are interpreted by the experimenters to indicate a stimulating effect of self-radiation as suggested by Gurwitsch's mitogenetic rays. It is clear that such an interpretation is far-fetched at present; but the results indicate, despite careless reporting, that there is an optimum number of eggs which lies well above the minimum at which, under certain conditions at least, growth is favored as compared with that shown by eggs isolated into similar amounts of sea-water.

HETEROTYPIC CROWDING IN TISSUE CULTURES

We have noted above the case of fibroblasts growing in plasma only when under the close influence of leucocytes, as an instance of the direct effect of different kinds of tissues grown together upon the ability of the one to grow at all; there is also evidence that differentiation is stimulated or accelerated by the presence of two sorts of cells in close association. Thus Ebeling and Fischer (1922) combined a ten-year-old strain of fibroblasts grown in pure culture with a two months' strain of epithelium which had been similarly grown in pure culture. After the two had been grown together for some time, the epithelium became rounded into a sort of epithelial glandular tissue lying within a supporting network of fibroblast cells. Champy (1914) and Drew (1923) have reported somewhat similar results from combining these two kinds of tissue cells into one culture.

GROWTH-PROMOTING SUBSTANCES

The possibility of growth being promoted by small amounts of obscure chemical substances is indicated by the well-known work upon vitamins in connection with the growth and well-being of man and certain other mammals and of birds. The exact application of the facts developed in connection with work on vitamins with animals at the level of group life with which we have been dealing is at present unknown, since practically no work has been done upon the vitamin relations of the invertebrates and little upon those of the lower vertebrates. From the work upon the higher vertebrates we know that vitamins A and B are both growth-promoting substances whose absence from the diet leads to serious disturbances and finally to death, and whose presence even in minute amounts promotes the normal metabolic processes which result in growth.

The possibility of growth-promoting substances being concerned with the physiological effects of groups of animals upon the individuals composing the group is further indicated by the work on "bios." The literature on this subject is voluminous and confused. Tanner (1925) presents an exhaustive review and bibliography of the researches from 1860 to 1924. "Bios" is the name provisionally given by Wildiers (1901) to a mysterious organic substance which he believed to be necessary for the proliferation of yeast cells. After approximately a quarter of a century of work upon the subject, Tanner summarizes the situation regarding bios as follows:

"One group of investigators denies the existence or need on the part of the yeast plant, of a substance like 'bios.' They feel that yeasts will grow without this accessory substance.

"Another group believes that 'bios' is necessary for the growth of yeasts. They are unable to secure growth of yeasts in pure solutions without it. Certain of these investigators have reported fractionation of 'bios' into components which are necessary to one another.

"A third group of investigators believe that yeast will grow in pure nutrient solutions without 'bios' but that the addition of a 'bios' containing substance may cause increased growth. Whether this acceleration in growth following the addition of a 'bios' containing substance is due to 'bios' or to some other factor in the preparate has not been satisfactorily established. In this connection it is well to point out that even a medium such as beer-wort which is rich in 'bios' may be improved by the addition of other 'bios' containing substances.

"A fourth group may also be recognized including those who have isolated 'bios' or substances having 'bios' properties."

Throughout his review Tanner's attitude is satisfactorily critical; and from a study of it, supplemented with certain of the original research reports, it seems to me that the evidence favoring the view that there is a growth-promoting substance which markedly stimulates yeast growth is too strong to be disregarded at the present time. Concerning whether the yeast cells are able to synthesize this substance from nutrient solutions lacking it, as has been claimed, the evidence is not yet so clear.

The same problem in a somewhat different guise is met with in the studies concerning whether inorganic substances taken alone are adequate for the growth of green plants. This question was most recently raised by Bottomley (1915 and subsequent papers) and Mockeridge (1920, 1927), who showed that certain complex organic substances, when partially broken down by bacterial action, stimulated the growth of *Lemna* and other water plants to a marked degree. Bottomley gave the name "auximones" to these substances which were effective in promoting growth for green plants. It soon became apparent that the green plant *Lemna* can grow and multiply for indefinite periods in a purely inorganic medium (Clark, 1924, 1926; Ashby, 1929), and Wolfe (1926) was led to the point of view that Bottomley's theory of the need of growth-promoting substances by green plants was completely refuted. Ashby made a more complete analysis of the problem (1929*a*) and has demonstrated that small amounts of organic substance obtained from fresh horse dung will increase the rate of growth of *Lemna* if present in only 0.2 parts per million and that the growth-rate is little affected by additions of this material beyond 2.0 parts per million.

The duckweed which Ashby used in these experiments had been growing for 6 months on a purely inorganic medium made up in glass-distilled water. Environmental conditions such as light, temperature, and pH were adequately controlled. The mean frond weight remained the same in control and experimental solutions; the mean frond number increased 42 per cent; the area of the fronds, 33 per cent; and there were roughly 80 per cent more chloroplasts in the cells of fronds treated with 0.002–0.02 grams per liter of dry extract, as compared with untreated fronds.

The addition of 0.2 parts per million of organic matter to a solution containing already 1,210 parts per million of mineral matter will significantly increase the growth of *Lemna*. The power of the extract is not affected by autoclaving; hence the effect is not due to an enzyme. The ash constituent of the extract does not increase the growth-rate, and increasing the nitrogen content by adding 0.003 grams per liter of KNO_3 did not affect the growth, while adding one-hundredth of this amount of nitrogen as organic matter did significantly increase the rate of growth. It seems clearly established by this work that, while Bottomley's "auximones" are not essential for plant growth, the addition of extremely minute amounts of organic material produces this effect by acting as a catalyzer.

The work with vitamins, tissue extracts, bios, and auximones indicates clearly that the presence of very small amounts of organic material may strongly affect the growth processes of organisms present. Relatively slight conditioning by the products of plant or animal metabolism produces marked results, which are not necessarily increased by increasing the amount of material present. One reason for the failure to recognize the presence and effectiveness of such compounds lies in the exceedingly minute minimal quantities necessary to produce maximal effects. Unless especial care is taken, traces of such materials will be present as contamination and will produce as great an effect as if more were added.

CHAPTER X

STIMULATING EFFECTS OF CROWDING ON THE RATE OF REPRODUCTION

In a preceding chapter we have seen that there is much support for the conclusion that crowding decreases the rate of reproduction among animals generally, with specific instances among the Protozoa; the Crustacea, of which *Daphnia* is an example; in the insect *Drosophila;* and among birds. We have also seen that there is at times an optimum crowding for the growth-rate, which does not necessarily coincide with the minimum population density. It is now necessary to examine whether the evidence that has been advanced demonstrates a similar optimum, at least for certain animals at some time in their life-cycle, in so far as their rate of reproduction is concerned.

The phenomenon which we are to discuss may deservedly be called "Robertson's phenomenon," since Robertson was most active in collecting evidence of its existence. He himself gave it the name of "allelocatalysis," which he defined (1924*a*) as meaning "the acceleration of multiplication by the contiguity of a second organism in a restricted volume of nutrient medium."

His announcement of the existence of this phenomenon naturally awakened an immediate interest among biologists generally and among students of the Protozoa in particular, many of whom have been unable to confirm its existence. Hence it becomes necessary to examine the development of the problem in an effort to evaluate the results of work centered on it.

Robertson (1921*a*, 1923, 1924*a*, 1924*b*) found that when two infusorians, *Enchelys* and *Colpidium* (later identified as *Colpoda*), are introduced into the same restricted amount of fresh culture medium, the early rate of reproduction following a period of readjustment, called the "lag period," is not merely double that shown when a single infusorian of the same species is similarly treated, but is some multiple in excess of this. He reports that he has obtained a rate of reproduction from two and a half to ten times that which might otherwise be expected. In his later work Robertson found this effect to be more marked when the transplants are freed from contamination with the parent culture medium by repeated washings. The increased rate of reproduction, which Robertson calls the "allelocatalytic effect," does not depend upon conjugation, because this does not occur within the conditions of the experiment. Robertson attributes the stimulation to the diffusion of some agent from the organisms into the culture medium, which accelerates their reproductive rate. When more than one organism is initially present, the concentration of this substance within the organism is higher; and the rate of multiplication is increased as a direct consequence.

Table VIII gives Robertson's allelocatalytic data from his 1921*a* paper, showing the effect of placing 2 *Enchelys farcimen* together in a single drop (0.08 cc.) of culture medium as compared with isolations of single individuals into the same amount. The figures given include all cases recorded in this paper, except those which Robertson says were run under conditions unfavorable for allelocatalysis but which were run and described to test out the conditions under which allelocatalysis might occur. Omitted cases include those in which the parent culture was over 3 days old; those in which one of the two was purposely killed; and those introduced into bacteria-free media.

Jahn (1929) calls attention to the fact that the averages of the generation time, during the first observation after isolation, in four much cited experiments of Robertson's, Nos. 238A, 240A, 237A, and 242A, show a variation of ±31 per cent, about a median of 9.1 hours; and that one pair of experiments, Nos. 310A and 311A, with 1 isolated individual and one culture of 2 individuals, shows a similar variation of ±20 per cent. He reasons that since Robertson has listed these experiments in several publications they must be typical, but that the acceleration in experiment No. 311A is less than the variation in 1-animal cultures and that therefore all Robertson's results must be questioned.

However, enough data has been cited in Table VIII to show that

TABLE VIII

Showing Robertson's Original Allelocatalytic Data

(The Data in the Last Column Are Based on Those of the Preceding Column, Restated To Show Directly the Effect of the Presence of a Second Organism in the Same Limited Amount of Medium)

Culture No.	No. of Animals Introduced	No. after 24 Hours	Ratio	Ratio per Single Original Animal
Normal Hay Infusion; Parent Culture 24 Hours Old				
15B	1	12	1:5	1:2.5
16B	2	60		
19B	1	16	1:5.6	1:2.8
21B	2	90		
36A	1	60	1:4.3	1:2.15
37A	2	260		
64B	1	24	1:4.2	1:2.1
65B	2	102		
55B	1	14	1:3.4	1:1.7
56B	2	48		
251A	1	14	1:2.7	1:1.35
252A	2	38		
251B	1	10	1:8.5	1:4.25
252B	2	85		
55A	1	14	1:4	1:2
55B	2	56		
Normal Hay Infusion; Parent Culture 48 Hours Old				
41A	1	4	1:7.5	1:3.75
42A	2	30		
41B	1	5	1:4.4	1:2.2
42B	2	22		
62A	1	8	1:2.5	1:1.25
63A	2	20		
62B	1	7	1:2.9	1:1.45
63B	2	20		
188A	1	2	1:4	1:2
189A	2	8		
188B	1	2	1:5.5	1:2.75
189B	2	11		

TABLE VIII—*Continued*

Culture No.	No. of Animals Introduced	No. after 24 Hours	Ratio	Ratio per Single Original Animal
Normal Hay Infusion; Parent Culture 48 Hours Old—*Continued*				
242A.............	1	8	1:3.9	1:1.95
243A.............	2	31		
242B.............	1	10	1:3.4	1:1.7
243B.............	2	34		
Normal Hay Infusion; Parent Culture 72 Hours Old				
23A..............	1	1	1:6	1:3
24A..............	2	24		
23B..............	1	2	1:3	1:1.5
24B..............	2	6		
191A.............	1	2	1:3	1:1.5
192A.............	2	6		
191B.............	1	1	1:6	1:3
192B.............	2	6		
202A.............	1	31	1:3.16	1:1.58
203A.............	2	98		
Normal Hay Infusion; Parent Culture Age not Given				
224A.............	1	8	1:3	1:1.5
225A.............	2	24		

they cannot be so easily dismissed. As will appear shortly, I hold no brief for the allelocatalysis theory as stated by Robertson; but it is only fair to note, as Robertson himself states (1927), that in order to be closely comparable, experiments such as these must have the same history and be run at the same time, a fact well known to experimental workers in this and other related fields of physiological zoölogy. It is therefore hardly fair criticism to compare a single allelocatalysis experiment run at one time with another run later. Fortunately, we do have a statistical method of analysis which takes into consideration the biological sensitiveness of paired experiments, that is, "Student's" method (Student, 1925; Fisher, 1925). Applying this method to the results listed in Table VIII, we find that there are 128 chances in 10,000 of random sampling giving as large differences in the ratios per individual animals as those given in the first eight comparisons listed, which is well within the range of statistical significance. It will be noted that these include all the experiments cited by Robertson (1921*a*) that were run in normal hay infusion with the parent culture 24 hours old. If one extends the inquiry with Student's method to include all the cases listed in Table VIII, one finds that the results are almost statistically perfect; that under no conditions would random sampling be expected to give the results shown in that table.

But the fact that the results are statistically significant does not necessarily mean that they support a given hypothesis. It merely means that in the case of Robertson's experiments here cited, *Enchelys* divided significantly faster when 2 individuals were placed together in a drop of culture medium, as compared with 1 individual of similar history similarly treated.

Robertson thought at first (1922) that old *Enchelys* culture fluid contained no substances that were toxic to the Infusoria isolated into it from young cultures; but as evidence accumulated, he changed his opinion to conclude (1924) that "the ultimate cessation of reproduction in old cultures is attributable also to the accumulation of a product of growth, and possibly of the same product that was originally responsible for the acceleration." The presence of an accelerator Robertson finds indicated by: (*a*) the accelerative effect due to the addition to the culture medium of a small proportion of culture fluid which was previously inhabited by the infusorian *Enchelys;* (*b*) the increase of reproductive rate when the volume of fluid is reduced so that less of the accelerator passes into the fluid and more remains within the organism; and (*c*) the fall of the reproductive rate when one of two recently divided individuals is isolated into fresh culture medium, in comparison with the reproductive rate of the other left in the original medium.

Originally Robertson did not believe this phenomenon to be due to a heavier growth of bacteria, although the results are said to depend on the presence of an optimum amount of bacteria in the culture medium. Later, after the evidence furnished by Cutler and Crump showed that the growth of *Colpidium* is much more facilitated by the "contaminating bacteria" which they found ordinarily associated with this infusorian than it is by *Sarcina*, which they employed as a food organism, Robertson (1927) considered the possibility that *Colpidium* and other ciliates are restricted in their food to certain species of bacteria for which they modify the nutrients available, thus existing in a sort of symbiosis with their food species. However, he concludes that whether the allelocatalytic effect originates with the ciliates or with associated food organisms, it remains a mutually accelerative effect of contiguous organisms upon their reproductive rates.

Robertson shows that his results are not due to the carrying-over of twice the amount of parent culture medium when 2 organisms, in place of 1, are transplanted; for in his experiments infusorians washed with medium similar to that into which they are subsequently subcultured show the effect even better than unwashed animals. Neither is the effect due to mere numbers, for recently divided or dividing individuals, if transferred together, act as single animals.

To explain the observed results, Robertson advances the following hypothesis (1923). During nuclear division each nucleus retains the charge of autocatalyst with which it was provided, and adds to it during the course of nuclear synthesis. At each division the autocatalyst is shared between the nuclear substance and the surrounding medium in a proportion determined by its relative solubility and by its affinity for chemical substances within the nucleus. The mutually accelerative or allelocatalytic effect of contiguous cells is due to each cell's losing less of the autocatalyst to the medium because of the presence of the other. According to this hypothesis, the autocatalytic effect of growing colonies, whether attached or detached, is due to the same cause.

Fischer's work (1923), in which he found that fibroblasts grow *in vitro* only when tissue cells are numerous and close together, can be interpreted as giving supporting evidence to the allelocatalytic hypothesis. Burrows (1924), growing cancer cells *in vitro*, found a similar stimulation; this and other similar evidence from tissue culture has been reviewed at some length in the preceding chapter. Similar evidence from the field of general bacteriology will be presented in chapter xv under the heading proposed by Churchman (1920), "Communal Activity of Bacteria."

On the other hand, Cutler and Crump (1923), in cultures of *Colpidium*, failed to find the allelocatalytic effect when the volume of the medium was reduced; and Greenleaf (1924), using Paramecia and *Pleurotricha*, in a brief note records his failure to confirm Robertson's work. Peskett (1924, 1924*a*) observed such stimulation to division in but 3 cultures of yeast out of 128 examined.

Robertson (1924*a*, 1924*b*) re-examined the problem in the light of results published up to that date, and explained the lack of success of other workers as being principally due to their failure to wash the organisms before transfer. His reasoning here (1927) is as follows: Presumably, if contiguous animals can affect each other's growth without the occurrence of conjugation, they must do so through the agency of some soluble substance which they emit and which is transferred from one organism to another through the medium which they inhabit. Presumably also, this soluble substance must be very abundant in the thickly inhabited cultures from which subcultures are usually prepared. This leads, of necessity, if Robertson's reasoning is sound, to the removal of the parent culture medium from the animals by washing before transfer if allelocatalysis is to result. Robertson reports (1924*a*) that allelocatalysis increases with progressive removal of *preformed* catalyst until a maximal effect is reached just before its total removal. In a preceding paper (1924) he presented evidence that washing not only removed adherent autocatalyst but also washed out the accumulation of this hypothetical substance from the living cells themselves.

Robertson also emphasizes (1924*b*) an earlier statement that in comparing the reproductive rate care must be taken to estimate the population some time before it has attained maximal density. The end result of introducing a second individual is to reduce the rate of reproduction, since the final maximum is the same in all cases and is independent of the size of the seeding.

Peskett (1925, 1925*a*) returns to the problem of possible allelocatalysis in yeast. Apparently he used very careful technique, washing his transplants and plotting a growth curve based on a number of examinations at different phases of the culture. He finds that in 46 cultures with synthetic media whose volumes ranged from 2 to 146 cu. mm. he cannot attribute any differences observed to changes in volumes, as would be required if Robertson's hypothesis of allelocatalysis holds for yeast. With 57 cultures, containing bios and varying from 0.5 to 362 cu. mm., he observed a tendency for more rapid growth in the cultures with the larger volumes. In only 2 cases out of 8 near the minimal volume did he find evidence of acceleration in the cultures. The statistics on mortality of the cultures show that those started with 2 cells usually have a higher viability but that the percentage of difference is not great. He concludes that allelocatalysis does not occur in the yeast studied.

Cutler and Crump (1925) also repeated their experiments with *Colpidium*, thoroughly washing their animals before transfer. They again failed to obtain evidence of the stimulation demanded by Robertson's hypothesis. It may be that their technique differs sufficiently from that of Robertson to account for some difference in results, since the infusorians which Robertson washed were frequently injured in the process while Cutler and Crump found no deleterious effect from their washings.

Calkins (1926) gives the results of 60-day experiments with *Uroleptus*, in which 1, 2, 3, and 4 animals were introduced into 1 drop of fresh medium, respectively. The division rate was found to be inversely proportional to the initial number of individuals inoculated.

Returning to this problem in 1927, Robertson states that all of the data and conclusions concerning allelocatalysis in Infusoria which have been issued from his laboratory in recent years remain valid, save that they may apply to the associated food organism and not to the Infusoria themselves.

Robertson, in further considering the possible reasons for the discrepancy between his own work and that of his critics, stresses the need of being sure that the animals isolated to test for allelocatalysis should have the same division rate and that this effect is shown only when more than one living cell is present and that the presence of a second cell which has died does not affect it. He particularly criticizes the work of Cutler and Crump and of Peskett because of the high death-rate in their subcultures, not only among the inoculum but also, as Cutler and Crump have shown, among the animals produced in the subculture itself. He also criticizes the artificial medium used by the latter investigators, and presents evidence to show that their synthetic medium is not well buffered above pH 7.8, which is the optimum region for the growth of *Enchelys*. Data are also presented showing the great difference in viability of different Infusoria from a crowded culture, and he suggests that a part of the discrepancy between his own work and that of Cutler and Crump is due to differences in the protozoan-bacteria ratio on account of the different techniques employed.

Greenleaf (1926) reported in full his experiments along this line, using *Paramecium aurelia*, *P. caudatum*, and *Pleurotricha lanceolata*, inoculated in from 2 to 40 drops and *Stylonychia pustulata* in 2 and 5 drops of culture medium, to determine the influence of volume of culture medium on the rate of cell division, and the influence of cell proximity on this rate, and to investigate the factors concerned with the lag period. In the experiments covering the first point, Greenleaf reports the results by means of giving the average per diem rate carried through a period of 5 days. Under these conditions, in 4 experiments only, out of 27, were 2 drops found to be more favorable for reproduction than 5 drops. Similarly, in only 4 experiments out of 27, did the 5-drop cultures give a higher rate of reproduction than the 20-drop cultures. In the comparisons between the 20- and the 40-drop cultures, the former gave a higher rate in 8 experiments; the latter in 9 cases; and 1 was a tie.

Again, in his investigation of the effect of 2 animals present in a given volume, 2 or 5 drops, he used the average per diem division rate based on 5-day periods and found that the single animals divided the more frequently in both cases, although he did find that "there was less difference in the division rates of the animal carried alone and the two carried together in the 5-drop series than in the 2-drop series. This indicates that in the larger volume the depressing effect of the presence of 2 animals is less marked than in the smaller volume." He did not attempt to carry this further to find the relative effect of a further increase in volume.

Greenleaf's work is chiefly to be criticized because of his taking average division rates for a period of 5 days instead of following the actual division rate per day—the more so since Robertson had previously called attention to the necessity of such care and Peskett (1925) had used particular care to examine his cultures at fairly frequent intervals in his later studies upon the rate of growth in yeast cells.

In keeping with his other results, Greenleaf found that isolating infusorians into 5 drops of culture medium causes a higher rate of reproduction if the cells are re-isolated after 24 hours than if they are allowed to stand for 48 hours in the same medium.

Myers (1927) worked with washed *Paramecium caudatum* and found that "increasing the numbers of individuals present in a given volume does not increase the rate of reproduction either from the beginning of the cultures or after the first fission." In his work he used 2–16 drops of culture medium, measuring from 0.1 to 0.8 cc. His observations were usually made three times per 24 hours, at intervals of 6, 6, and 12 hours.

Petersen (1929) justly comments on a part of this work of Myers, showing that his system of taking observations at 6-, 6-, and 12-hour intervals may mask significant results since an indicated interpretation of his data shows that he does not regard changes of from 6.0 to 8.4 hours before the first division time as being significant, although the higher number represents an increase of 40 per cent. Of course, the technique used may require greater differences before significance is reached.[1]

[1] In the light of this criticism supported by tabular evidence, it is hard to understand Jahn's (1929) difficulty which causes him to assert that Petersen has selected certain data from two sets of Myers' experiments. Actually she used all the data published by Myers concerning 16-drop cultures which bore on this point, and lists all three sets of these experiments both in her summarizing table of Myers' work and in the accompanying discussion. From other remarks concerning Petersen's work, Jahn seems to have overlooked the fact that she was interested in contrasting the effects obtained by testing for allelocatalysis in small and in relatively large volumes of medium, from which came her particular attention to the results reported by Myers from his 16-drop

Yocum (1928) reported results resembling allelocatalysis from his work with *Oxytricha.* He isolated unwashed individuals to 4 and to 10 drops of sterile medium when 100 drops equal a cubic centimeter. After 24 hours he found that when all cases were averaged, the division rate in the smaller volume exceeded that in the larger by 14 per cent, and that the average in the smaller volume in each of his series was greater than in the larger volume by over 10 per cent. He interpreted this as meaning that the smaller volumes reach a high concentration of an autocatalyst earlier than do the larger volumes.

The work of Petersen (1929) chronologically belongs at this point; but it will be examined in more detail shortly, after considering the work of Jahn (1929), who investigated the relation between density of population to the growth rate in *Euglena* sp. *Euglena* was chosen for its ability to grow in autotropic media. Single isolations were abandoned on account of excessive evaporation. Of 22 cultures started with washed sister-cells isolated into 3 or into 6 drops of medium, 5 cultures were counted at the end of 10 days. Three of these showed larger numbers in the smaller volume, one showed the reverse, and the other had about equal numbers. No data are given for the intervening period.

Thereafter mass cultures were run in which the ratio of the initial volume of individuals to medium ranged from 1:383,300 to 1:3,833,-000. Results are recorded as obtained by a series of dilutions of from 0.005 to 0.55, 0.066 to 0.66, 0.089 to 0.0894, 0.081 to 0.73, 0.14 to 0.555, and 0.42 to 9.5 thousands per cubic centimeter of culture fluid. Initial counts were made in from 1 to 4 days in the different series. In Series 2, 3, 5, and 6[1] the initial count in the more concentrated medium was made approximately 1–3 days before that in the least crowded series; hence it is hard to judge initial conditions.

Only in Series 1 do the recorded data show the initial count made at the same period in the different experimental lines; and here the

---

cultures. It may be, as Petersen suggests, that the method of recording data chosen by Myers does not allow his results to be used as evidence for allelocatalysis. If that be true, there are similar reasons for thinking that, in so far as the 16-drop cultures are concerned, they cannot be used as evidence against this interpretation.

[1] The time of making the initial count in Series 4 is not given.

results at the end of the first day's exposure clearly bear out the author's conclusions that his mass experiments with *Euglena* offer no evidence of any allelocatalytic effect. As would be expected, from all of the work to date, Robertson's included, after the first few days Jahn found a significant difference in rate of increase, with a more rapid rate in the less dense populations, which he is inclined to ascribe to the combined effect of relatively more food and relatively less wastes. The effect of mass of *Euglena*, which is capable of growing in autotropic media, may well differ from that of holozoic organisms, particularly since their CO_2 relations diverge so decidedly.

The work which throws most light on this tangled problem is that of Petersen (1929). Petersen used *Paramecium caudatum* in her experiments. They were grown in a boiled-hay medium which was bacterized 24 hours before using. Animals were isolated by means of a drawn-out pipette. Any attempted washing was done by re-isolation. Throughout her experiments 24 drops equaled 1 cc.

The usual method of comparing individuals of the same clone was not followed in this work because it was found early in the investigation that there may be a marked difference in the division rate of animals from the same clone. Even the division rate of sister-cells showed a difference. Her method, as finally worked out, was to isolate 2 animals with approximately the same rate and time of division. These were allowed to divide until 4, or 8, or 16 cells were produced, according to the needs of the experiments, at which time all the descendants of one animal were subjected to one set of conditions while all from the other were placed under the conditions with which comparisons were to be made. Because of the re-isolation so early in the life of the subculture, the lag phenomenon was avoided.

Under these conditions, four sets of experiments were run, as follows: (1) cell proximity in small volumes of medium; (2) variation in volume of medium; (3) cell proximity in large volumes of medium; and (4) conditioning of medium, both large and small volumes. Petersen's results will be given in some detail because of their importance.

Preliminary experiments in 2 drops of medium gave some slight indication of a stimulating effect upon the rate of reproduction when 2 organisms, instead of 1, were present; and two series of experiments were run to investigate this tendency. In the first series the animals were washed once; in the second they were isolated unwashed. Examinations at 12, 18, 24, 36, and 48 hours failed to reveal any indication of acceleration due to the presence of 2, as compared with 1 animal, in the original subcultures, either with the washed or with the unwashed cultures. Again, contrary to the report of Robertson, there was no speeding up of the division rate after the first fission. Further experiments, made with small volumes after Petersen had found that a different relation exists when animals are subcultured by one's, by two's, and by four's in larger volumes, still gave the same negative effects. With these small volumes the work of Petersen is in agreement with that of Myers and of Cutler and Crump rather than with that of Robertson.

Because of the evidence of retardation in the early experiments, Petersen thought that the volumes used might be entirely too small. Acting on this lead and using 20 drops of culture medium (24 drops equal 1 cc.), Petersen transferred 1, 2, and 4 Paramecia, which had been washed three times, into this volume of bacterized medium, with the results shown in Table IX. In these experiments there is evidently an acceleration in the early division rate of the subcultures, associated with an increase in the number of animals introduced. Further experiments shown in Tables X and XI demonstrate that this is not an accidental result.

The question as to whether the accelerative effect is species-specific was apparently answered for this particular organism by the fact that in 5 sets of experiments similar to those reported above which were found to be noticeably infested with a small ciliate after the first 24 hours of isolation, the rate of division was nearly or quite independent of the initial seeding. Similar results were obtained with the only other set of contaminated cultures entering into this series of experiments.

So far, these experiments of Petersen's have shown that with small volumes, $\frac{5}{24}$ cc. or less, there is no acceleration with the introduction of a single individual in the initial subculture, but that with

TABLE IX

Showing Division Rate of Sets of 4 Paramecia Transferred Singly, in Two's, and in Four's, to 20 Drops of Bacterized Medium. Washed Twice in Bacterized and Once in Alkalinized Medium
(Data from Petersen)

Transfer	One			Two			Four		
Hours	0	24	48	0	24	48	0	24	48
	4	5	9	4	5	16	4	8	18
	4	4	9	4	5	10	4	5	12
	4	5	9	4	6	13	4	6	16
	4	4	11	4	5	20	4	7	16
	4	4	9	4	6	10	4	6	14
	4	8	16	4	8	17	4	8	24
	4	6	12	4	8	24	4	8	24
	4	7	14	4	8	16	4	8	16
	4	7	16	4	14	24	4	14	24
	4	9	10	4	8	12	4	10	14
Total	40	59	115	40	73	162	40	80	178
Mean division rate	......	1.47	2.87	......	1.85	4.05	......	2.0	4.45

TABLE X

Showing the Effect of Double Washing and Transfer of Paramecia to 20 Drops of Alkalinized Bacterized Medium
(From Petersen)

Transfer	One			Two			Four		
Hours	0	24	48	0	24	48	0	24	48
	4	6	21	4	8	30	4	8	38
	4	7	12	4	8	16	4	10	36
	4	6	25	4	7	34	4	8	37
	4	8	23	4	8	30	4	8	52
	4	6	22	4	8	34	4	8	40
	4	5	28	4	5	34	4	7	42
	4	7	25	4	8	47	4	9	56
	4	6	9	4	8	22	4	8	33
	4	7	18	4	8	32	4	8	38
	4	6	16	4	7	18	4	8	20
	4	6	14	4	8	16	4	8	23
	4	5	22	4	8	17	4	8	24
	4	5	14	4	6	18	4	8	20
	4	6	28	4	7	34	4	7	40
Total	56	86	277	56	104	382	56	113	499
Mean division rate	......	1.53	4.94	......	1.85	6.81	......	2.01	8.80

large volumes, $\frac{20}{24}$ cc., such acceleration does occur; with still larger volumes we begin to see that the effect is in reality a relationship between the mass of animals and the mass of culture media. Inoculations similar to those made above, but made into 40 drops of medium, showed in 48 hours some evidence of acceleration of division rate with an initial seeding of 4, but with this volume 2 individuals reacted as did 1. Whatever the process of conditioning the medium,

TABLE XI

SHOWING THE EFFECT OF DOUBLE WASHING AND TRANSFER OF PARAMECIA TO 20 DROPS OF UNALKALIZED BACTERIZED MEDIUM
(From Petersen)

Transfer	ONE		TWO		FOUR	
Hours	0	24	0	24	0	24
	4	14	4	16	4	16
	4	16	4	16	4	16
	4	15	4	16	4	16
	4	20	4	28	4	30
	4	18	4	16	4	28
	4	14	4	24	4	30
	4	19	4	22	4	26
	4	21	4	26	4	26
	4	16	4	27	4	30
	4	18	4	23	4	31
	4	19	4	24	4	26
	4	21	4	26	4	30
Total	48	211	48	264	48	305
Mean division rate	....	4.39	....	5.5	....	6.35

2 were no more effective in 48 hours with 1.67 cc. of medium than was a single individual, while 4 were significantly more effective than either. These results are exhibited in detail as Table XII.

When the numbers produced in 20-drop cultures initially containing 1 and 4 individuals are examined by Student's method for determining the statistical value of paired experiments, one finds, if all of Petersen's comparable data are considered, that there are 2 chances in 10,000 of obtaining so great a deviation from the mean by random sampling. When the similar data for the 40-drop cultures are similarly considered, the probability, even from this short series of experiments, is shown to be slightly less than 5 chances in a 100, which is usually considered to suggest statistical significance.

Petersen approached this problem from another angle. Experiments were carried on to determine whether or not the division of an animal in a limited amount of medium would accelerate the rate of division of individuals isolated into the same volume. Both washed and unwashed Paramecia were so tested. First, washed animals were

TABLE XII

PARAMECIA WASHED THREE TIMES AND TRANSFERRED SINGLY, IN TWO'S, AND IN FOUR'S, TO 40 DROPS OF ALKALINIZED, BACTERIZED MEDIUM

(From Petersen)

Transfer	One			Two			Four		
Hours	0	24	48	0	24	48	0	24	48
	4	8	16	4	8	16	4	8	16
	4	8	16	4	8	16	4	8	16
	4	8	16	4	8	16	4	9	19
	4	4	4	4	4	4	4	4	4
	4	4	4	4	4	4	4	4	4
	4	4	4	4	4	4	4	4	8
	4	8	16	4	8	16	4	12	18
	4	8	16	4	8	16	4	10	16
	4	8	16	4	9	16	4	10	16
	4	8	32	4	8	32	4	16	40
	4	8	22	4	8	32	4	8	24
Total	44	76	162	44	77	172	44	93	181
Mean division rate	......	1.75	3.69	......	1.75	3.69	......	2.11	4.11

isolated into 2 drops of bacterized medium. At the end of 12 hours the animals were removed, and the slides in which fission had occurred were noted. Five sets, consisting of 8 animals each, each set being the fission products of 1 animal, were isolated singly into the drops in which fission had recently occurred; and comparable isolations were made into those drops where animals had lived for the same length of time but without division. There was no significant difference in division rate at the beginning or at the end of the experiments, on washed or unwashed individuals.

Petersen says: "This is not what would be expected on the basis of Robertson's work. His reports indicate that acceleration of division rate should appear after the first division in a limited volume, due to the liberation of an autocatalyst from the nucleus to the pericellular medium at the time of division.

"Another attempt to accelerate division rate by the use of conditioned medium was made with cultures of large volumes. Cultures of 20 drops in which an animal had lived and divided were inoculated with isolated unwashed animals and their division compared with that of other comparable animals also unwashed, which were isolated to 20 drops each of fresh bacterized medium.

"Animals isolated into the 20 drops in which one individual had divided and lived (d) and animals isolated into 20 drops of fresh bacterized medium (n.d.) at the end of 18 hours gave the following numbers respectively.

d............4, 2, 4, 4, 2, 2, 3, 3, 2, 3, 2, 2, 2, 2, 2, 2—41
n.d...........1, 1, 1, 2, 1, 1, 1, 1, 1, 1, 1, 1, 2, 1, 1, 2—19

"Thus indications are that an animal living and dividing in 20 drops of medium conditioned it in such a way that an acceleration in the division rate of other individuals isolated into it is marked.

"Apparently in the small volumes, one individual is able to condition the medium sufficiently to cause rapid multiplication; and therefore when one animal is transferred to a small volume of conditioned medium, its rate of division is not faster than that of a similar individual transferred to unconditioned medium. But in large volumes, one individual is not able to condition the medium immediately; and when, therefore, one animal is introduced into a large volume of conditioned medium, division takes place faster than in the case of a similar individual introduced into unconditioned medium.

"Greenleaf and Myers in work with conditioned medium used only small volumes, 2–5 drops. Both report a depressed division rate in the conditioned medium, which they interpret as due to the increased toxicity of the old conditioned medium."

The importance of Petersen's work lies in the fact that for the first time in the development of this phase of the problem of the effect of initial numbers on the rate of reproduction in relatively small subcultures, she has been able to obtain both positive and negative results at will, depending on her manipulation of volume relations. Her work and discussion make it impossible for one to lump together all volumes in testing for the possible stimulating effect of the presence of more than 1 cell in a limited amount of medium.

Here we shall have to let the discussion of Robertson's phenomenon rest for the time being, pending the accumulation of further data. So far, the evidence demonstrates that reproductive rate does not always depend on the number of the original transplants; that, in fact, under many conditions and with many organisms it has not yet been shown to have any positive correlation. All are agreed that in the end the rate of reproduction falls sooner in small subcultures seeded with more than one organism and more rapidly than it does when a single individual is isolated. However, the repeated experience of Robertson, supported now by Yocum's work in so far as it covers the same ground, and by the work of Petersen with larger volumes of culture medium, demonstrates that true acceleration of the rate of reproduction may occur associated with the introduction of more than one organism into a limited amount of medium.

When the very establishment of the phenomenon is a matter of such great difficulty, it is to be expected that the explanation of the phenomenon, when it does occur, is still uncertain. At present we cannot accept Robertson's hypothesis of the action of an autocatalyst; and while there is evidence for the production of some conditioning agent, some *X*-substance, which renders the culture more favorable to growth, more light is needed, particularly concerning the rôle of bacteria in the phenomenon among other possible factors, before much progress can be made toward a solution of this aspect of the problem. Perhaps when we have examined the relations of masses of larger animals to survival under adverse conditions, a subject to be taken up soon, we can better understand the complexities of the present problem.

An effect strikingly similar to Robertson's phenomenon has been described by Chapman (1928), though not with that in mind. In his work upon the effect of a limited environment, in this case whole wheat flour, upon the number of confused flour beetles, *Tribolium confusum*, that will develop, Chapman found, as was stated earlier (chap. vii), that with these animals and with this medium it is possible to demonstrate that the total population per gram of flour arrives at a constant level of 43.97 ± 2.88. Chapman's table recording the population level attained at successive intervals when 2, 4, 8, 16, 32, and 64 beetles are placed in 32 grams of flour for each lot is repeated here as Table XIII.

TABLE XIII

SHOWING THE NUMBER OF BEETLES PER GRAM OF FLOUR WITH INITIAL POPULATION VARYING FROM 2 TO 64 PER 32 GRAMS. THE RESULTS ARE GIVEN IN TERMS OF BEETLES PER GRAM OF FLOUR

(From Chapman)

DAYS	GRAMS					
	32	32	32	32	32	32
0.	0.062	0.125	0.25	0.5	1	2
11.	1.15	5.5	9	16	27	42
25.	2.5	12	19	20	30	38
54.	11	27	32	30	43	55
69.	28	31	30	29	46	47
81.	30	44	42	43	53	53
104.	40	43	38	42	52	44
123.	45	43	39	50	50	46
139.	42	47	42	44	50	45

Chapman is interested in showing by these figures, as by those cited in chapter vii, that, regardless of the initial seeding, a point of equilibrium is reached after which the population remains relatively constant. His medium is distinctly advantageous in this respect, for with the Protozoa which we have just been considering, after the similarly constant point of equilibrium is reached, the population declines on account of the changes induced in the culture medium. In flour beetles, as we have said before, this decline can be controlled by changing the culture medium.

Our own interest is in examining the effect of numbers in the initial population upon the early rate of reproduction. The relationships existing for the 11- and 25-day periods shown above are given in Figure 13, which has been drawn from the data just given. The figure shows immediately that for these first periods in the development of the population the greatest increase comes with an initial population of 4, while the least rapid rate of reproduction comes with the smallest number of individuals present.

This indicates, as did the data of Petersen, that there is an optimum initial population-medium relationship, which, for certain

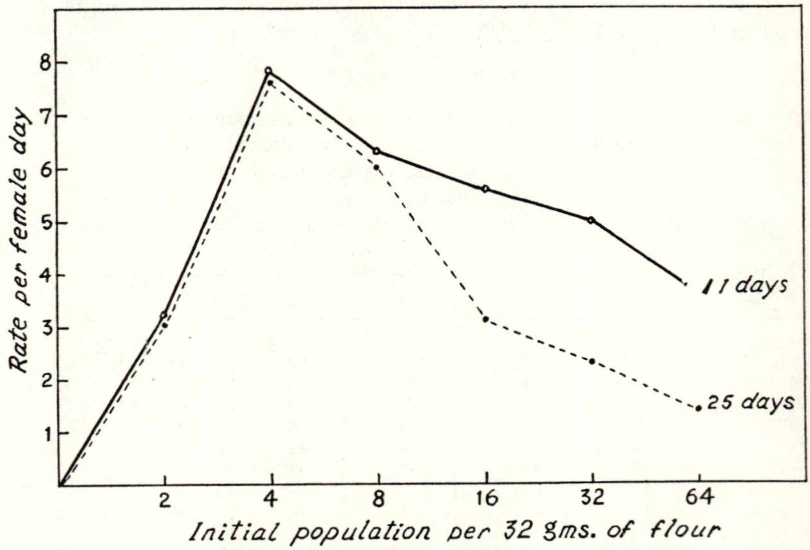

FIG. 13.—Showing the relation of initial density of *Tribolium* population in 32 gm. of flour to rate of reproduction. Recalculated from data reported by Chapman.

volumes of medium, is different from that given by the smallest population present. The fact here seems as clear as could be expected from an experiment not designed to test this particular point. More recently T. Park, working in this laboratory, has confirmed Chapman's results (unpublished data). The explanation is even less apparent than in the case of Robertson's phenomenon in the Protozoa.

CHAPTER XI

EFFECT OF CROWDING ON SURVIVAL AND OXYGEN CONSUMPTION

Studies on the causes and effects of animal aggregations begun in 1912 and carried on intermittently thereafter were summarized (Allee, 1920) for the water isopod *Asellus communis*, three species of land isopods, and the ophiurid starfish *Ophioderma brevispina*. The results showed that, in general, bunching in these species is most prevalent under adverse conditions and when there is no means of satisfying the normally positive thigmotactic reaction. Of the single factors tested, the tendency to collect in bunches is most strongly encouraged in *Asellus* by the breeding reaction, in the land isopods by the amount of moisture present, and in *Ophioderma* by the amount of light.

All of these animals had shown in preliminary experiments a lowered rate of metabolism immediately following the formation of the aggregation, as measured by their oxygen consumption or carbon-dioxide production. When isolated and bunched animals stand for long periods of time, the effect on the metabolic rate is reversed. Both come to have lower rates than at the beginning of the experiment, but the decrease is much greater with isolated than with bunched animals. In the land isopods this decrease is accompanied by a greater loss of water by the isolated individuals. The conclusion was drawn that "under laboratory conditions the formation of aggregations serves to make these animals more quiet and in the long run proves to be what is usually called an adaptive reaction." This early statement remains a fair summary of the evidence since collected in this laboratory.

In an earlier chapter attention was called to the fact that land isopods aggregate into fairly compact masses in the absence of water and separate when water is added. Experiments on two species, *Oniscus asellus* and *Cylisticus convexus*, were run to test the effect of bunching on changes in water content as indicated by weight of the animals (Allee, 1926). When the isopods were placed on a moistened filter paper with maximum moisture present with which the animals would remain bunched, the aggregated isopods increased in weight 3.8 per cent from water intake while their isolated fellows were increasing 9.7 per cent.

Similar experiments with bunches and single individuals placed under desiccating conditions showed that in a typical experiment isolated isopods lost water three times as fast as did a group of 10 isopods gathered into a close bunch. Further, all 10 of the bunched animals in this typical experiment were alive with a loss of weight of less than 16 per cent after 7.45 hours, when the last of the isolated individuals was found dead. The last 2 isopods to die showed a water loss at this time of about 44 per cent. Sixty per cent of the isolated isopods were found dead after 4.38 hours' exposure.

These observations show a definite survival value of a group of animals even at the low level of social integration existing among land isopods. It may be remarked again that, outside the breeding season, as these were, isopods collect in bunches due primarily to non-social, individual tropistic reactions, and that almost their only social attribute is tolerance for other animals and their products within a limited space.

STARFISH AUTOTOMY

The essentials of the situation among the ophiurid starfish are similar, but the mode of expression differs (Allee, 1927). Ophiurids have the practice of fragmenting their arms under certain conditions. Early experiments showed that there is a greater tendency to practice autotomy with isolated than with bunched *Ophioderma*. The tests to be considered were run in connection with respiration studies to be described later. Two groups of 8 each were placed in two large bottles each containing 8 liters of sea-water. Eight similar starfishes were isolated into flasks of about 1 liter each. The water was usually changed daily and never left longer than 48 hours. No attempt was made to supply food.

The following schedule of numerical symbols was adopted: 0, perfect arms; 1, tip gone; 2, end gone; 3, one-third arm gone; 4, one-half arm gone; 5, two-thirds arm gone; 6, mere stump left. Individuals were removed from the bottles when they showed fragmented arms yielding on the foregoing scale a total value for the five arms of a single starfish of 15–20, regardless of the distribution of effects between the different arms. Table XIV gives the mean time in days before each lot was discarded. The data at hand are faulty

TABLE XIV

SHOWING THE TIME IN DAYS IN LONG RESPIRATION EXPERIMENTS BEFORE THE DIFFERENT LOTS WERE DISCONTINUED BECAUSE OF DEATH OR THE FRAGMENTATION OF ARMS

Experiment No.	First Bunch	Second Bunch	Mean	Isolated	Difference	Glass*	Duration in Days
1	19.6	20.6	20.1	16.6	3.5	....	21
2	13.5	13.3	13.4	12.5	0.9	....	14
3	5.0	5.0	5.0	5.75	0.75	....	5
4	9.0	9.25	9.13	6.13	3.0	....	8
5	9.63	9.56	9.6	6.2	3.4	9.75	11
6	7.75	7.5	7.63	6.5	1.13	7.5	8
Mean	10.75	10.87	10.81	8.85	....	....	....
Mean of last two only	8.69	8.53	8.62	6.35	....	8.63	....

* The *Ophioderma* in this column were isolated into liter flasks containing glass rods bent into various shapes.

in that, after respiration experiments were stopped, there were frequently some individuals that had not yet reached the degree of fragmentation necessary for removal. In some cases these were observed, though under altered conditions, until autotomy had progressed past the arbitrary dead line; but in others this was not done, and the last day of respiration tests was recorded for use in this table. This procedure markedly favors the isolated individuals since there were fewer of them so treated. The length of time the experiment ran depended mainly on the temperature, since fragmentation proceeds more rapidly at high temperatures.

In every instance except one, the mean survival time was greater for the bunched animals than for those isolated under similar conditions and at the same time. In four of the six comparisons the difference is large enough to be of significance. Student's method of statistical evaluation shows that for all six experiments there are 49 chances in 1,000 of random sampling yielding so great a variation from the mean in either direction.

The one exception where the isolated individuals showed a greater survival time is instructive. For some reason one or two of the animals in each of the bunches of this experiment died soon after one of the daily inspections, disintegrated and polluted the whole liquid, and caused the death of the remainder. Such an extreme catastrophe could not happen with the isolated individuals.

The survival of the starfishes isolated into liter flasks containing small heaps of variously bent glass rods is also instructive. Here the starfishes came to rest on or among these glass rods just as in nature they rest among eelgrass blades. It will be noted that for the two tests made, the survival is comparable with that of grouped animals rather than with the starfishes isolated into bare containers. It appears as though the satisfaction of the thigmotropic appetite has approximately the same survival value whether the satisfaction comes from contact with the piles of glass rods or the group of individuals of the same species.

RESPIRATION STUDIES WITH ISOPODS

The effect of aggregation upon the aggregants can be followed more closely by studying the effect upon the rate of respiration. Several such studies have been made with different animals, some of which will be summarized.

Preliminary experiments upon the two species of land isopods mentioned above, indicated that soon after these isopods aggregate their rate of respiration is decreased, as compared with similar animals isolated for a similar time (Allee, 1926). This tendency appears after the isopods have been bunched for 5 minutes, and extends at least through the first hour of bunching. When the bunches and solitary individuals are compared after standing in the laboratory for a longer period of time, the isopods taken at random from the bunches are giving off carbon dioxide the more rapidly.

Oxygen consumption of two other species of land isopods was

tested in manometer respirometers such as were described by Krajnik (1922). The results obtained support the conclusions reached in earlier studies with carbon-dioxide production, that is, that with land isopods the recently bunched individuals are carrying on respiration at a less rapid rate than when recently isolated. In exact ratios, the rate of oxygen consumption with recently bunched and recently isolated *Armadillidium* in two tests were 1:1.486 and 1:1.368, while with *Tracheoniscus* the same ratios were 1:1.214 and 1:1.344.

The *Armadillidium* tests allow other comparisons. Since groups of isolated or bunched individuals were set away in the dark for approximately 24 hours, one can compare their physiological condition near the beginning of this period with that at the end and also make cross-comparisons.

In general, the determinations show that, under the conditions of the experiments, there is a very marked decrease in oxygen consumption after 24 hours' isolation and starvation—70 and 65 per cent, respectively, in two sets of experiments. There is a similar but less pronounced reduction when the animals are bunched and starved—in that case 31 and 29 per cent, respectively.

At the end of 24 hours the bunched isopods are uniformly using more oxygen per unit weight than are the isolated individuals. The observed ratios were 1:1.64 and 1:1.48. All the differences mentioned are statistically significant, since the least difference in means is still over eleven and a half times their combined probable error.

Experiments extending over 50 hours showed that the bunched isopods kept their higher rate of respiration during this period as compared with similar isolated individuals. The aggregated individuals exhibited a greater range in rate of oxygen consumption during this time, probably due to the greater difference between repose and occasional activity. They also were much less likely to move about than were the isolated individuals. In all these experiments neither set of isopods was fed, but the water content was manipulated so that there was no essential difference in weight, as shown by random weighings.

Calibrations show that oxygen consumption for the *Armadillidium* isolated for about 1 hour is at the rate of 219.6 cc. per hour and kilogram; for those isolated for about 24 hours it is 77 cc. per hour and kilogram; for those bunched approximately an hour, 160.5 and for those bunched for about 24 hours, 130.2 cc. per hour and kilogram. Similar calculations for the *Tracheoniscus* show a mean oxygen consumption of 263.5 cc. per hour and kilogram when isolated less than 2 hours, and of 196.7 cc. when bunched about the same time.[1]

STARFISH RESPIRATION

The marine ophiurid starfish, phylum Echinoderma, stands well removed from the land isopod, phylum Arthropoda, in the evolutionary scale and in habitat. The oxygen consumption of isolated and of bunched *Ophioderma* was tested by Winkler's method.

A typical experiment was made on three groups of 8 animals each, selected so that there would be the same size relations in each group. The members of each lot were momentarily dried on filter paper and weighed in a known amount of sea-water. The lot weighing most was placed together in a bottle holding approximately 8 liters (actually 8,500 cc.); the lot weighing next most was separated, and each individual was placed in an Erlenmeyer flask of about 1 liter capacity (actual average 1,143 cc.). The third and lightest group of 8 animals were placed together in a second bottle like the first. Appropriate blanks were run on both sizes. All bottles and flasks were fitted with rubber stoppers, which carried tubes for drawing titration samples without contact with air and with a minimum disturbance. The containers were gently reversed twice just before sampling to insure an even mixing of the sample, and other due experimental precautions were taken. Determinations were made 4 hours after the start of the experiment to determine initial relations. Later, tests were made at 24-hour intervals, except that during the middle course of some experiments they were made at 48-hour intervals.

[1] Hyman (1923) found the oxygen consumption of the anterior pieces of *Planaria dorotocephala* to be at the rate of 260 cc. per hour and kilogram. Krogh (1916) lists a rate in calories equivalent to 393.6 cc. per hour and kilogram for *Apis mellifica*, and of 72 and again of 115.2 for *Musca*. The highest rate listed in Krogh's tables for crustaceans, all gillbreathing, is of 6.528 cc. per hour and kilogram. The values given above for land isopods are subject to a calibration error of about 5 per cent.

The longer tests were terminated piecemeal, as different individuals fragmented their arms to such an extent that comparisons were no longer fair. The records show the rate of oxygen consumption, as well as the total oxygen consumption, of twelve groups of aggregated *Ophioderma* during their entire starvation period, and of six lots of similar size and number which had been isolated until they, too, died of starvation and confinement. In addition to the corrections for the slight difference in amount of water available for each animal, the crude data were further corrected for differences in weight, since tests had shown that, within the size limits used, oxygen consumption is directly related to the size of the animal, although the greatest observed spread between the total weight of 8 isolated animals and either of the accompanying bunches was only 0.4 gm.

Six tests were run to determine the initial relations in oxygen consumption of isolated and bunched individuals. When these were examined at the end of the first 4 hours, it was found that on the average the groups of larger individuals, totaling 48 in all, had consumed 208 cu. mm. of oxygen per individual of 1.25 gm. moist weight. The 48 smaller animals making up the second bunch had consumed 197 cu. mm. when reduced to the same standard weight; while the 48 isolated starfishes had consumed 379 cu. mm. per standard individual of 1.25 gm. Here we see again that at the start of the experiment the isolated individuals are consuming oxygen at a much higher rate than are their aggregated fellows when the only known difference between them is that the latter have been allowed to collect in bunches. When these results are examined statistically by Student's method, we find that there are 66 chances in 10,000 of random sampling yielding such a wide deviation from the mean. Needless to say, such results are statistically significant. This initial relationship continued in most cases past the first 12 hours, and sometimes past the 24-hour sampling, but had usually disappeared before 48 hours.

Two of these initial sets of experiments and four other similar ones were followed through with starving animals until terminated by death or by fragmentation of the arms. The mean results in terms of total oxygen consumption of the surviving members of each lot are summarized in Figure 14, which in *A* plots the mean rate per hour per living individual in the isolated (solid) and aggregated state (broken line). The broken line is drawn as the mean of the two

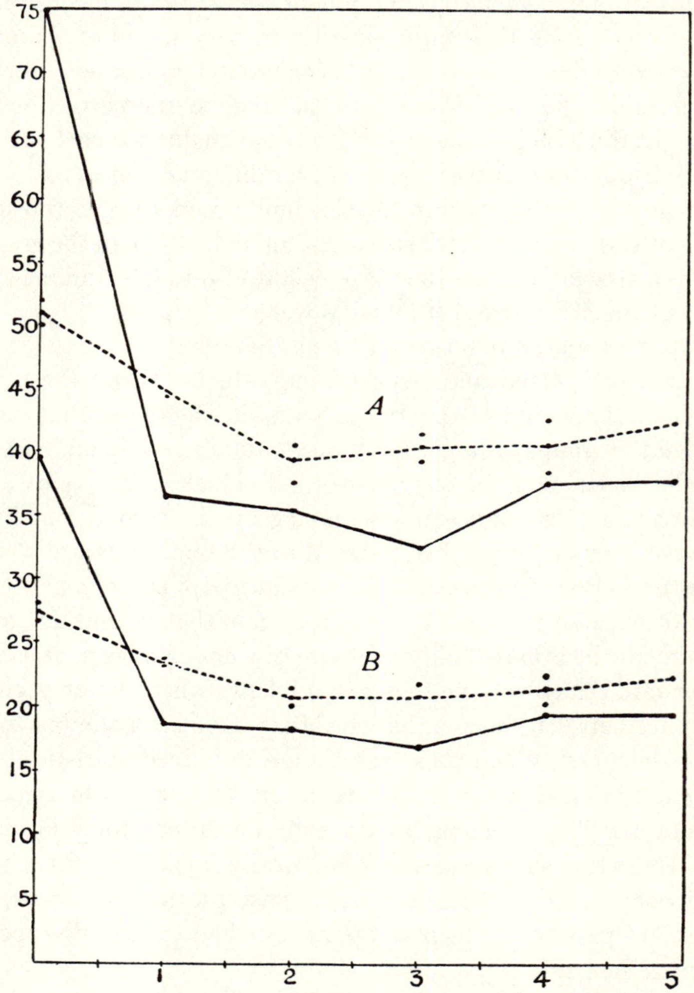

FIG. 14.—Showing, *A*, the mean rate of oxygen consumption per hour at the beginning and during successive fifths of the starvation period; each space on the ordinates represents 1 cu. mm. oxygen. *B*, The mean oxygen consumption; each space on the ordinates represents 100 cu. mm. of oxygen. The broken line gives the respiration of the bunched individuals.

series of bunches which were run simultaneously. The means for each of these are indicated for each point. The graph starts with the rate given during the first 4 hours and proceeds with that given during the different fifths of the experiments. This means of charting is used because of the varying length of the experiments to be summarized, all of which were physiologically the same length, in that they were terminated when the starving individuals being tested had reached approximately the same degree of depression as measured by death or by fragmentation of the arms.

Graph *B* shows similarly the total mean oxygen consumption for each group. The initial amount indicated is that which would have been given had the animals continued to respire for a period equaling the others in length but at the rate given in the first 4 hours of experimentation. Again the solid line represents the isolated individuals and the broken line the bunched animals, and again the spread of the means for each series of bunches is indicated. This form of presentation shows at a glance the more rapid consumption of oxygen in initial stages, particularly on the part of the isolated individuals, in contrast with the reduced rate later. The increase in the rate of respiration near the close of the experiment is due to the greater consumption of oxygen in the process of autotomy and to the inclusion of some cases in which decay of fragmented parts of arms may have occurred.

The isolated animals have a rate of oxygen consumption 186 per cent above that of the groups during the first 4 hours. Later this mean rate of consumption falls for both, but more rapidly for the isolated than for the bunched individuals, so that, when the entire course of the experiments is considered, the rate of use of oxygen by the isolated animals is only 83 per cent of that of the accompanying groups. The differences of these means have about the same statistical value as that noted for the initial 4 hours of respiration.

The foregoing statement and graph does not show the extent of the difference between the oxygen consumption of the isolated and the bunched animals, since, on the average, the former died off more rapidly than did the bunched individuals. The results reported here are in terms of the mean oxygen consumed by the animals still liv-

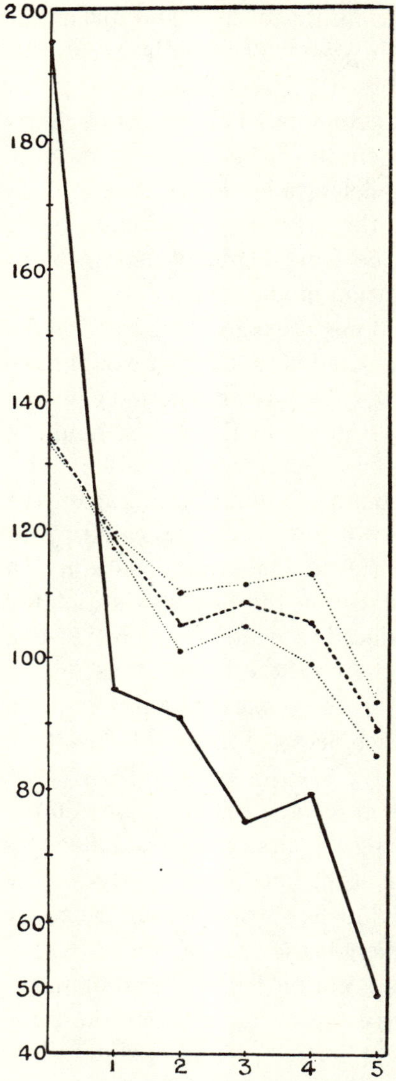

Fig. 15.—Showing the total oxygen consumption of isolated (solid line) and bunched starfish (broken lines) in the different fifths of the starvation experiments. The heavy broken line gives the mean of the two bunches shown by the lighter lines. Each space on the ordinates represents 1 cc. of oxygen.

ing; hence, in the later stages of the tests we are comparing the mean rates of the most hardy of the isolated individuals with that of a large bunched group that has been subjected to less rigid selection.

When the differential autotomy is considered in connection with oxygen consumption, it is found that the effect of isolation is much more marked than when only means of survivors are compared. These relations are shown in Figure 15, which gives the total oxygen consumed, corrected for weight of the different animal groups. Again the initial value is that which would have been consumed had the initial rate held for a period equaling the others shown. It is added only for the purpose of graphic comparison. The preceding types of analysis showed that after the initial period the rate of oxygen consumption remained approximately constant at their respective levels for both bunched and isolated individuals. On the other hand, the complete data show a decided falling-off in oxygen consumption near the end of the experiments, due to the decrease in numbers of animals present. The effect of the differential length of life under experimental conditions is shown in the earlier and greater reduction in oxygen consumption in final stages of the isolated rather than of the bunched individuals.

The possibility of consuming a similar amount of oxygen in the same time was the same for the lot of isolated as for the bunched *Ophioderma*, had they been given equal treatment. The conclusion is obvious that when similar *Ophioderma* are isolated without food in clean glass receptacles of approximately equal volume per number of individuals present, their rate of oxygen consumption, following a period of initial stimulation, is significantly depressed, and their expectation of remaining intact, or of living, is less than if they are allowed to aggregate.

FACTORS CONTRIBUTING TO GROUP PROTECTION

The question immediately arises as to the type of beneficial influence exerted by the grouped individuals upon each other. There has not been opportunity to investigate this aspect of the problem thoroughly. Certain data have been collected that are of decided interest, and these will be presented not as a final answer but as a possible guide to the ultimate solution of the problem.

The earlier students of the effects of crowding, who uniformly found only deleterious effects, supposed that there is some harmful secretion given off by the crowded animals. Similarly, workers who have reported beneficial effects of increased numbers of individuals in relation to volume of the containing water have postulated some beneficial secretion which is preservative in action. Drzewina and Bohn, in their series of studies upon this subject, in which they used a wide range of aquatic organisms, suggested that the observed beneficial effects are due to a hypothetical chemical which is secreted by the mass of animals in sufficient amounts to have an autoprotective effect. When the crowded conditions produce harmful results, they postulated another chemical substance which is autodestructive.

Inasmuch as these *Ophioderma* live naturally among eelgrass where they crawl over the blades, and since the formation of dense aggregations is greatly retarded if not entirely prevented by the presence of eelgrass in the stock aquaria, it seemed possible that the observed effects might be due to physical rather to chemical factors, at least in this case. In other words, it seemed possible that in the absence of suitable non-ophiurid materials the animals might themselves serve as substitutes for certain elements of the normal physical environment.

This possibility was tested by placing glass rods of different lengths, bent into different shapes, in some of the Erlenmeyer respiration flasks and using these for isolation chambers. The glass

TABLE XV

SHOWING THE INITIAL RELATIONS IN OXYGEN CONSUMPTION OF *Ophioderma* ISOLATED INTO FLASKS CONTAINING IRREGULAR HEAPS OF GLASS RODS, IN COMPARISON WITH BUNCHED INDIVIDUALS AND WITH THOSE ISOLATED INTO PLAIN FLASKS

(The Results Are in Terms of Cu. Mm. of Oxygen Consumed per Mean Individual of 1.25 Gm. Weight)

Experiment No.	Bunches Mean	Glass	Difference	Isolated	Difference	Temperature (Centigrade)
1.	192	280	88	421	141	22–24
2.	120	155	35	176	21	23–24
3.	226	229	3	522	293	23–22
4.	220	410	190	397	−13	24–24.5
5.	290	269	−21	507	238	24–24
6.	166	291	125	253	−38	25–25
Mean. . .	202	272	70	379	107	. . . .

rods formed a loose irregular pile, over and through which the starfish could crawl or against which it could come to rest. Four such flasks, with their accompanying controls, were added to the series already described. Such experiments were begun only after the conclusions outlined above were clearly indicated; they comprised six sets of 4-hour tests of initial respiration and two of the long-time respiration experiments. The tests were made similarly in every respect to those already described; the flasks containing glass rods were placed alongside those with isolated and bunched individuals with the behavior of which they were to be compared.

The results of exposure for 4 hours under these conditions are shown in Table XV. Here the respiratory rate is calculated in terms of 1.25 gm. per individual and for the actual amount of water present after the glass rods were added. The table gives a comparison of the initial respiration of 24 *Ophioderma* isolated into flasks containing glass rods, with that of 48 isolated into plain flasks, and with 96 divided among the accompanying bunches.

The *Ophioderma* associated with the clean glass rods consumed more oxygen per individual in 5 out of 6 cases than did their associated bunches. The mean difference of 70 cu. mm. is beyond the statistical border of significance, since there are 85 chances in a 1,000 in this case and 112 in the next, that random sampling would give so great a deviation from the mean. On the other hand, the

TABLE XVI

SHOWING FINAL MEAN RELATIONS IN OXYGEN CONSUMPTION OF *Ophioderma* SEPARATED UNDER THE CONDITIONS SHOWN FOR THE HEADING FOR TABLE XV

(The Results Are in Cu. Cm. of Oxygen Calculated to Equal Weight for Each Series)

Experiment No.	First Bunch	Second Bunch	Isolated	Isolated with Glass Rods	Weight (Gr.)	Temperature (Centigrade)
5	14.050	14.103	11.914	17.926	1.34	21–25
6	7.820	7.809	5.220	10.811	1.21	21–24

individuals associated with the glass rods consumed less oxygen than the accompanying animals isolated into plain flasks, in 4 out of the 6 cases, with a mean difference of 107 cu. mm. Thus the initial effect of the presence of the heap of glass rods with the isolated individuals was to produce a rate of oxygen consumption intermediate between that of the bunched and the other isolated individuals. The position of the mean, nearer to that given by the bunched individuals than to those isolated in plain flasks, while suggestive, is not significant statistically.

The results in the two total respiration experiments which contained this feature are given in Table XVI. Here the presence of the glass rods was accompanied by an even greater total oxygen consumption than that of the bunched individuals and, as was shown in Table XIV, the autotomy effects are more closely related to those of the bunched than to those of the other isolated animals.

We have, then, indications that the provision of an opportunity for physical contact with the lifeless glass rods produces effects similar to those given by the bunching of live starfishes. Obviously, more work is needed at this point before we can come to a definite conclusion. Other preliminary experiments indicate that the presence of irregular heaps of paraffined glass rods or of paraffined rubber tubing tends to prevent autotomy of arms of individuals as compared with other starfishes isolated into plain dishes.

NATURAL AGGREGATIONS OF ASELLUS

The fortunate discovery of a group of gigantic aggregations of water isopods, *Asellus communis*, gave opportunity for further respiration experiments on these under natural conditions. An account of some aspects of these aggregations has already been given, describing the methods of formation of such groups. For our present purposes it is necessary to remember that these aggregations occurred at the downstream end of culverts and on the downstream side of an overflow across a sand roadway which divided an extensive cat-tail swamp at the head waters of Dune Creek, in the Indiana dunes region. At the lower edge of this overflow, just before it widened out into the lower swamp, great masses of *Asellus* collected in winter and early spring about willow shrubs, old cat-tails, or in depressions where they might find a lodging.

As was shown in the analysis in a preceding chapter, the formation of these aggregations was conditioned to a large extent, perhaps completely, by the interaction of the tropisms of the individuals with environmental factors. In turn, the accumulation of animals was sufficient to affect decidedly the water surrounding and penetrating them. This effect was shown by the loss of 60 per cent of the oxygen normally present in the stream. In one observation the stream above the main bunch had 6.37 cc. of oxygen per liter, while a collection from the midst of this large aggregation had but 2.61 cc., a loss of 3.76 cc. of oxygen. Similarly the pH of the stream was lowered as much as 0.2 of a pH unit in extreme cases, and averaged 0.09 of a unit less than normal when all the twenty-three tests were considered. These changes were brought about when the clusters were located in a stream 15–18 cm. deep where the surface water was moving at the rate of 25 cm. a second.

Field tests were run to determine the effect of aggregation upon the rate of oxygen consumption of the aggregated isopods. For this purpose, a respiration chamber was made by firmly attaching a short piece of snugly fitting rubber tubing to a small wide-mouthed bottle of about 14 cc. capacity. The rubber tubing was long enough to extend beyond the bottle mouth about the length of the bottle neck. A second similar bottle inserted into this rubber collar until the mouths of the two met tightly made a respiration chamber of 27.3–30 cc. capacity. The rubber tubing served as a collar to hold the two bottles firmly bound together, and came in contact with the water only to a very slight extent. At the end of the respiration period the contained isopods were all shaken down into the first bottle. The other was removed, closed with its glass stopper, and used as an isopod-free collection sample. The agitation necessary to shake the isopods all into one bottle must have mixed the contained water fairly thoroughly.

A sample collected in this manner was treated according to the usual Winkler's method, except that only one-fourth of the usual amount of the reagents was added. Titration was carried on in the field, using a standardized sodium-thiosulphate solution of about 0.002 normality. Blanks were run with each set of five respiration tests. These tests were always run in parallel series—one composed of animals taken from the aggregations; the other containing the same number of isopods of similar size that were scattered about singly when collected, and which had not been aggregated recently if at all. Ten male isopods, or 20 of the smaller females, were the standard number used at one time in each respiration chamber. The respiration tests of lots from the aggregation and those that before the tests had been scattered singly were, with one exception, run simultaneously. In this exceptional case, the results obtained may be affected by the greater amount of oxygen present in the water when the isolated animals were tested.

Care was taken to collect the water for the tests from the same depth at the same point in the stream, and in as nearly the same manner as possible. After the respiration chambers were closed, they were placed side by side in running water and were thus exposed to identical conditions during the respiration period, which usually lasted for 1 hour. The isopods from each series were brought into the laboratory, where their moist weights were determined. The results obtained are summarized in Table XVII. The figures on oxygen consumption and weight are mean results from the number of groups indicated in the first column, or from the same number of isolated individuals as were tested in the groups.

TABLE XVII

SHOWING THE AMOUNT OF OXYGEN CONSUMED BY GROUPS OF 10 MALE ISOPODS DURING THEIR FIRST HOUR AFTER REMOVAL FROM A LARGE BUNCH, AS COMPARED WITH A SIMILAR NUMBER OF ANIMALS THAT WERE ISOLATED WHEN COLLECTED

Number Groups	Temperature (Centigrade)	Bunched Isopods			Scattered Isopods		
		Cu. Mm. O_2 Used	Weight (Grams)	O_2 Calculated to Equal Weight	Cu. Mm. O_2 Used	Weight (Grams)	O_2 Calculated to Equal Weight
6	5	5.7	1.338	....	19.9	1.326	....
5	12	28.9	1.1055	36.6	42.9	1.402	....
5	9	31.4	1.330	....	41.9	1.187	46.9
4	15	33.2	1.337	....	48.0	1.252	52.8
5	13*	64.0	1.245	....	80.3	1.16	86.2

* Last test ran 2 hours.

Twenty-five comparable tests were run with or without correcting for weight. The set of 10 isopods that had previously been scattered, in all cases but two, used more oxygen during the respiration period than did the set of the same number taken from the aggregation and started at almost the same time. Always the total amount of oxygen consumed by all the members of one series of animals that had been scattered was greater than that used by the series from the aggregated isopods run at the same time.

Without correcting for weight, the mean difference in oxygen consumption in these five different series shows that those that were solitary when collected used 12.33 cu. mm. more oxygen during the first hour's exposure in a respiration chamber than those that came from the aggregation. Application of Student's statistical method shows that there is less than 1 chance in 100 of getting so great a deviation in either direction from the mean in random sampling.

Such results are statistically significant. If observed differences in weight are considered, the calculated difference in oxygen consumption between the bunched and the isolated isopods is still greater.

A group of tests were run to find the respiration rate of the bunched isopods as compared with those scattered among the vegetation in the upper swamp. These latter animals had not aggregated; nor had they been exposed recently to the stimulus of a strong current. The tests were made as were those preceding, except that females were used because they were so much more numerous than males in the swamp. Since the females are much smaller than the males, 20 were placed in each respiration chamber, where 10 males had been used in the other tests. The rate of oxygen consumption was slightly higher in the isopods taken from the aggregations than that of the similar females from the vegetation of the upper swamp, even when the observed figures are corrected for differences in weight; but the results as observed are without statistical significance. Before calculation for weight differences there are 8 chances in 10 of securing so great a difference by random sampling. More work is needed at this point, but the results indicate that the females of the aggregations are in more nearly the same physiological state as those scattered in the upper swamp than were the males from the aggregations like those scattered in the current below the swamp but above the aggregations.

The rate of oxygen consumption of these females taken from the aggregations approaches that of the males run at approximately the same temperatures. The males tested on the afternoon of April 15, at a temperature of 15° C., consumed, on the average, 33.2 cu. mm. of oxygen per 10 males. The 20 females tested 2 days later at 17° C. used, on the average, 30.2 cu. mm. of oxygen. They weighed 1.1853 against 1.3765 grams for the males. If they had continued to consume oxygen at the same rate per unit weight and had weighed as much as the males, they would have used 35.1 cu. mm. of oxygen in the same time.

If the comparison is carried a step farther, we find that these females do not have a rate of oxygen consumption comparable with that of males scattered through the stream just above the aggregations. Using values calculated from equal weights, we find that the latter consumed 52.8 cu. mm. of oxygen as against 35.1 for the aggregated females. If the male isopods in the upper swamp have about the same rate of oxygen consumption as the females there, a point unfortunately not directly tested, we can reconstruct their respiration history as follows:

The isopods in the upper swamp, on being swept off their feet, struggle against the current and are borne along clutching at straws as they pass. Sometimes an isopod clasps another and both are carried along head over heels until they are deposited in some depression with a sandy bottom, from which they start upstream only to be carried down again. Such isopods have their rate of respiration greatly increased as a result of their exertions. If they are carried on down so that they lodge among the grasses below or among their fellows who have already lodged there, they again take on about the same rate of respiration they had shown in the upper swamp.

COMPARISONS AND CONCLUSIONS

Like the land isopods and the brittle starfish, *Asellus* is not usually regarded as a social species. These isopods are frequently found in greater abundance in one place than another; but outside the breeding season, environmental analyses usually reveal significant differences which may condition differential distribution; and, as we have seen, these natural aggregations originate largely through the reactions of individuals to their environment, rather than through social impulses. Great aggregations of these animals in nature have hitherto escaped notice; and, so far as I am aware, no other great natural aggregation at the low level of group organization here existing has been analyzed to find the physiological effect of aggregation upon the aggregants.

The laboratory experience outlined above had previously shown that, conditions being otherwise favorable, land isopods and the brittle starfish, *Ophioderma brevispina*, have their rate of oxygen consumption increased as a first effect of aggregation. Later the rate of oxygen consumption is lowered as compared with individuals isolated under similar conditions. The respiration relations were not obtained at the beginning of the aggregations in the case of the isopods in the field. Neither was it possible to arrange to make respiration tests without disturbing the aggregations as was done in the laboratory. One must use care in interpreting results based upon the oxygen consumption given during the first hour following the considerable disturbance occasioned by the extraction of 10 isopods from the midst of an aggregation as compared with that of 10 isopods for the first hour that they are placed together in a small respiration chamber after having been previously scattered.

It must also be noted that the isopods from the aggregations and those collected while still separate did not behave similarly in the respiration chamber. The former tended to gather in the restricted space furnished by the necks of the bottles forming the respiration chamber, where they were well shaded by the rubber tubing that held the two bottles together. Those collected while scattered tended to remain so, and were more likely to be found exposed to the brighter light at the ends of the bottles. Since all lots tried behaved uniformly in this regard, and since previous location also made a difference in the oxygen consumption, the differential behavior within the respiratory chamber may be taken as one of the factors in determining the respiration relations observed. The exact weight that should be given this factor has not been determined. Such behavior might result in the shaded, more aggregated isopods using less oxygen during the respiration period than would the unshaded scattered individuals, as was found to occur.

As a result of experience with respiration of laboratory aggregations, it seems probable that the differential respiration observed is due to a difference in muscular tonus between the aggregated and the isolated animals. If this idea is applied, we have the reasonable assumption that isopods taken from the upper swamp far enough away so that they are undisturbed by the currents set up by the overflow are under approximately the same muscular tension as are those taken from the dense aggregations, and that both are under less muscular tension than are the isolated iospods that have recently been exposed to the full impact of a current which in many instances has been carrying them hurtling along over a sandy stream bed. The evidence at hand allows no suggestion as to whether the respiration results observed are due to muscular tension acting as a direct result of the previous position of the isopods or whether they are due to the difference in muscular tension resulting from the differential behavior, which is in turn conditioned by their previous position.

The possibility that the increased carbon-dioxide tension and the decreased oxygen supply within the aggregation may have reduced the oxygen requirements of these animals is negatived by the fact that the rate of oxygen consumption of isopods from the center of the aggregations was slightly, but not significantly, higher than that of those taken from the upper swamp where the oxygen tension and pH were approximately that of the main current above the large clusters.

The results obtained from the study of these enormous aggregations of isopods in nature demonstrate that such masses can affect environmental conditions markedly, even in flowing water. Obviously, such groupings would have still greater effect in quiet water or on land. As we shall soon see, such aggregations may have survival value in the laboratory, due to the modification they produce in their environment. No tests were made of the possible survival value of the isopod aggregations in nature, but the fact that they do alter their environment to a measurable degree extends the possibility of the application of the laboratory studies on the survival value of masses of animals. As in the case of the laboratory groups of land isopods and of *Ophioderma*, these water isopods in nature under adverse conditions tend to use other isopods in place of inanimate elements of their physical environment, when the former are present in considerable numbers and the latter are lacking in their usual ratio of abundance in proportion to the numbers of animals.

In conclusion we may recall the statement at the beginning of this chapter, which these later studies have served to confirm, that under laboratory conditions the formation of aggregations serves to make these animals more quiet, and in the long run proves to be what is usually called an "adaptive reaction"; and, so far as we have information, the same results come also from aggregations under natural conditions.

CHAPTER XII

PROTECTION FROM TOXIC REAGENTS

Shortly after the publication of the preliminary announcement which the preceding chapter restates, Bohn and Drzewina (1920) published the first of a brilliant series of studies, apparently not yet concluded, which opened for investigation another phase of the subject that has yielded much evidence concerning the possible survival value of groups of animals for the individuals composing the groups. The general results obtained may be summarized in their own words (1926):

"Nous avons recherché l'intervention du facteur *masse* dans les réponses de divers organismes vis-à-vis de multiples agents nocifs du milieu extérieur. Dans une masse M d'eau, on introduit une masse *m* d'un être vivant (*m* étant égal ou inférieur à M/100), cette masse résiste à un agent nocif déterminé (substance chimique, par exemple); mais une masse plus petite, *m*/10 ou *m*/100, ne résiste pas; tout se passe comme si la masse de matière vivante exerçait, vis-à-vis d'elle-même, un effet protecteur (auto-protection)."

MASS PROTECTION FROM COLLOIDAL SILVER

Drzewina and Bohn worked with colloids of the heavy metals, particularly with colloidal silver. We have spent much time in this laboratory, also, studying mass relations of animals when exposed to colloidal silver;[1] and if the following report and discussion is

[1] The stock suspension of colloidal silver was made as follows, from directions furnished by Dr. Terry-McCoy of the Department of Chemistry of the University of Chicago: Dissolve 4 grams of commercial dextrine in 100 cc. of water and then 4 grams of pure caustic soda. Dissolve 3 grams of silver nitrate in 20 cc. of water and add to the dextrine-soda solution. The precipitate of silver oxide that is formed is gradually reduced by the dextrine, the color changing to reddish brown. After 20–30 minutes, add 100 cc. of 90 per cent alcohol and stir. Allow the mixture to settle for 15–20 minutes, and then pour off the liquid from the particles of silver at the bottom. Add about 200 cc. of water and the silver particles will generally disperse; however, if this is not the case, shake or stir until an even suspension of silver results.

based largely upon our own work (Allee and Schuett, 1927), it is only because I am more familiar with this, and by no means because it is more important than work in this field done by Drzewina and Bohn in following out their original discovery.

It is easy to demonstrate, as Drzewina and Bohn state, that within limits, the greater the mass of animals, the better the protection. Thus with *Planaria dorotocephala*, twelve sets of 2 each, exposed at room temperature to 10 cc. of water containing 10 drops of the standard suspension of colloidal silver, showed the beginning of head disintegration in from 4.5 to 10 hours; while ten similar lots, each containing from 10 to 75 worms, all lasted over 36 hours in the same volume and concentration. A species of Cladocera and the isopod *Asellus communis* showed similar, but less marked, group protection.

Similar tests, in which the animals were exposed to the action of dilute suspensions of colloidal silver for considerable time and were then washed and transferred to water to which they were accustomed, yielded such results as are listed in Table XVIII. With the worms, the results are summarized in terms of survival after exposure, with added information concerning the number of worms that were able to crawl either with or without stimulation with a camel's hair brush. Other data at hand show that the protective effect of the group was greater than is indicated by this summary, but the summarized account is sufficiently conclusive. With the brittle starfish, *Ophioderma*, tests were made concerning the righting ability of the animals after exposure to colloidal silver. In these tests, after exposure, the washed animals were transferred to separate finger bowls each containing 250 cc. of well-aërated sea-water. It will be noted that the mean righting time of the isolated animals is given in minutes while that of the animals tested in groups of 5 is given in seconds.

More cases with different concentrations could be given, but the result is the same in all. This agrees entirely with the experience of Drzewina and Bohn, who used a wide range of animals; and with that of Bresslau, using different toxic reagents with ciliates. Yet the possible menace of crowding, so frequently given as the only result of confining many animals within a small space, may easily be demonstrated even here. With the *Ophioderma*, for example, if the glass finger bowls were covered with glass plates during the exposure to colloidal silver, thus stopping free gaseous exchange, the grouped animals were in much worse condition than were their isolated fellows.

A priori, one would expect exactly such a result, since in each case there is much less of the toxic substance present per individual

TABLE XVIII

SHOWING THE RESULTS OF EXPOSURE TO SUSPENSIONS OF COLLOIDAL SILVER FOLLOWED BY TRANSFER TO WATER SIMILAR TO THAT TO WHICH THE ANIMALS WERE ACCUSTOMED

All were run at room temperature.

Animal	Cc. of Water	Drops Colloidal Silver	Number Animals	Time Exposed to Colloidal Silver (Hours)	Number That Survived	Number That Were Moving	Number That Died
Planaria maculata	15	12	6(3)	2	11	2	7
	15	12	1(75)	2	72	66	3
	5	1	5(3)	26	9	0	6
	5	1	1(25)	26	25	23	0
P. dorotocephala..	10	2	40(1)	7	15	1	25
	10	2	3(25)	7	75	63	0
Dendrocoelum lacteum..........	5	1	5(3)	26	3	1	12
	5	1	1(24)	26	18	18	6
					Righted	Mean Time	
Ophioderma brevispina........	50	2	48(1)	14.5–16	41	30 min.	0
	50	2	12(5)	14.5–16	59	38 sec.	0

composing the larger groups than when from 1 to 3 animals are placed in the same amount of the same concentration of toxic substance (Pieron, 1921). Drzewina and Bohn consider this possibility (1921*c*, 1928) and dismiss it as inadequate because of their experiments upon the effect of reconditioned toxic solutions and upon the relation of volume of the toxic material to its effect upon exposed animals. They summarize their conclusions repeatedly in some such words as the following (1921*d*):

"Bref, si chez diverses espèces que nous avons examinées jusqu'ici (Paramécies, Colpodes, Stylonichies, Stentors, Hydres, Convoluta, Glossiphonies, têtrads de Grenouille) tout se passe comme s'il y avait émission rapide d'une substance *ou* de substances assurant une défense; chez *Polycelis* c'est le contraire:[1] on assiste non pas à une *auto-protection* mais à une *auto-destruction.*"

They do not believe that the greater resistance of the group is due to a more rapid using-up of the toxic substance, and cite two types of experiments to support their contention (1921). In the first type they expose some hundred of *Convoluta* in one suspension and 2 each in the accompanying dishes. Even if the former receive 10 drops of colloidal silver to 25 cc. of water and the isolated animals have only 2 drops for the same volume, they find that the large group survives after the others are disintegrated. This is a fairer test than if the same amount of colloidal silver were added in both cases, but still it must be remarked that the ratio of colloidal silver per individual is about ten times greater in the case of the 2 individuals, when compared with the larger group.

We have made some exact tests to cover this point, and I present the summary of our results in Table XIX. The tests reported show clearly that if a group of 25 *Planaria* are placed in the same volume and the same concentration of colloidal silver as are isolated individuals, the former survive the experience in good condition while the majority of the latter succumb, and almost all the others are severely affected. Further, if similar planarians are placed at the same time in the same volume of water, to which has been added sufficient colloidal silver to make the concentration per individual equal to that given the isolated worms, the bunched animals all succumb to a treatment that left 37.5 per cent of the isolated individuals alive, even if most of them were strongly affected. And finally, if the groups are placed in proportionate volume and proportionate concentration of colloidal silver, their condition approaches that of the isolated individuals. This evidence shows that the greater protection which the mass furnished when exposed to colloidal silver was due to the smaller amount of the

[1] Against KCl (W. C. A.).

toxic substance which each individual had to remove from suspension in order to lower the strength below the threshold of immediate toxicity. The mechanism for this removal will be discussed later.

Another type of experiment was run with *Ophioderma* to determine whether the protective effect of the bunch was due to the fact

TABLE XIX

ANIMALS USED: *Planaria dorotocephala*, 15–25 MM. LONG, EXPOSED TO ACTION OF COLLOIDAL SILVER FOR 7 HOURS, THEN WASHED AND TRANSFERRED TO WATER TO WHICH THEY WERE ACCLIMATED, AND EXAMINED ON THE DAY FOLLOWING THE EXPOSURE

	No.	Percentage
I. 40 planarians isolated into 10 cc. of water plus 2 drops colloidal silver		
Normal, crawling when stimulated	1	2.5
Less than half disintegrated	2	5.0
Half and less than two-thirds disintegrated	4	10.0
Two-thirds but not all disintegrated	8	20.0
Wholly disintegrated	25	62.5
II. 150 planarians in 6 lots each in 250 cc. water plus 50 drops colloidal silver		
Normal, crawling when stimulated	0	0
Less than half disintegrated	0	0
Half, but less than two-thirds disintegrated	18	12
Two-thirds but not all disintegrated	52	35
Wholly disintegrated	80	53
III. 71 planarians in three lots each in 10 cc. of water plus 2 drops of colloidal silver		
Normal, crawling when stimulated	63	88.8
Normal, not crawling when stimulated	1	1.4
Head started to disintegrate	7	9.8
Head completely disintegrated	0	0.0
IV. 75 planarians in three lots each in 10 cc. of water plus 50 drops of colloidal silver		
All wholly disintegrated.		

that the bunched animals were exposed to a smaller concentration of colloidal silver than were the isolated animals. Since the area of the surface would affect the aëration and so affect survival values, the larger numbers were placed in crystallization dishes with a diameter of 24 cm., while the smaller bunches and the isolated animals were in the usual finger bowls with a diameter of 10 cm. The essentials of the set-up were:

One bunch of 10 animals with 25 cc. of sea-water and 1.5 drops of colloidal silver each, and with a surface exposure of 45 sq. cm. per animal.

One bunch of 5 animals with 5 cc. of sea-water and 0.3 drops of colloidal silver each, and with a surface exposure of 15.6 sq. cm. per animal.

Five isolated animals with 25 cc. of sea-water and 1.5 drops of colloidal silver each, and with 78 sq. cm. surface exposure.

A second set was exactly similar, except that the animals were exposed to 2, 0.4, and 2 drops of colloidal silver per individual.

The exposure in a typical experiment lasted 15.5 hours, after which the animals were washed as usual and transferred to fresh sea-water for observation for the following 24 hours. At intervals records of numbers that righted after having been turned over, and other evidences of activity, such as spontaneous motion of tube feet, were recorded.

All of the animals bunched in the 10 cm. finger bowls recovered sufficiently to right themselves within 24 hours, despite the fact that 4 of the 10 were much corroded by the action of silver that had settled to the bottom of one of the finger bowls. In both experiments, the isolated animals fared next best, although they made a much poorer showing than did the bunched individuals that were exposed to a much lower amount of colloidal silver per animal; only 6 of this set of 10 righted. The bunched animals, exposed to the same volume and amount of colloidal silver as those isolated, ranked a poor third, which may be due to the decreased surface exposure or to some of the better-known ill effects of crowding.

These experiments indicate either that the decreased amount of colloidal silver or the reduced volume of sea-water present per individual, or both, markedly favored the survival of the animals bunched in the small amount of water present in the finger bowls. Drzewina and Bohn emphasize the latter factor, and accordingly experiments were run to test out this point. Contrary to the experience of Drzewina and Bohn (1921*a*), the toxicity with *Ophioderma* was found to depend on the concentration of the colloidal silver rather than on the volume to which the animals are exposed.

It is worth noting that the protective action of the bunch appears only in the more toxic suspensions, where apparently the toxic strength per individual was more nearly reached even with the bunched animals. Drzewina and Bohn apparently would interpret such an observation as meaning that the autoprotective substance is secreted more rapidly under more toxic conditions.

RECONDITIONED SOLUTIONS

The second and more convincing experiment used by Drzewina and Bohn as evidence in favor of their view that the group protection is furnished by some sort of autoprotective secretion, rather than by exhaustion of the toxic substance, is as follows (1921*a*):

"Nous décantons la solution où depuis 24 heures séjournent une cinquantaine d'embryons (tadpoles of *Rana fusca*) et dont la teinte révèle la presence du colloïde; nous y ajoutons le même nombre de gouttes que la veille, 1 par exemple, et nous y plaçons deux embryons neufs du même âge. Ceux-ci survivent, alors que des individus témoins, placés dans une solution neuve à 1 goutte de collargol, succombent, comme c'est la règle pour les isolés. Il semble ainsi que, attaquées par le colloïde, les larves émettent, *rapidement*, une substance (ou des substances) qui a pour effet de les protéger."

Wholly similar experiments were run with the brittle starfish, *Ophioderma*, to test for the presence of the postulated autoprotective secretion. Colloidal suspensions in which *Ophioderma* had been exposed were reconditioned by 3 hours aëration with room air in order that the new lot might not suffer from low oxygen tension, and were filled to the original volume with distilled water. The original suspension had been made with 1 drop of colloidal silver for each 25 cc., and the same amount was added to this reconditioned water. Tests showed that the aëration of freshly prepared suspensions for this length of time caused a color change but did not markedly affect the toxicity.

Three lots, each consisting of one group of 5 bunched animals and 4 isolated individuals, were placed severally in 50 cc. of such a reconditioned suspension, 9 hours after the previous experiment had closed. These were run simultaneously with three similar lots in freshly prepared suspensions of 2 drops of colloidal silver in 50 cc. of sea-water. All conditions were similar. The temperature was 18.5° C. The results are summarized in Table XX.

If Drzewina and Bohn were correct in thinking that the colloidal silver is not removed from solution on exposure to the animals, there should have been an excess accumulation of this substance in the twice-used water; but after again conditioning the used substance, it was necessary to add the same amount, 1 drop of colloidal

TABLE XX

SHOWING RIGHTING TIME OF *Ophioderma* AFTER AN EXPOSURE OF 17 HOURS IN RECONDITIONED AND FRESH SUSPENSIONS OF COLLOIDAL SILVER

All gave the righting reaction.

Bunch			Isolated		
Number	Mean Time	Spread	Number	Mean Time	Spread
Reconditioned Suspension					
15	73 sec.	10 sec.–12 min.	12	25 sec.	5–118 sec.
New Suspension					
15	20 sec.	7–75 sec.	12	25 sec.	10–65 sec.

silver per 25 cc. to this suspension, in order that it might be as deeply colored as was a fresh suspension made up with that amount of the colloidal silver. There was only sufficient colloidal silver left in this twice-used suspension to discolor the liquid. Either the animals or the aëration, or both, had removed the greater part of the silver; and it can be readily demonstrated that the effect was not wholly due to aëration. The water thus treated was distinctly sirupy, and evidently held organic matter, received from the two lots of *Ophioderma* which it had contained.

The effect of this twice-reconditioned suspension was tested as before. Comparison tests were run with newly prepared suspensions, and the results are summarized in Table XXI.

There is no evidence here of the presence of an autoprotective secretion in the sea-water, protecting animals from the action of the colloidal silver; nor is there evidence for the presence of an active autodestructive agent, which Drzewina and Bohn also postulated to explain certain cases in which the greater mass of animals exposed to KCl die more rapidly than do the solitary individuals. In the presence of an active autodestructive secretion one would expect the mass to die more rapidly than do the isolated individuals, which is not the case.

TABLE XXI

SHOWING RIGHTING TIME OF *Ophioderma* AFTER AN EXPOSURE OF 19 HOURS IN TWICE-RECONDITIONED AND IN FRESH SUSPENSIONS

Bunch			Isolated		
Number	Mean Time	Spread	Number	Mean Time	Spread
Reconditioned Suspension					
15	46 sec.	10–221 sec.	11*	3.45 hrs.	17 sec.–22 hrs.
New Suspension					
15	34 sec.	10–175 sec.	12	1.5 hrs.	10 sec.–14 hrs.

* One failed to give the righting reaction.

There is good evidence that the secretion of slime or other organic matter into the suspension does remove the colloidal silver and so render the solution less toxic. It is particularly significant that, after having been twice used, the suspension required as much colloidal silver as was used in a fresh suspension to bring it to the same color. Similar results were obtained from other experiments.

More recent experiments in this laboratory, the results of which are unpublished as yet, show clearly that groups of goldfish live longer in a given volume of a given concentration of colloidal silver than do isolated goldfishes similarly isolated into the same amount of the same concentration of colloidal silver suspension; and that, under these conditions, the group removes significantly more silver from suspension than do the isolated individuals. The experiments do not yet show conclusively whether this will account for all of the observed group protection.

SPECIFICITY OF MASS PROTECTION AGAINST COLLOIDAL SILVER

There are two types of specificity possible: the protective secretion may have no other function, and so be specific in that sense; or it may be limited in protection to the species producing it. The first type of specificity will be discussed later. The latter aspect was investigated by placing one animal in a restricted volume of water containing a number of animals of a different species. All our observations show that the protective action of the mass is not limited

TABLE XXII

SHOWING THE NON-SPECIFICITY OF THE PROTECTION AGAINST COLLOIDAL SILVER

Animals	Water (Cc.)	Colloidal Silver (Drops)	Time to Death (Hours)
Many *Cladocera*, 1 *Asellus*	12.5	8	Over 36
1 *Asellus*	12.5	8	4.5
Many *Cladocera*, 1 *Planaria*	12.5	8	Over 36
1 *Planaria*	12.5	8	5.5
50 *Asellus*, 1 *Planaria*	12.5	8	Over 36
1 *Planaria*	12.5	8	5.5
Desiccated parotid gland, 2 *Planaria*	10	10	Over 36
2 *Planaria*	10	10	7.5
Snail slime, 2 *Planaria*	10	5	Over 36
2 *Planaria*	10	5	Less than 18
Snail slime, 2 *Planaria*	10	5	Over 36
2 *Planaria*	10	5	Less than 16
1 *Physa*, 2 *Planaria*	10	5	Over 36
2 *Planaria*	10	5	11

to a given species. This is what would be expected if the fixing of the colloidal silver in some manner is the principal element in the protective action. The results of typical experiments are summarized in Tables XXII and XXIII.

As these tables show, such diverse organisms as *Cladocera*, *Asellus*, pond snails, pond leeches, *Dendrocœlum*, and even pond moss, if present in quantity, markedly protect planarians from the toxic action of colloidal silver. Even the actual presence of living organisms is unnecessary; snail slime without the snails protects efficiently apparently by adsorbing the colloidal silver. The slime becomes densely colored as the water becomes lighter. Suspensions in water of desiccated parotid glands of sheep exhibit similar adsorptive phenomena and have similar protective value.

A still more severe test of the specificity of the protection was made by placing recently killed *Asellus* in glass dishes containing 10 cc. of water plus 2 drops of colloidal silver, and introducing with these dead isopods one *Planaria dorotocephala*. At the end of 7 hours' exposure, 20 planarians isolated into similar volumes of the same concentration showed, 6, one-third disintegrated; 13, one-half disintegrated; and one, wholly so. The 5 worms isolated into suspensions containing dead isopods showed 1 intact though bloated; 3, heads disintegrated; and 1 with both head and posterior end

TABLE XXIII

SHOWING COMMUNITY PROTECTION AGAINST COLLOIDAL SILVER

Two drops of colloidal silver in 5 cc. pond water; exposure 13 hours, then washed and placed in pond water. Temperature 21° C.

Animals	Crawling	Lived	Died
23 *Planaria maculata*	23	23	0
10 isolated *P. maculata*	0	1	9
25 *Dendrocœlum*, 1 *P. maculata*	1	1	0
30 *Glossophonia*, 1 *P. maculata*	0	1	0
26 *Asellus*, 1 *P. maculata*	1	1	0
25 *Segmentina*, 1 *P. maculata*	0	1	0
Pond moss, 1 *P. maculata*	1	1	0

disintegrated. Thus the presence of the recently killed isopods decidedly protected the otherwise isolated worms from the toxic action of the colloidal silver. This, taken with all the other evidence at hand, furnishes convincing proof that the protective action of the mass against colloidal silver extends beyond the species limits.

We have shown in the preceding pages that, within limits and other conditions being equal, there is greater protection the greater the mass of the animals present when exposed to the same amount of colloidal silver in the same volume of water. Further, the protection is largely, and perhaps completely, furnished by the fixation of the toxic substance by the mass of animals, so that each escapes receiving a lethal dose; while, with the same concentration, isolated individuals receive a stronger dose of the toxic substance. The colloidal silver may be differently fixed in different animals; but with those that secrete slime, like planarians, the colloidal silver is adsorbed on the slime. With other animals observed, it may be removed by adsorption on the surfaces of the animals themselves. Finally, as would be expected from this mechanism, we have demonstrated that the protection furnished by the mass, is, at least to a considerable extent, independent of the species present.

Our experiments do not support the hypothesis that group protection among these aquatic animals is furnished by the rapid production in the presence of a toxic agent, such as colloidal silver, of a more or less mysterious autoprotective secretion. It is true that the production of slime by planarians actually serves as an autoprotective agent, not only in fixing colloidal silver in these experiments, but very probably in protecting the planarians from sudden changes in culture or habitat water. But slime production cannot be regarded as a specific autoprotective secretion, either in the sense that it is used for no other purposes, for obviously it plays many other rôles in the economy of slime-producing organisms, or in the sense that slime is limited in its protecting power to the one species producing it, since the protection furnished by aggregations of mixed species is easily demonstrated.

OTHER CASES OF GROUP PROTECTION

Bresslau (1924) found that Protozoa give off a substance which he calls "tektin," a mucin-like body, which is given off in greater abundance when the animals are stimulated by heat, pressure, methylene blue, iodine, etc. The tektin, when given off, takes up water rapidly and exhibits strong surface activity, adsorbing foreign particles readily. Bresslau tells of putting 2 cc. of liquid from a culture of Infusoria (either *Colpidium* or *Paramecium*) containing many individuals in one dish and a similar amount from the same culture, but with few individuals, into another. Into each he introduced 1 cc. of 1 per cent solution of methylene blue. Both produced the tektin and adsorbed this poison: the culture with the many animals so much more completely that the possibility of surviving the toxic action of the poisonous material was greatly increased.

Carpenter (1927) approached this problem from a wholly different angle. Without referring to the work of Drzewina and Bohn, she attacked the problem of the toxicity of metallic salts on fishes on account of her interest in stream pollution. Using lead nitrate, she found that the fatality curve is described by the equation $K = 1/t \log 1/\text{conc.}$, where K is a constant dependent upon the toxic substance employed and t is survival time for any given concentration (conc.).

When a number of minnows (*Leuciscus*) were killed successively by placing them in the same 500 cc. solution of N/10 lead nitrate, the survival time was found to be prolonged with each successive fish until the eighth had a survival time approximately double the standard figure for that concentration and size of fish. The actual survival times reported are, in order, 73, 89, 92, 93, 94, 120, and 130 minutes. Carpenter makes the deduction that the solution was successively weakened by the abstraction from it of a certain proportion of the lethal substance by each fish tested and that the lethal efficacy was thus progressively reduced.

The lethal efficacy of the solution varied in inverse proportion to the actual size of the fish exposed and directly according to the amount of lead (as Pb) required to kill the fishes. If the original solution be relatively weak, a single fish will remove an important fraction from a small amount of solution, and will therefore show a longer survival time than a similar fish isolated into a large volume of solution of similar strength. This is in keeping with Huxley's observations (1922) on the action of mercuric chloride on the gill tissue of *Mytilus*. Carpenter concluded that the lead salt causes death, due to the formation of a film over the gills and skin of the fish, by an interaction of mucus and the metallic ion, which causes death by suffocation. If insufficient lead ions are present, the film is shed; and the solution being thus freed from its toxic element, recovery ensues.

An attempt was made to establish the amount of lead which enters into combination with the mucus, and the amount, if any, which enters the blood-stream. Two large minnows were placed together in 1–5 liters of distilled water, to which was added sufficient lead nitrate to supply 6.21 mg. of lead. Immediately after death, the bodies were transferred to fairly dilute acetic acid for 4 minutes until the gills were quite clear. After washing the bodies in distilled water and combining both liquids, these were found to carry 4.62 mg. of lead. The original solution was found, by similar analysis, to have 1.8 mg. of lead after the death of the fishes. Adequate analysis of the fishes' bodies showed no lead had penetrated. Carpenter points out that the overplus of 0.21 mg. found in the analyses as compared with the original solution makes the experiment inconclusive in some respects, but that it does show clearly that the fishes remove toxic substances by adsorption, just as do our results with colloidal silver.

In later work (1930) Carpenter finds that the survival time of a number of North American fishes is directly influenced by the value of the ratio of the volume of solution to mass of individual fish so long as the solution is of constant molar concentration. When the concentration and volume are constant, the survival varies inversely according to the mass of the fishes. For equal masses of animals, the survival varies directly with the ratio of the volume of the solution to the molar concentration. The results upon survival of exposing groups of fishes to lethal concentrations and volumes of lead salts as compared with isolated fishes, are not yet clear, but the later members of a processionary series killed in the same solution survive longer than do the earlier members on account of the using up of the harmful agent.

Preliminary experiments in our own laboratory in which the survival of groups of 10 goldfish was compared with that of the same number of individuals isolated into the same volume and concentration of lead nitrate have revealed no significant differences when solutions of from N/10 to N/200 were used. The technique used in these experiments differed from that of Carpenter; hence we are not ready to criticize her results. But our experiments do show that the group effects clearly exhibited with colloidal silver are not so readily demonstrated with lead nitrate.

Pawlow (1925) came at the same general problem from still another angle. Led on by Ostwald's basic work on adsorption of toxic salts, Pawlow worked out theoretical formulas to express the relation between toxicity and adsorption on animals. These show that if the active toxic substance has a direct effect upon the organism, then the survival depends on the volume of the medium for any given concentration of the toxic agent and also upon the density of the population. The same relationship holds if the active lethal agent works through capillary attraction or by division among phase boundaries. These relationships are recognized as forming a biological aspect of the law of mass action.

In experiments on the brine shrimp, *Artemia salina*, exposed to reduced or to increased salt solution, Pawlow found that the theory given above was verified. The mean life-duration varied from 32.1 to 6.1 in different salt concentrations when the volume was quadrupled, while halving the numbers of animals present decreased the survival between 6.7 and 15.2 per cent. Similar relationships were found for the amphipod *Gammarus pulex* when exposed to certain concentrations of NaCl.

Stimulated to similar studies by their interest in the implications of the work of Gurwitsch on mitogenetic rays, Frank and Kurepina (1930) have demonstrated mass protection for sea-urchin eggs from the action of KCN, both as regards the rate of development and the percentage surviving. They report that 10^{-5} N solution exerted the following effects:

Number of eggs	1 in each 13 drops	68 in 3 drops
No cleavage	85%	20%
2 blastomeres	15%	40%
4–8 blastomeres	0%	40%

Another type of protection has already been shown to be operating, in at least certain instances, with aquatic forms (chap. vii) and is of sufficient importance to be repeated here in some slight detail. The discovery of this factor is a result of the work of Fowler (1931), who has examined the resistance of *Daphnia pulex* to a great many chemicals. Usually he found the groups more resistant than were isolated individuals run in the same volume of the same solutions. When he pushed the analysis of this phenomenon, particularly with $CaCl_2$, he found that the group was consuming less oxygen per individual for a given exposure time than were the accompanying single individuals. Further, when the experiments were run in such a way that the CO_2 was absorbed as soon as produced, the survival of the animals in the groups became the same as that of the isolated animals. When he re-examined his original data in the light of these findings, it was discovered that the cases in his early work in which the singles lived for a longer time than the group were those in which the dilution of the chemical was such that acclimatization, rather than direct resistance, was an important factor. In such cases, as Child has repeatedly pointed out (e.g. 1915), the animals with the higher rate of general metabolism have the greater power of acclimatization, and hence of survival. On the other hand, whenever the individuals tend to produce conditions which depress their rate of metabolism and are exposed to solution strengths lethal within a relatively short time before acclimatization becomes a factor, such depression has survival value.

Fowler's work also demonstrated that when *Daphnia* are exposed in dilute solutions of sodium or potassium hydroxide, the grouped individuals lived significantly longer than did similar isolated animals, and that under these conditions the hydroxide in the solution surrounding the group was reduced in strength as compared with that surrounding relatively isolated animals. Eight experiments with NaOH showed that 2 animals in 20 cc. solution in 3 hours reduced the hydroxide from a mean concentration of 0.00094 N to 0.000764 N. Under the same conditions 20 *Daphnia* reduced the same amount of the same solution to 0.000606 N, a difference of almost 30 per cent, with a statistical probability of 0.002. Similarly significant results were obtained with KOH solutions. Here the carbon dioxide given off by the animals reacts with the hydroxides to form carbonates and water, thus weakening the toxic concentration of the medium. The groups produce carbon dioxide more rapidly than do animals isolated into the same volume of medium, and hence decrease the toxicity more rapidly and survive longer.

The problem of the number of animals present in relation to their survival when exposed to toxic agents presents a somewhat different aspect when applied to insects placed in containers into which toxic gases have been introduced in concentration just sufficient to kill only a part of the isolated individuals with the exposures employed. Here the production of autoprotective materials presents a distinctly different problem from that found in aquatic animals. Discussion of experimentation in this connection will be reserved until chapter xiv, which deals with aggregations and insect survival.

GROUP PROTECTION FROM HIGH TEMPERATURES

Not all adverse conditions in nature are the result of the presence of toxic materials in aërial or watery solutions where the types of protection we have just been considering might operate. One other adverse condition is that furnished by the physical factor of high temperature. Concerning the ratio between the mass of exposed animals in relation to volume of medium with respect to this physical factor, Robertson (1921) records observations upon the Australian infusorian *Enchelys farcimen.*

A temperature of 30° C. prevents subcultures of this protozoan from multiplying, and the isolated individual almost inevitably dies. But shade temperatures of 30° C. are known in pools of South Australia in which wild *Enchelys* live. Wild infusorians brought into the laboratory are similarly killed by such temperatures if isolated; but if the culture slides are populated by 20–30 individuals, they can successfully resist exposure to temperatures of 33–34° C. for as long as 7 days in succession without apparent injury or abnormality. Also, a similar number of individuals put into fresh hay infusion at this temperature survive and multiply, while isolated individuals perish.

Typhoid bacilli can be rendered abnormally susceptible to hydrogen peroxide by a degree of heat which is just sublethal; and the growth inhibition so caused can be neutralized, as in bacteria naturally susceptible to peroxide, by the accumulation of certain products of bacterial growth (Burnet, 1925*a*). These peroxide-neutralizing materials are more rapidly produced the larger the colony of bacteria and the more vigorous their growth. This subject will receive further attention in chapter xvi.

Drzewina and Bohn (1926, 1928) report similar results with spermatozoa. Although these will be considered at greater length elsewhere, it should be stated that both of these observers interpret the greater protection which they find to be given by the presence of greater numbers, to the conditioning of the water by some sort of autoprotective exudate. Robertson brings his observations into line with his general theory of the extrusion of an autocatalytic substance, while the others regard mass protection from high temperature as an excellent example of the effect of an autoprotective secretion.

PROTECTION FROM ULTRA-VIOLET RADIATION

Another type of experimentation which should yield critical tests of the efficacy of group protection and of its mechanism should be found in treatment with radiant energy applied as ultra-violet radiation. Hinrichs (1927) has reported that *Arbacia* sperm are less affected in concentrated than in dilute suspensions (obviously a mass protection); and Petersen has obtained similar results in this laboratory in radiating Paramecia (unpublished). Accordingly, tests (Allee, 1928) were made concerning various aspects of survival value of groups as compared with isolated planarians, when exposed to the full spectrum of a quartz mercury-vapor arc, with temperature controlled during the exposure. The results collected demonstrated two facts:

In the first place, the presence of products of cytolysis produced by exposure to ultra-violet radiation are more harmful than beneficial to those worms that have been exposed to the action of ultra-violet radiation or to those that have not been exposed. Similarly, water containing products of metabolism or exudates given off by the animal, either in its usual laboratory culture or when exposed to ultra-violet radiation, shortens rather than lengthens the survival of other worms isolated into it.

In the second place, the exposure of a number of worms in a limited amount of water with limited exposed area gives much greater protection for the individuals than if they had been exposed singly or in pairs. Some of the implications of these results deserve consideration and will be discussed in the order just given.

When the survival of worms in various sorts of worm-conditioned water is compared with that of similarly treated planarians in aërated well-water, the worms in the conditioned water are found to die more quickly. The difference in the survival periods is usually small, and the periods themselves are highly variable for different lots of worms. Eleven such group comparisons are possible with the data at hand; and in these, nine cases definitely favor the well-water, one favors the conditioned water, and the other is recorded as a tie but with more careful observation it would probably have been placed definitely in the majority column. The main inquiry was concerned with the possibility of there being a protective value in the conditioned water, and there is no doubt of the negative answer given by the experiments. Rather, the converse is indicated though the evidence is not complete concerning the possibility that with different water and with smaller proportions of the conditioning matter, such conditioned water may prove advantageous to the worms placed in it.

With regard to the definite protection furnished when many worms were exposed to ultra-violet radiation in a limited amount of water and with a limited surface area, the experimental evidence indicates that this protection is due to some sort of interference with the penetration of injurious rays or to some other biophysical effect of numbers, rather than to the presence of some exudate or exudates released as a result of the radiation. As has been stated above, water containing such exudates produced harmful rather than beneficial results.

In the experiments where the worms were more densely crowded (60 to 1 cc. of water), the protection was obviously connected with the "shading" which Hinrichs mentions in her studies on the radiation of sperm suspensions. In the less dense groups (15 to 1 cc.), "shading" in the usual sense is less obvious; but exposure at this density also resulted in definite protection to the group-exposed worms, which suggests that some other factor or factors may have been operating. Protection furnished by the presence of so few worms may be an illustration of the phenomenon called by Drzewina and Bohn "catalysis by contact." Similar shading would undoubtedly have resulted from the exposure of isolated worms in water conditioned by the pressure of products of cytolysis. The experiment was not tried, since the only information to be gained would have been the relative value of the protection and the toxicity of such conditioned water.

It is worth noting that all the members of the group exposed together were benefited by the fact that they were together. In such circumstances it might have happened that certain animals at the surface would take the brunt of the harmful rays. Their presence might serve to protect the other members, and the group might thus be of value to the species if not to all the individuals composing it. In one experiment, where the subsequent history was taken for all the exposed animals, the interval before the first effects of exposure was observed in the 30 animals exposed as a group ranged from 27 minutes to 32.8 hours. Their final cytolysis ranged from 8.25 to 106 hours. The same conditions were observed for the 26 worms exposed in pairs in from 1 minute to 23.5 hours, and from 3.4 to 84 hours, respectively. Thirteen (50 per cent) of those exposed as pairs were visibly affected before the end of the first hour after exposure, while 6 (20 per cent) of the grouped animals were similarly affected in the same time. Six of those radiated in pairs were dead before the first of the group died.

I am not prepared to generalize widely from these experiments with the relatively large and highly pigmented *Planaria dorotocephala* that mass protection from ultra-violet radiation is always due solely to some sort of physical interference rather than to the possibly protective action of some exudate. With small organisms such as sperm of sea-urchins or with Paramecia, which are more translucent, the mass protection may be due to the latter factor, as Hinrichs suggests. Her observations show, however, that even with such minute organisms, the fact of physical interference is significant.

Single planarians, and to a greater extent massed planarians, would be more nearly analogous to sperm or protozoan aggregates than to a suspension in which distribution is fairly equal in three dimensions and each organism is surrounded by a medium of approximately uniform consistency. In the latter case, it is conceivable that a closer equilibrium between cells and medium must be maintained than with larger forms. This subject is one for experimentation rather than a priori discussion.

In general the experiments on the group resistance of *Planaria* to ultra-violet radiation forward our understanding of mass relations of individuals in two ways: First, the results emphasize the fact that the phenomena of possible protection of individuals by chemical exudates, which has been demonstrated for many animals exposed to different situations, is not universal. Second, they show that even when the massed animals are known to produce chemical exudates which are harmful, the massing may still have survival value through the changed physical conditions which it produces.

The work reviewed in the present chapter makes clear that group protection from toxic agents may operate in diverse ways. There may be group protection solely as a result of the distribution of toxic material among so many individuals that no one receives a lethal dose, or the toxic substance may be adsorbed on the slime which is often produced in copious quantities under the stimulation of the abnormal situation. The survival value of the group may be due to the depressed physiological condition obtaining among its numbers, which favors increased survival when the animals are exposed to strongly toxic solutions which kill actively metabolizing animals more rapidly than those whose rate of metabolism is lower. The group may act in a purely physical manner by altering the electrical conditions, or against light and ultra-violet radiations by a simple shading phenomenon. The evidence to date does not exclude the possibility of protection from active toxic conditions by the group conditioning the medium through some protective secretion, although this kind of explanation should be adopted only on positive proof. While it is clear that group protection is a fact, it is also certain, as might have been expected, that there are many ways by which groups accomplish this protection, and that more than one may be acting in the same case.

CHAPTER XIII

RESISTANCE TO HYPOTONIC SEA-WATER

We have just seen that when exposed to a variety of toxic agents, with the mass of animals in optimal relation with the volume of the medium and other conditions being favorable, groups of animals will survive longer than will equal numbers of single individuals when each is isolated into the same volume of the same medium to which the group is exposed. Our experience with colloidal silver and with several other reagents indicates that such protection is largely, and at times probably completely, furnished by the fixation of the toxic substance either directly by the mass, or by slime and other products given off, so that it is either removed or distributed among so many individuals that each receives a sublethal dose. If the animals are closely aggregated, only those on the outside receive the full impact of the toxic agent, so that those within are protected by another type of mechanical action of the mass. Such an attack on the problem of the protection of the individual by the mass, while demonstrating the fact of the protection, does not furnish critical evidence concerning the production of an autoprotective secretion other than slime, which in these cases presumably owes its protective action to its adsorptive power.

A better opportunity to obtain critical evidence is furnished when the toxic properties of the solution are due to the absence of easily measured materials rather than to the presence of some toxic substance added to the water. Such conditions are fulfilled when marine animals are placed in hypotonic sea-water. In concluding a note on the effect of exposing the marine turbellarian *Convoluta roscoffensis* to hypotonic sea-water, Bohn and Drzewina state (1920):

"Toutes conditions égales d'ailleurs, les individus isolés sont infiniment plus sensibles que les individus groupés, comme si le fait d'être groupés constituait pour eux une protection. Le contraste est souvent saisissant. Soient deux verres de montre, dont l'un contient, dans l'eau deluée à 75 pour 100, quelques individus, et l'autre plusieurs centaines de *Convoluta;* les premiers sont cytolysés en quelques heures, les derniers après plusiers jours."

Elsewhere they state clearly their belief that the observed protection is due to the secretion of an autoprotective substance by the mass in greater quantity in proportion to the available volume of water than is possible for the isolated animals.

Lapicque (1921), in discussion, made the obvious objection that the introduction of differing numbers of marine animals into the same quantity of hypotonic sea-water would have a differential effect upon the salt content and that the difference in survival might

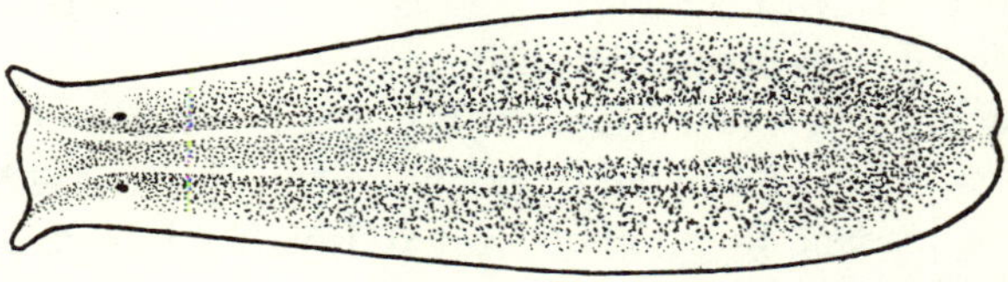

FIG. 16.—*Procerodes wheatlandi,* dorsal view. The posterior end forms a muscular sucker by means of which the animal attaches to the under side of rocks near the low tide line. These animals are usually found in considerable numbers on a given stone, if at all.

be due to this direct action of the greater mass of animals. He rightly thought there should be a quantitative determination of the salt content at the end of the experiment.

Drzewina and Bohn (1921*b*, 1928) replied by calling attention to the small size of the *Convoluta* worms, which are about 3 mm. long. They calculate that the amount of salt such worms would carry would not affect sensibly the salt concentration of the solution, but have not refuted this criticism by experiment.

At Woods Hole I have had an opportunity to test this situation with two main objectives: First, the question of fact involved: Is there a greater protection furnished by the presence of large numbers of animals in hypotonic sea-water when compared with fewer similar animals in the same volume of water? And, second: What is the mechanism of the protective action of the group, if it be found?

For these studies a turbellarian, *Procerodes wheatlandi* (Girard), Figure 16, was selected as representing a form taxonomically somewhat related to *Convoluta*. These animals are normally subjected to hypotonic sea-water when a heavy rain occurs at low tide.

These small worms reach a length of about 5 mm. and are about 1 mm. in width. In nature they are found in abundance on the lower sides of stones in small tide pools, near or below low tide line. They are not abundant in deeper water. Usually they were taken from the protected sides of stones that were firmly located on a sandy substratum; evidently they cannot stand the full sweep of the waves. They were usually present on a given stone in considerable numbers, if at all. As the water went stale in the laboratory, if there were numbers of worms present they would collect on the surface film in shaded areas in dense aggregations. Here, as in the field, they did not occupy all the apparently optimal space.

Appropriately safeguarded experiments showed that these worms will survive exposure to equal amounts of tap-water the better (*a*) if they are present in numbers; (*b*) if isolated into tap-water in which other living *Procerodes* have previously been exposed; and especially (*c*) if exposed in tap-water in which other *Procerodes* have died and disintegrated in whole or in part, even when such a conditioned medium is boiled or filtered. As is to be expected, these worms live longer if the salinity of the tap-water is increased by as much as 0.05 per cent above a minimum of that value. With well-washed worms it is possible to demonstrate protection when the tap-water contains exudates from living or dead *Procerodes* in which the salt concentration, as measured by the amount of chlorides present, is not the determining factor. The method used consists in titrating with N/100 silver nitrate, using a 1 per cent solution of potassium chromate as an indicator.

The first experiments (Allee, 1928) demonstrated that the protection was due neither to sea-water contamination nor to the leaching-out of electrolytes in the proportions in which they exist in sea-water. They did not exclude the possibility that the protection may have been given by the leaching-out of electrolytes in some other proportion from that found in sea-water. Such an explanation of the observed protection is the most simple and obvious one to be advanced. The possibility of its operation was tested at the first opportunity.

For this purpose experiments were repeated, using the technique suggested but with the additional precaution of checking the electrical resistance of the different solutions to which the worms were exposed, at the beginning and at intervals during the progress of the survival tests (Allee, 1929).

This method of determining the amount of electrolytes present measures all electrolytes, instead of depending on the well-known relation between the amount of chlorine and the total salt concentration in sea-water. The experiments to be reported fall into two main groups: those in which the water had an initial resistance of about 1,900 to 2,500 ohms, and those in which the initial resistance was 5000 to 6,500 ohms. The experiments reported in 1928, as nearly as can be told from chlorine titrations and survival values, belong to the former level. Preliminary tests showed that with hypotonic sea-water at the greater dilutions, and under the conditions of these experiments, an initial difference of about 1,000 ohms is needed to affect significantly the survival time of these worms.

All the comparisons made concerning the effect of homotypically conditioned medium upon the survival of *Procerodes* are summarized in Table XXIV, which gives the results of nine separate sets of experiments, each of which consisted of from three to six independent sets of tests. In the results down to those of Experiment 71 the initial resistance of the conditioned water was regulated by the addition of distilled or of tap-water to the conditioned medium. In the later experiments the resistivity was controlled by dialysis.

Direct comparisons indicate that dialyzed *Procerodes* culture medium is less effective than is similar dialyzed culture medium which also contains the water extract of dead *Procerodes* worms. These results are in keeping with those obtained in 1928 with the methods then in use, and in all probability represent the true state of affairs.[1]

In all, 640 worms were used, which were divided equally between *Procerodes*-conditioned water and fresh water to which enough dilute

[1] Castle (1928) in his work on the life-history of *Planaria velata* similarly found that in placing 50 small fragments of the flatworm in 10 cc. of distilled water some will survive. If the same mass of the same number of small fragments is placed in 200 cc. of water, usually all will die. With small volumes, more than 75 per cent will survive. He attributes this protective action of the mass primarily to the rapid conditioning of the medium by the products from disintegrating pieces.

sea-water had been added to bring it to an equal initial resistivity. These worms showed a mean survival of 18.38 hours longer in the worm-conditioned water than in the controls. Each group of experiments considered singly gave positive results, although some individual pairs did not. Despite the variability in procedure used in the different experiments, the combined results show a statistical significance of 0.03 when considered as nine paired experiments.

These results are graphically summarized in Figure 17, which shows two sets of histograms. That marked *A* gives in black the

TABLE XXIV

SHOWING THE EFFECT OF *Procerodes*-CONDITIONED WATER UPON SURVIVAL OF *Procerodes* ISOLATED INTO EXTREMELY HYPOTONIC SEA-WATER

EXPERIMENT	RESISTANCE IN OHMS				NUMBER TESTED	SURVIVAL IN HOURS		
	Conditioned Water		Dilute Sea-Water			Con-ditioned Water	Dilute Sea-Water	Differ-ence
	Start	Late	Start	Late				
41–43	1,916	1,358	1,897	1,445	56	88.16	28.44	59.72
43–45	2,100	1,438	2,100	1,480	60	59.97	37.03	22.94
46–48	2,350	1,780	2,350	1,760	60	58.1	44.05	14.05
49–51	2,400	1,930	2,400	1,888	56	20.45	11.16	9.29
52–54	2,000	1,320	2,000	1,735	60	10.05	7.52	2.53
71	5,150	3,550	5,150	4,000	58	37.18	11.97	25.21
71*a*	5,150	3,550	4,250	3,400	58	37.18	13.23	23.95
73	6,200	4,700	6,200	4,500	118	12.06	7.75	4.31
73*a*	6,200	4,700	5,100	4,000	114	11.30	7.88	3.42

percentage of worms surviving at the indicated hourly intervals when placed in tap-water with an initial resistivity of from 5,000 to 6,550 ohms. Just above, in the shaded blocks, is given the added percentage of survival resulting from the presence of sufficient sea salt to decrease the initial resistance to from 1,890 to 2,400 ohms. The upper clear blocks give the added survival resulting from the presence of water extract of *Procerodes* worms with the same initial resistance as the hypotonic sea-water.

Graph *B* shows similarly the survival in tap-water with an initial resistivity of from 5,000 to 6,550 ohms, in hypotonic sea-water made by adding dilute sea-water to pond or distilled water, bringing it to a resistivity of from 5,150 to 6,550 ohms, and finally, at the top,

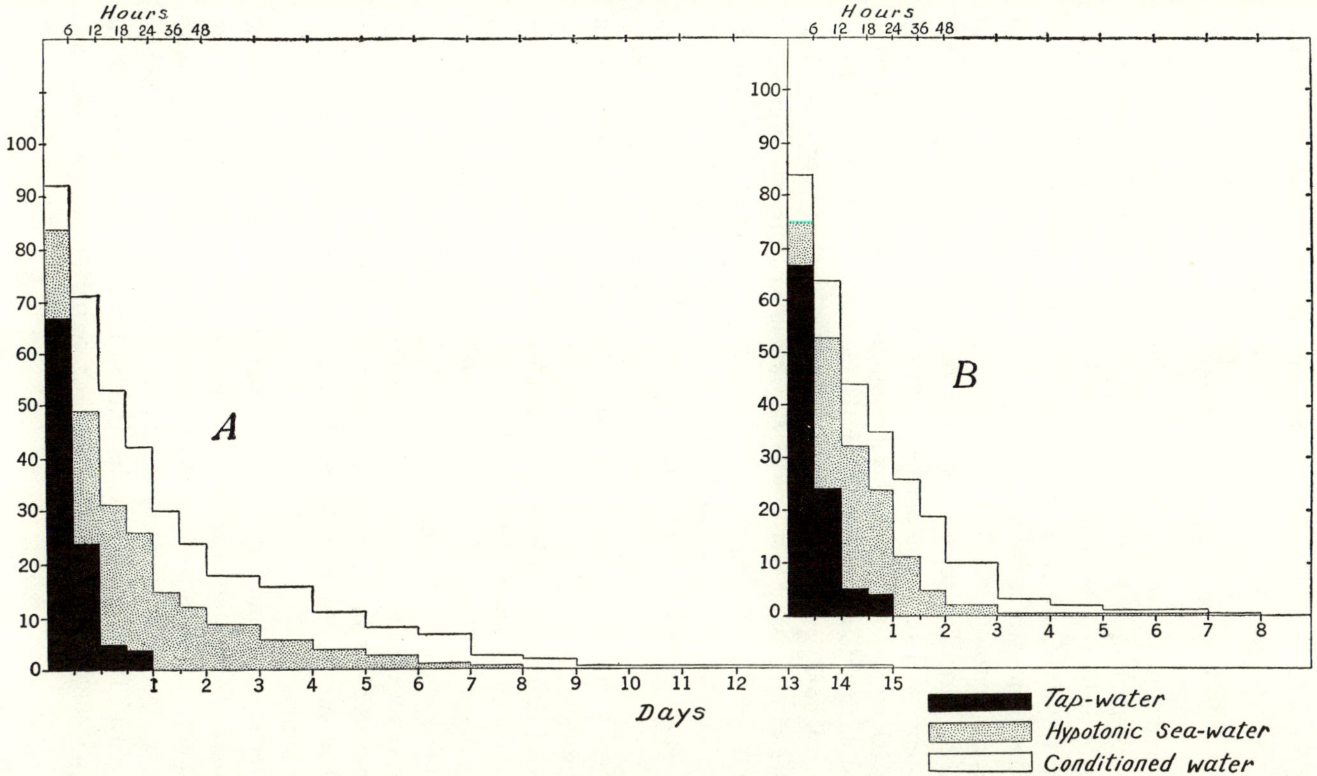

FIG. 17.—Showing the percentage of survival at different time intervals of *Procerodes* isolated into homotypically conditioned medium (total height), hypotonic sea-water of the same initial resistivity (black plus stippled), and tap-water (black).

the survival in dialyzed *Procerodes*-conditioned water having the same initial resistivity.

HETEROTYPICALLY CONDITIONED WATER

Having demonstrated in two successive years that *Procerodes* may be protected, at least to some extent, from the harmful effect of extremely hypotonic sea-water by being isolated in water conditioned by exudates or watery extracts of other *Procerodes*, and that whatever the protective device may be, it is not a direct result of an increase in total electrolytes present, it becomes of importance to inquire carefully concerning the specificity of this protection. Evidence will be presented here concerning the species specificity only, since to date no work has been done to attempt the analysis of possible functional specificity. In the course of this study the effects of the following kinds of media were tested against very dilute sea-water with the same resistivity: *Paramecium* culture medium, water extract of *Planaria maculata* (a fresh-water turbellarian) and of *P. maculata* culture media, and water extracts of a marine amphipod. The effect of each of these will be discussed in the order given.

Paramecium culture medium.—The solution used was the clear brownish surface liquid from a hay infusion which contained a vigorously growing and almost pure culture of *Paramecium.* The general situation can be presented most clearly by describing one experiment in some detail and by adding summaries of all. In Experiment 74, hay-infusion liquid, such as has just been described, containing many Paramecia was boiled and dialyzed against running tap-water until it showed a resistance of 6,450 ohms at 20° C. Tap-water was brought to the same resistivity by adding 0.25 per cent sea-water. Another lot of tap-water was brought to 5,450 ohms resistance.

One hundred and fifty *Procerodes* were isolated, 50 into each of the modified tap-waters and 50 into the dialyzed *Paramecium* culture medium. At the end of 6 hours the water in the more dilute tap series had changed from 6,450 to 3,350 ohms and that in the culture medium had fallen from the same initial level to 3,550 ohms. Both were changed to 6,550 ohms in their respective media. After another 6 hours, this had fallen to 4,950 ohms for the modified tap-water and to 5,750 ohms for the culture media. Again the liquids were changed to appropriate solutions, each at 7,400 ohms at 19° C.

After 24 hours the water from worms just beginning disintegration in the dilute sea-water showed 4,850 ohms resistance, while that from worms in the same condition in the culture medium showed

TABLE XXV

SHOWING THE SURVIVAL TIME, IN HOURS, OF *Procerodes* ISOLATED INTO 1 ML. EACH OF *Paramecium* HAY INFUSION AND OF DILUTE SEA-WATER OF THE SAME OR GREATER RESISTIVITY

MEDIUM	WORM NO.	SURVIVAL TIME IN HOURS			RESISTANCE RANGE IN OHMS
		Mean	Maximum	Minimum	
Control I	1– 9	18.05	36.0	3.5	
Culture	11–19	21.85	38.5	6.0	
Control II	21–29	15.75	34.0	3.5	
Control I	31–40	12.70	34.0	3.5	
Culture	41–50	27.95	58.5	11.5	
Control II	51–60	22.50	58.0	11.0	
Control I	61–70	17.00	40.0	8.5	
Culture	71–80	28.30	40.0	8.5	
Control II	81–90	19.20	64.0	3.5	
Control I	X1–10	16.85	34.0	7.0	
Culture	X11–20	25.93	49.3	7.0	
Control II	X21–31	18.50	41.0	8.5	
Control I	X31–40	11.40	17.0	4.0	
Culture	X41–50	25.20	58.5	8.5	
Control II	X51–60	20.50	41.0	8.1	
SUMMARY: Control I	49	15.20	40.0	3.5	7400–3350
SUMMARY: Culture	49	25.85	58.5	6.0	7400–3550
SUMMARY: Control II	49	19.30	64.0	3.5	6150–3550

4,950 ohms. Again the surviving worms were changed to water having a resistance of 7,200 ohms for each solution. Ten hours later both showed a resistance of 4,475 ohms.

The approximate survival time was obtained for 49 worms exposed singly to 1 ml. each of the two solutions whose history has just been given. It will be noted that in no case was the resistivity less in the *Paramecium* culture medium than in the accompanying control solution. The survival times are summarized in Table XXV, in which Control I is used to mean the modified tap-water with the same resistivity as the culture medium, while Control II designates the less-dilute salt solution.

The summary shows that the worms in the culture medium lived, on the average, 25.85 hours, while those in the very dilute sea-water with the same initial resistivity lived only 15.2 hours. The difference of 10.65 hours, when examined by Student's method, shows a statistical probability of 0.0714, when considered as 5 pairs of tests as listed above, but when considered as 49 individual pairs, the probability becomes 0.00002, which is clearly significant.

As convincing as is this experience that *Paramecium* culture medium with equal or less total electrolytes than accompanying dilute sea-water has a survival value for marine *Procerodes* isolated into it, the same series of experiments furnished still further evidence that such is indeed the case. In the series labeled "Control II" the initial resistance was 1,000 ohms less, and therefore a less-dilute solution of sea-water than Control I. This water was renewed each time the others were changed, and was kept at least 1,000 ohms more concentrated at these times. It was never found to be less concentrated than was the culture medium, and only once to have the same concentration. The worms in this less-dilute sea-water lived longer than did those in Control I by an average time of 4.1 hours, a difference which is not statistically significant.

The worms in the culture medium showed a mean survival time of 6.55 hours greater than did those in the more concentrated Control II. When individual pairs are considered, this has a probability of 0.02, and hence is statistically significant. Further experiments with *Paramecium* culture medium and with water extracts and culture medium of *Planaria maculata*, and water extracts of marine amphipods found in close association with *Procerodes* in nature, give essentially the same results and show that heterotypically conditioned fresh water has protective value for *Procerodes* isolated into it as compared with the survival when isolated into hypotonic sea-water with the same electrical resistivity.

While this general relation holds, there was not necessarily the same degree of protection from each type of solution. Whether this is due to the conditions under which the experiments were run, or whether it is an inherent property of the different heterotypically conditioned solutions, is not yet apparent. At any rate, the findings to date are summarized in Figure 18, which gives the experience from comparable experiments run simultaneously. The vertical axis shows the percentage of worms surviving at any given time; the horizontal axis gives the time elapsing since the beginning of the experiment. One finds that there is a marked drop in the percentage of survivals within the first 48 hours and that thereafter the curves tend to flatten, reaching extinction in from 16 to 20 days.

The mean survival for the whole group is not plotted on this chart, but practically coincides with the graph for *Procerodes*-conditioned water, except that it continues just above the base line until after the 15-day mark. The graph for hypotonic sea-water having the same initial resistivity runs below the lowest graph on the chart at all points, except that at the 2- and 3-day periods it is very slightly above this lowest graph.

These results show clearly the lack of species specificity in the protection of *Procerodes* against the lethal effect of hypotonic sea-water; and remind one, in this respect at least, of the results obtained by Allee and Schuett (1927) that protection from such toxic substances as colloidal silver also lacks species specificity. The protection against colloidal silver appears in a large part to lack function specificity. Whether or not the present protective mechanism also has other and general functions is not clear at the present time, although on general grounds one would be inclined to think that such would be the case.

POSSIBLE FACTORS CONTRIBUTING TOWARD THE OBSERVED SURVIVAL VALUE OF CONDITIONED SOLUTIONS

The facts recorded are plain. The lethal effect of the fresh water is clearly less for solutions that have been conditioned by the presence of living organisms, when compared with hypotonic sea-water having equal initial resistivity. The exact source of this conditioning is not yet clear. In some of the *Procerodes*-conditioned media the survival value is due to exudates from the living worms; in others, where water extracts were prepared, the survival value may

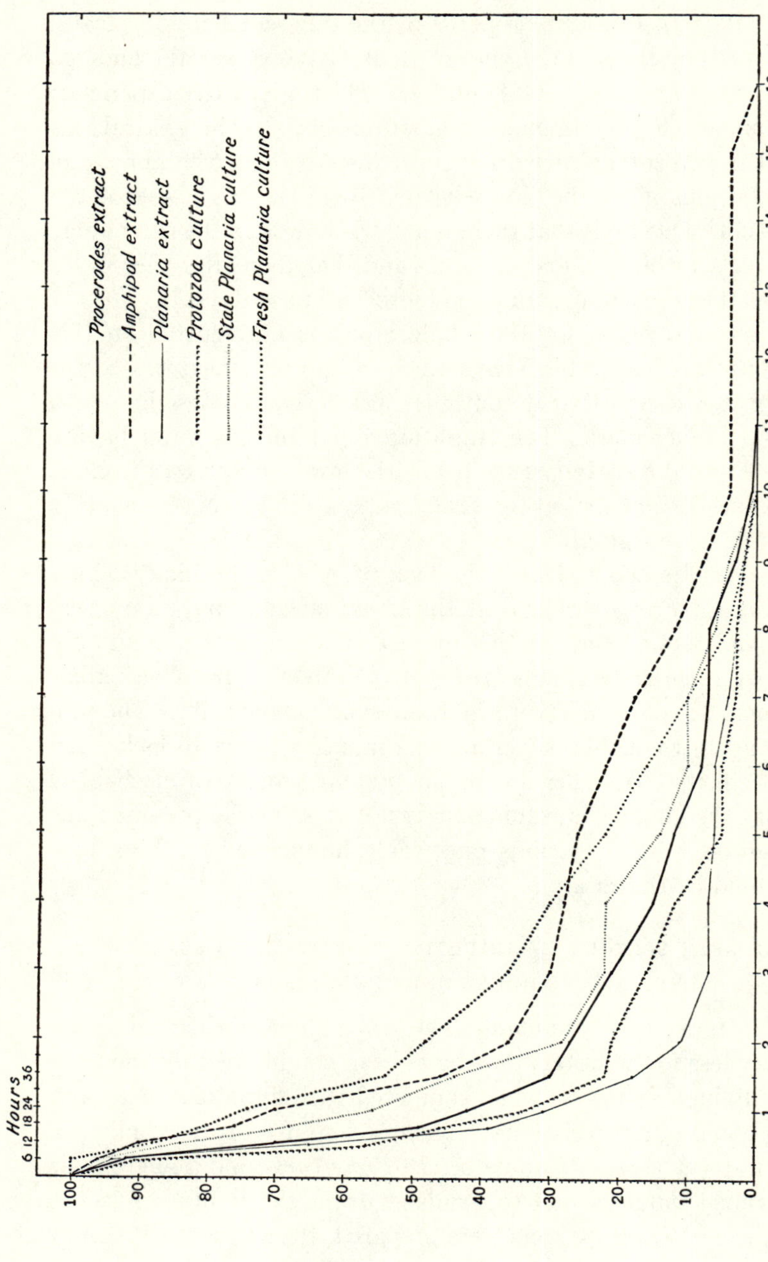

FIG. 18.—Showing the percentages of *Procerodes* surviving isolation into different types of animal-conditioned water with the same initial resistivity.

have been in part due to the action of bacteria, although this is practically ruled out in those experiments in which the solution was given a fair pasteurizing treatment in the formation of the extract, which was later boiled vigorously. The bacterial growth in such solutions must have been slight, at least in the early history of the isolations.

With the *Paramecium* culture medium the rôle played by the different elements entering into the conditioning of this medium could be told only by experimentation. The Paramecia were growing in a hay infusion which contained the water extracts of the hay as well as the products of bacteria and of Protozoa other than the Paramecia. All these must be considered as possible conditioning factors, since the work above has clearly shown that the protection is furnished by heterotypic as well as by homotypic conditioning.

It has been possible to begin experimental analysis of the factors leading to increased survival in these biologically conditioned solutions. To date these indicate that the results are not primarily due to the depressing action of the conditioned medium, in so far as these are mimicked by dilute alcohol. They are not due to differences in pH. Gelatine suspensions do not confer similar benefits. The colloids present do not mask the amount of electrolytes in solution.

Preliminary tests have been run to determine whether the protective substance would be adsorbed on animal charcoal. These gave results of sufficient interest to deserve being presented in outline. Animal charcoal was added to the dialyzed culture medium. The whole was stirred, boiled, filtered twice with suction and once without, and, after redialysis to 6,050 ohms, a standard survival experiment was set up with hypotonic sea-water controls at 6,050 ohms and 5,050 ohms. The resistance decreased in all solutions, as usual, during the progress of the experiment. The solutions were changed twice in the first 20 hours. As usual after standing, the culture water had less electrolytes than did either of the accompanying solutions.

Under these conditions the 50 *Procerodes* isolated into the treated culture medium lived a mean time of 11.03 hours, while those in hypotonic sea-water with the same initial resistivity lived 9.21 hours.

The difference, 2.72 hours, has a statistical probability of 0.0564, which is too great for much significance to be attached to the observed difference. These values are to be compared with the results shown in Table XXV, which were obtained immediately preceding the present experiment. That test showed a difference in survival of 10.65 hours, with a probability of 0.00002.

When comparison is made between the survival in the charcoal-treated medium and the less-dilute sea-water which had an initial resistivity of 5,050 ohms, the difference in survival time is 2.53 hours, with a statistical probability of 0.072. These values are to be compared with a survival difference of 6.55 hours and a probability of 0.02 in the preceding experiment with the same type of culture water similarly dialyzed but not treated with charcoal.

These results indicate that the charcoal did remove a part of the actively protective substance, and suggests the need of further work at this point. Perhaps more extended adsorption would have a still greater effect. Another possibility must be mentioned, although here, too, the evidence available at the present time does not warrant the drawing of definite conclusions. I refer to the possibility of the differential leaching of materials in raw and in biologically conditioned water. Early in the analysis of the results here summarized, indications appeared that worms isolated into water conditioned by living organisms tended to lose electrolytes less rapidly than did their controls isolated into hypotonic sea-water with the same initial resistance. Accordingly, the resistance records were examined for comparable experiments. The different exposures were for varying lengths of time, and hence show a marked variation in the change in resistance from that given at the beginning of the experiment. The initial resistance of the observed solutions stood at different levels in the various experiments, but always at the same level in both the experiment and the control for any given experiment. The mean difference in the whole series of experiments shows that there was 4.8 per cent less change in the resistivity of the biologically conditioned media than in their accompanying hypotonic sea-water controls. Despite the many known elements of variation, this has a statistical probability of 0.00068.

One must be cautious in interpreting this lessened change in electrical resistivity in the animal-conditioned media as being the fundamental cause of the longer survival of the worms isolated into such media, because in general we have seen that much greater initial differences in resistivity have relatively little effect upon survival, and also because gelatine suspensions showed a smaller resistivity change than did the controls, and yet had no protective action.

There is a possibility that if the worms lose contained electrolytes less rapidly into the surrounding medium in the presence of materials from other organisms, organic materials necessary for the continued well-being of the worms, which are not measured by present methods, may also escape less rapidly, as Robertson assumes to happen in the case of allelocatalysis in Protozoa; and that the observed protection may be due to the retention within the organism of a certain substance or substances necessary for continued existence, rather than to the presence of a definite protective material in the treated solutions.

We are here dealing with some chemical difference in the medium unanalyzed as yet, rather than with some sort of physical protection of which different types have been described (Allee, 1920, 1926, 1927; Drzewina and Bohn, 1928). The material behaves in some aspects like that which Banta and Brown (1929) find affecting the percentage of males in crowded cladoceran cultures, which they attribute to the accumulation of excretory products; in others, like that which Petersen reports as conditioning *Paramecium* cultures so as to allow higher division rates, especially since the latter has been shown not to be species-specific. We are probably dealing here with the sort of protection which Drzewina and Bohn originally postulated in their communication of 1920, quoted at the beginning of this chapter.

CHAPTER XIV

RELATION BETWEEN DENSITY OF POPULATION AND INSECT SURVIVAL

We have already seen that the insect *Drosophila* reproduces less rapidly the greater the population density (chap. vii), and that, on the other hand, the confused flour beetle (*Tribolium*) during initial stages in cultures of flour reproduces more rapidly if the population for given environments be larger than the minimum (chap. x). In later chapters we shall find that crowding exerts certain physiological effects upon aggregated insects which are expressed in morphological changes. These facts concerning mass relations between non-social insects are the more interesting in view of the high state of social organization to be found at times in this group. Unless we can find a strong substratum of generalized co-operative survival values among the subsocial insects, we shall have difficulty in demonstrating a connection between the phenomena associated with relatively slightly integrated animal aggregations and those connected with the more closely knit social communities.

For the present we desire to consider the relations between numbers of non-social insects and their survival in the face of unfavorable conditions. Three lines of evidence exist upon this point: (1) the data from laboratory and field experiments concerning the relation between numbers of insects present and their survival in the presence of toxic reagents; (2) studies on the relation between population density and longevity in *Drosophila;* and (3) the more usual type of natural-history observations showing that the survival value of crowding in the presence of predators holds for insects as well as for the larger animals, for whom it has been more frequently reported.

MASS PROTECTION FOR GRASSHOPPERS

Two workers have independently reported mass-survival values for insects exposed to poisonous substances. Deere, for the majority of his experiments in this laboratory, used different species of grasshoppers, working upon the meadow grasshopper, *Xiphidium fasciatus,* more than on any other single species. Experiments were run upon the effects of ether, carbon tetrachloride, ethylene chloride, ethyl alcohol, and hydrocyanic acid. Two types of experiments were carried out. In one set the insect containers had a direct volumetric relation to the number of individuals per bottle, so that if 5 insects were used, the container had five times the volume of that containing but 1 insect; and if 10 were used, the container had ten times the volume of that employed with only 1 animal. In such experiments the volume percentage of the toxic gas was kept constant.

In the other type of experiment, all tests were run in the same size bottle regardless of the numbers exposed. In these, too, the volume percentage of the toxic agents was kept constant. The amount of gas used was calculated to produce narcosis in most cases, without death necessarily following.

In all experiments the group survived longer; but, as might be expected, there is relatively greater group resistance when the volumes are identical than when they are proportional to the number of insects. Similarly, single individuals have been found to have a somewhat longer survival when placed in a smaller bottle rather than a large one, when each has the same volume percentage of toxic gas. The results of one of the constant volume experiments are given in Table XXVI.

The effect of exposure of these animals may be summarized as follows: In the group of 30, 82 per cent survived for 24 hours; in the group of 20, 90 per cent survived; in the group of 10, 90 per cent survived; while of the 10 singles only 60 per cent survived for the 24 hours of the experiment. In control lots under similar conditions, but without the ethylene chloride gas, 97 per cent of the group of 30 survived, and all of each of the other lots lived for the period of the test. Apparently the group of 30 suffered somewhat from the ill effects of crowding, but not enough fully to mask the survival value of the group when exposed to this amount of narcotic gas.

The *Xiphidium* have cannibalistic tendencies, which caused several experiments to be discarded. When *Melanoplus* were used, they showed greater initial activity but otherwise gave similar results.

Ethyl alcohol differed from the other vapors used in that the effect on the animals was much more gradual, extended over a longer period of time, and stimulated them to greater activity.

TABLE XXVI

SHOWING THE RELATIVE SURVIVAL OF GROUPS AND ISOLATED INDIVIDUALS OF THE GRASSHOPPER *XIPHIDIUM*, EXPOSED TO EQUAL VOLUMES OF 27.5 VOLUME PER CENT OF ETHYLENE GAS, AUGUST 17, 1928; BAROMETER 746, TEMPERATURE 24° C.

All were placed in 240 cc. bottles supplied with 66 cc. of the gas.

Number of Animals	Minutes until First Effect	Minutes* until Quiet	Condition after 12 Hours		
			Apparently Normal	Affected	Apparently Dead
30	25 (few)	50 (3)	24	3	2
20	35 (several)	45 (few)	17	1	2
10	25 (few)	45 (few)	9	....	1
1	25	37	1	....	....
1	24	44	....	....	1
1	13	43	....	1	....
1	27	....	1	....	....
1	29	49	1	....	....
1	12	63	....	....	1
1	7	58	....	1	....
1	18	....	....	1	....
1	24	41	....	....	1
1	25	60	....	1	....

* Removed at this time to ordinary air.

The period of recovery was also longer with this reagent. The difference in recovery between the group and the single individuals was also greater with alcohol than in the case of the other vapors used.

The outstanding effect noted in the case of insects exposed to HCN was the lack of stimulation when the insects were first exposed. *Xiphidium* in particular remained motionless from the time the gas reached them. Muscular activity could be induced, if at all, only by most vigorous shakings. This would loosen their attachment from the bottle or from whatever object to which they might be clinging. Those that were fully narcotized would fall to the bottom without movement; others might show slight movements of the feet, particularly if these came into contact with a solid object, even when there was no other visible movement. The recovery period from HCN was also shorter than from exposure to the other gases used; but otherwise the results were similar, particularly in that the group showed greater survival value than isolated individuals.

Such results from different numbers of insects exposed to the same volume of the same concentration of toxic gases are according to expectation from the experience previously recorded with aquatic organisms exposed to various toxic conditions. The apparent explanation is that, when the toxic gas is distributed among a number of organisms, no one individual is as likely to receive a dose sufficient to cause death as if it were exposed singly to the same quantity of the same gas. The explanation of the findings when groups and singles are exposed to volumes in proportion to the numbers present, indicating that there is group protection in proportion to the numbers present even in the face of similar conditions for each individual, is not clear, and further work is needed at this point.

It seems probable that under such conditions of crowding, even with equal space and an equal amount of the toxic gas present per individual, the group may cause a greater concentration of carbon dioxide in its neighborhood, or by some other effect may very well cause these animals to have a lowered rate of metabolism. Under these conditions, and with the concentration of poisonous substance at the right strength, one would expect the members of the depressed group to survive longer than would the more active single individuals. This suggestion is in line with the susceptibility work of Child (1915) and the resistance of *Daphnia* to various chemical reagents, as developed by Fowler and discussed in a previous chapter.

Experimental results similar to those of Deere have been obtained by Bliss (personal communication) under natural conditions. In his field studies Bliss found that the camphor scale insect (*Pseudoanidia*) has a greater natural death-rate the greater the density of the population on a given twig. When exposed to adverse conditions, such as are represented by spraying with an oil spray, the opposite conditions prevail, and there is decidedly greater survival the more dense the population. In making these observations, due care was taken to consider only those insects located on trees in comparable situations and on comparable parts of the trees. Bliss concludes that death from these sprays is not due to suffocation, as has been supposed, but is due to the taking-up of some poisonous material by the insects; and that when they are present in greater numbers, the toxic material is divided between more insects, no one of which is so likely to receive a lethal dose as if the population were less dense.

POPULATION DENSITY AND LONGEVITY IN DROSOPHILA

The relation between the density of population of laboratory cultures of wild-type *Drosophila* and their life-duration were less expected. Pearl and Parker showed in 1922 that statistical analyses of data accumulated in other studies indicated greatest longevity from bottles originally stocked with from 35 to 45 of the wild stock per bottle. In 1923 the same workers reported the results of an experiment made to test out this question of an optimal population. The data on the length of life of 12,382 individuals showed that the optimal density of population, when longevity is taken as the criterion, is not found in the minimal populations but lies in the region of initial densities of from 35 to 55 per 1-ounce vial, and the increase in length of life from the lowest density is at a much more rapid rate than is the decline of duration of life after the optimal density is passed.

A more complete analysis of the problem is given by Pearl, Miner, and Parker (1927). In this experimental work the flies were kept in 1-ounce vials stoppered with cotton plugs and held at 25° C. The bottles were examined daily, the dead flies were removed and their age recorded, and the living flies were at the same time transferred to fresh bottles of newly prepared food. In the first experiments banana agar was used as food, but similar results were obtained with a synthetic medium.

The extent of the experimental data may be visualized from the following statement of the numbers used in one experiment. One hundred and fifty vials were started with an initial population of

1 pair each; similarly 80 vials contained originally 2 such pairs; 50 vials contained 3; and 40 vials contained 4 pairs each. Thirty vials were started with 5, and 30 more with 6 pairs each, and 20 vials contained an initial population of 15 flies, or 7.5 pairs. Ten vials were started with each of the following populations: 20, 25, 35, 45, 55, 65, 75, 85, 95, 105, 125, 150, and 200. In another experiment the initial densities per 1-ounce vial were: 5, 25, 50, 75, 100, 200, 300, 400, and 500.

The results from the first experiment are given graphically in Fig. 19, which shows the mean duration of life of wild-type *Drosophila* at different densities of population. The second experiment yielded similar results and demonstrated that with the higher densities there is relatively slight effect upon longevity of marked increases in original density.

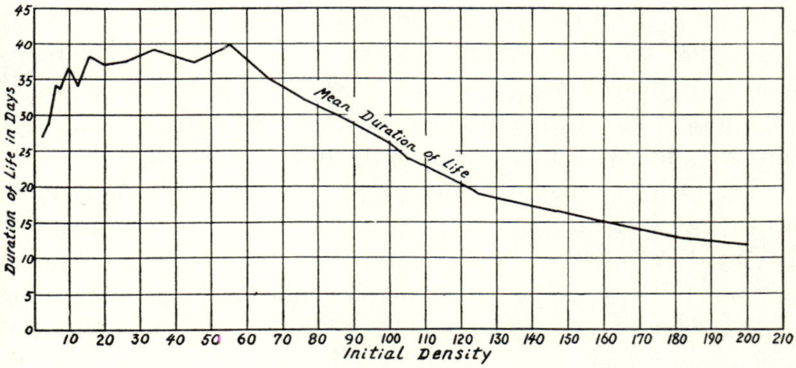

FIG. 19.—Showing the mean duration of life in *Drosophila* with relation to the initial density in 1-ounce bottles. (From Pearl, Miner, and Parker.)

Pearl, Miner, and Parker sum up their experience with this sort of experiment as follows (1927):

"The rate of mortality of *Drosophila* is profoundly influenced by the density of the population, that is, by the number of flies together occupying a limited universe in which volume of air, volume of food, and area of food surface are constant.

"There is an optimal density of population for *Drosophila* under the conditions of these experiments. This optimal density falls somewhere in the region of 35–55 flies per one ounce bottle containing 8 cc. of food substrate. At densities of population above and below the optimum, the specific death rates are higher at all ages than they are at the optimum."

Such results as these make one fairly bristle with questions, of which only a few can as yet be answered. Pearl and his associates have reported on investigations concerning the effect of changes in density during the progress of the experiments. Since an initial population of about 35 flies per vial was found to lie near the optimum under the conditions used, a number of vials were set up with this density and apparently were followed according to the practice of the preceding experiments until the sixteenth day, at which time the survivors were etherized and the tips of the wings of half of them were clipped, after which they were returned to their proper bottles. Appropriate tests showed that the wing-clipping as practiced did not significantly affect life-duration.

Vials with an initial population of 200 were treated in like manner up to the sixteenth day, when they were etherized and counted. In part of such bottles the population was then brought back to its initial density of 200 by adding marked flies which up to that time had lived under the optimal conditions furnished by an initial population pressure of 35 flies per ounce vial. Others were brought to the original population density of 200 per ounce bottle by adding flies surviving from stocks started from that density. The results of such treatment are shown graphically in Figure 20, which gives survival curves starting with 100 flies at 16 days of age, and the effect of previous history of crowding upon the duration of life after that age.

In commenting upon these results, the experimenters state that here, as before, flies subject to an initial density of 35 per ounce bottle survive, on the average, about double the time those live that are subjected to an initial density of 200 for the same size vial and with the same food supply. They continue:

"When flies which have lived the first 15 days of their lives under the conditions implied by an initial density of 35 are on the 16th day of their age submitted to a density of 200, and live out the remainder of their lives under the conditions implied thereby, their average duration of life is reduced in these experiments from the 34 or 35 days which it would have been had they stayed in the bottles with an initial density 35, to 22.83 ± 0.19. This result shows that

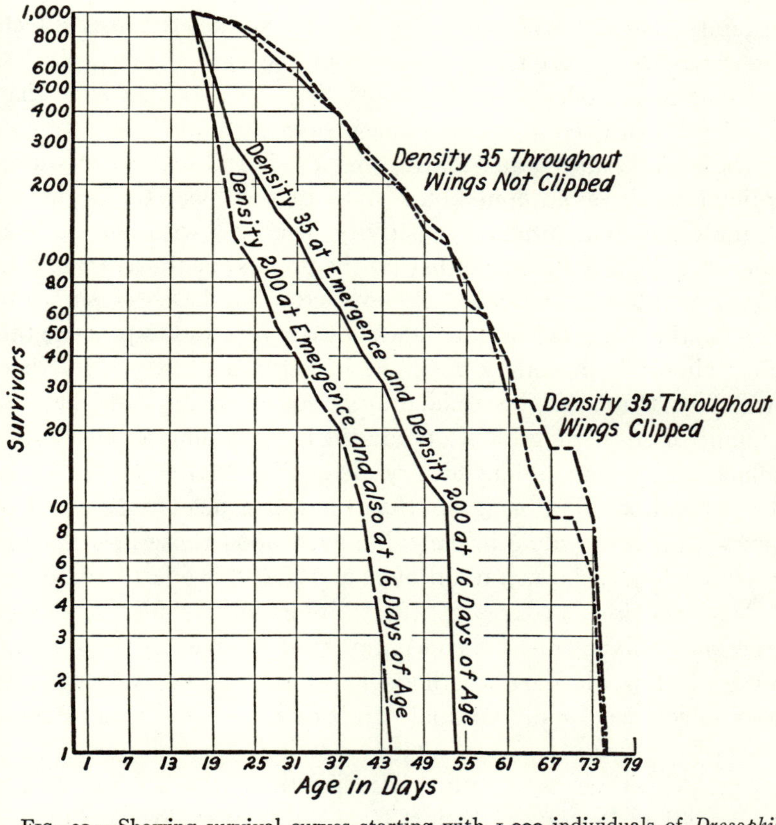

FIG. 20.—Showing survival curves starting with 1,000 individuals of *Drosophila* at 16 days of age which had been subjected to different densities of population up to that time. (From Pearl, Miner, and Parker, 1927.)

crowding produces a heavy increase in mortality even though it occurs as late as 16 days of age.

"Flies which have lived the first 15 days of their lives under the conditions implied by an initial density of 200, and then at the age of 16 days are again subjected to a density of 200, have a significantly shorter duration of life than do their companions in the same bottles who spent the first 15 days of their lives in bottles of initial density 35. The difference is 22.83 ± 0.19 minus 19.71 ± 0.17, which is 3.12 ± 0.25 days. This difference is more than 12 times the probable error. It may be taken as probable to the point of practical certainty that excessive crowding in early life deleteriously affects the survivors of 16 days of age so that they are significantly less resistant to the effects of heavy crowding again at that time than are flies which lived at optimal densities in early life."

These *Drosophila* data indicate that the harmful effect of supra-optimal densities of population is most marked near the beginning of adult life. This original sensitivity decreases with age, so that older flies are less affected either by the original or the average density of population. Even so, excessive crowding has been shown to increase the death-rate at later ages of these flies and to produce this effect almost immediately upon the increase in density. "Furthermore," as Pearl and his collaborators state, "it appears that the amount of shortening of life produced by crowding at any age is influenced by the previous history of the flies relative to density of the population. This suggests that there is a deleterious effect of supra-optimal crowding in early life even upon those flies that do not immediately die as a result of it, and that this effect endures for at least the first 15 days of life." Unfortunately for us, Pearl's experiments have not been concentrated at the part of the survival curve which would show whether suboptimal crowding will have the same effects that supra-optimal densities have been demonstrated to have.

These experiments on *Drosophila*, like those of Chapman on the confused flour beetle, are distinct from the great majority of crowding experiments in that they are carried on in a non-liquid environment, whereby the temptation to postulate the production of some *X*-substance responsible for the phenomena observed is much reduced. Pearl suggests that the biological explanation for his observations lies in an entirely different direction, but the matter has not been discussed publicly, even in general terms, by members of his laboratory. We know that in the case of effect of crowding on egg production in fowls, Pearl and his associates regarded the underlying stimulus to lie in the psychological field, and it is possible to regard the effects recorded here as expressions of some sort of stimulus physiology, such as Goetsch has recorded in his experiments upon the effect of crowding in retarding growth of rapidly swimming animals such as young tadpoles.

There are, however, other possibilities which must be considered. In vials stoppered with cotton there may be an increased CO_2 tension developed which may in part account for the observed effects. Then, too, the reduction in length of life at suboptimal densities may be an expression of the inability of the small populations present to gain control of "wild" organisms other than the food yeasts present in the cultures, while the supra-optimal density effects may be related at least in part to food shortage and to excess of excretion products. Obviously, the resolution of this situation into causal factors is not easy; but it is equally obvious that we have here one of the most suggestive of the phenomena yet presented, indicating the wide application and fundamental importance of the physiological effects of animal aggregations upon the aggregants.

MASS PROTECTION IN NATURE

A different aspect of the effects of numbers on insect survival is illustrated by the observations of Haviland (1926) upon the protective value of feeding aggregations of certain chrysomelid beetle larvae (*Coelomera cayennensis*). Haviland says:

"The larvae are thickly hairy and the last segment of the body is expanded into a strongly chitinized, shovel-shaped flange. The larvae feed in a compact mass on the upper surface of *Cecropia* leaves. Their heads are all directed inwards while the caudal expansions thresh to and fro on the outside of the circle. The chief enemy of these larvae is a carnivorous Pentatomid bug (*Phyllochirus*) which loiters on the outskirts of the throng awaiting the opportunity to impale a larva with his proboscis and drag it from its fellows. As long as the circle of shovels is unbroken, the bug stands little chance, for his stylets cannot penetrate their polished armor and he cannot reach the soft bodies beyond. But as the larvae feed they move outwards from the original center, the circle becomes wider, and sooner or later the enemy slips in between the defenses and secures a victim. According to my observations, however, the circle is not broken naturally until the larvae are full grown, in which case the loss of one or two individuals does not signify, for before the bug has digested his meal, the rest of the brood enter the pupal stage and so escape."

These observations bring us back again to the familiar biological level of the struggle for existence between predators and their prey. They are given here because it frequently appears that such masses of feeding larvae are more at the mercy of their enemies than if they fed alone, unless they have developed means of detecting the approach of enemies and transmitting the information through the mass or unless they have developed effective means of active group defense like that of the soldier caste of termites and ants.

CHAPTER XV

COMMUNAL ACTIVITY OF BACTERIA

Many of the relationships which we have found to hold for animals can also be studied with bacteria. The large amount of work done upon these micro-organisms and the excellent technique developed both in culturing and in making isolations cause the results obtained to be the more significant for our studies. In one way bacteria as presented here and spermatozoa as discussed in the following chapter may serve as test cases for us. The situation may be stated thus: We have been presenting certain evidence concerning the rôle of numbers of individuals present in relation to the physiology of each individual. The studies so far have dealt with a wide range of animals. If we turn to some totally different organisms, such as the bacteria, shall we find similar relationships prevailing there? If we do, and if we also find the same conditions holding for the spermatozoa, we shall have the greater reason for regarding the phenomena we are studying as being of universal biological significance.[1]

Buchanan (1918) has summarized the life-cycle of a bacterial culture into seven periods. These are illustrated in Figure 21, where each phase extends to the left of its respective numeral. This analysis of conditions in a bacterial culture is essentially like that known to obtain for many protozoan cultures (cf. Fig. 9) except that the analysis with bacteria has been more carefully made. The different phases will be discussed briefly here.

1. The *initial stationary phase* covers the time during which the number of introduced organisms shows no increase. This phase has been studied but little. If the bacteria are actively growing at the time of transfer, growth may continue on the new medium, indicat-

[1] Two summarizing books have appeared dealing with the material presented in the early part of this chapter, by Buchanan and Fulmer (1928) and by Henrici (1928). These assist a layman in this field to present a more seasoned summary than would otherwise be possible.

ing that the phase of the parent culture from which the inoculum was taken affects the duration of this initial stationary phase. This is the equivalent of the "lag phase" of Robertson (1924).

2. The *lag phase* is also called the *positive growth acceleration phase*, and is the period during which the rate of increase shows acceleration. This period, together with the preceding one, makes up the lag phase as so considered by many investigators working with bacteria and with Protozoa. Buchanan and Fulmer use it as meaning "the period elapsing between the beginning of multipli-

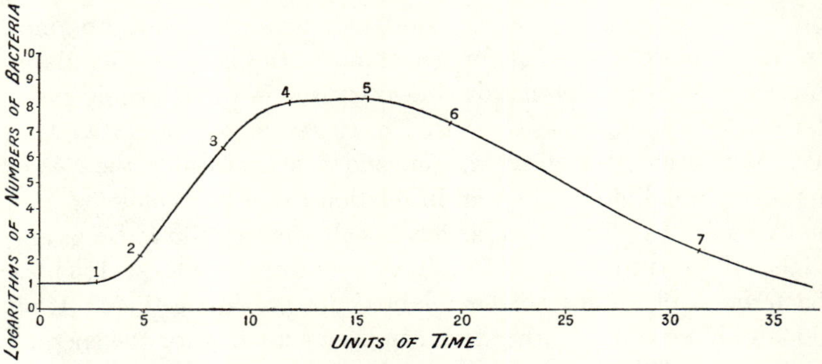

FIG. 21.—Showing the life-history of a bacterial culture. Figures on the vertical axis give logarithms of numbers of bacteria; those on the horizontal axis give units of time; those on the curve itself indicate the close of the respective growth phases. (Redrawn, with slight modifications, from Buchanan and Fulmer 1928 by permission of Williams & Wilkins.)

cation and the beginning of the rate of maximum increase per organism."

The phenomena associated with the lag phase interest us particularly because of the inherent implication that an increase in numbers or the continued occupancy of a new medium affect growth conditions. A number of theories have been advanced in explanation, of which the following may have significance in some cases (Buchanan and Fulmer, 1928).

a) The *essential secretion* theory is an application to the present phenomenon of a variant of Semper's suggestion that some *X*-substance must be necessary for growth. Here it takes the form that the transplanted organisms must give off some essential material into the new medium before maximal growth can occur. The fact that with bacteria, as we have already seen with Protozoa (chap. X), transfers from cultures during periods of maximal growth do not exhibit a lag period shows that such secretion is not always essential. Robertson's work (1924) for Protozoa and that of Valley and Rettger (1926) for bacteria indicate that it may at times have significance.

b) The *adaptation* theory is based on the idea that acclimatization to a new medium requires time. The evidence mentioned above shows that this is not always true; further, transfers from old cultures may show lag even though they are transferred to the same medium. Fulmer (1921) presents evidence that adaptation may be operative in the case of the transfer of yeast to a medium containing ammonium fluoride.

c) The *non-viability* theory which supposes that some of the organisms die off early may still hold for Protozoa, although it does not hold universally for bacteria, since culturing on agar plates reveals the lag phenomenon and non-viable bacteria would never be counted in this method.

d) The *agglutination* theory suggests that bacteria agglutinate on the plates, and the resulting count is of clusters of bacteria rather than of individuals. This may be true under certain conditions, but lag can be shown when agglutination has been avoided.

e) The theory of *recovery from injury* holds that exposure to the accumulated products of metabolism in the parent culture affects the different cells so that time is required to recover when placed in a new environment. Buchanan and Fulmer regard this as perhaps an "approximation of the truth, though probably not entirely adequate." If we cannot admit that the production of such toxic substances is universal, we may still think of the injury as being due to starvation in the deteriorated parent culture. Somewhat the same idea is expressed in the *inertia* theory, which holds that the transferred organism continues to grow for a time in the new medium as it would have grown in that from which it was taken. Buchanan's so-called *germination* theory of lag appears to be another statement of about the same concept.

f) Penfold (1914) and later Robertson (1924) regarded favorably the theory of *elaboration of essential chemical substances*. This theory assumes that some substance, *c*, may be required for growth which is not produced from *a* already in the solution, but from *b* which must be produced by the organism from *a*. Further, *c* must be present within the organism in optimal quantities before optimum growth occurs; and in a medium lacking the proper amount of *c* this substance diffuses out of the bodies of the non-growing cells and must be resynthesized before growth can take place.

The causes of the lag phenomenon may reside both in the medium and in the organism. They may differ in different conditions, and all the foregoing theories may operate in special cases.

In the light of the situation concerning Robertson's phenomenon among protozoa reviewed in chapter x, we are interested in the effects of number in the inoculum upon the duration of the lag period. Robertson (1922) defined "lag" in terms of the initial stationary phase and found no effect of numbers in the inoculum upon this phase preceding the first division. His "allelocatalytic effect," in which the rate of reproduction was much increased by the presence of a second organism in a limited amount of culture medium, belongs in the lag phase as used here, and the entire discussion of chapter x is pertinent at this point.

3. The *logarithmic growth phase* (see Fig. 21 again) is the period at which the rate of increase remains constant and at its highest value. During this phase the logarithms of numbers of bacteria plotted against time give a straight ascending line.

4. *Negative growth acceleration phase* covers the time when the rate of increase is falling, although the bacteria continue to increase in numbers. This phase sets in soon in bacterial cultures but comes more slowly in those of Protozoa. The causes contributing to its onset and development are those which later lead to the decline in numbers. These have already been discussed at some length in chapter vi. Briefly, those considered most important are the increase in concentration of harmful products of metabolism and the decrease in available food supply, which lead to cells entering the so-called "resting stage" or cause death.

5. The *stationary phase* is, as the name indicates, a time of practically no change in the numbers of the bacterial population, and, as is not indicated by the name, a period when the numbers are at a maximum. During this phase the multiplication and the death of cells are practically in equilibrium.

6. The *phase of accelerated death* covers the period when decrease in numbers begins slowly, and continues with increasing rapidity and leads into (7), the so-called *logarithmic death phase*, which covers the mid-senescent period of the cultures at a time when the rate of death sometimes remains constant. Henrici (1928) suggests the addition of (8), a final *phase of negative acceleration* in death-rate,

TABLE XXVII

SHOWING PENFOLD'S DATA CONCERNING GENERATION TIME IN MINUTES DURING THE FIRST 2 HOURS OF GROWTH OF BACTERIAL CULTURES

Dilution of Parent Culture	Experiment	
	A	B
1:100	44	45
1:400	48	47
1:1,000	52	48
1:10,000	91	56

since in many cases the death-rate decreases after a time and some cells remain alive after relatively long periods.

The form of the growth curve of a bacterial culture is influenced by such factors, among others, as temperature, composition of the medium, and the numbers, age, and previous history of the seeded bacteria. Of these, the effect of the size of the inoculum interests us particularly. Evidence clearly indicates that, within limits, the smaller the inoculum the longer the lag phase (Rahn, 1906; Penfold, 1914; Montank, *vide* Henrici, 1928). Penfold's data on the point is given in Table XXVII. Montank used yeast in his experiments. His results are summarized in Table XXVIII.

Inspection of these tables shows clearly that with the smaller seedings the combined initial stationary and lag phases are prolonged. With the yeast the prolongation is in proportion to the size of the inoculum. In these yeast cultures of Montank's, when the logarithmic growth stage is reached, growth is more rapid the smaller the seeding, except with the smallest amount used, where the rate of increase is distinctly less than with the next higher amount. This is in line with Robertson's findings for the effect of the size of the seeding upon rate of reproduction in certain protozoans in which

TABLE XXVIII

SHOWING MONTANK'S DATA ON THE INFLUENCE OF SIZE OF SEEDING UPON THE GROWTH CURVE IN YEAST

(Figures Are Given in Terms of Millions of Cells per Cubic Centimeter of Medium)

Hours of Incubation	Flask No.				
	1	2	3	4	5
0	20.9	2.0	0.2	0.02	0.002
2	23.8	2.2	0.2	0.02	0.002
4	37.8	3.1	0.2	0.025	0.003
6	58.9	8.3	0.3	0.02	0.003
8	105.5	19.6	1.0	0.02	0.004
10	129.0	31.6	3.8	0.024	0.004
12	161.6	54.5	11.8	0.04	0.005
16	174.7	91.5	44.8	0.26	0.005
20	176.2	118.0	89.3	3.0	0.006
24	181.0	124.9	123.7	18.3	0.007
36	183.6	144.1	138.3	114.6	0.289
48	186.0	156.7	146.7	135.0	1.089
72	185.1	165.6	162.2	160.0	93.264
96	195.0	173.5	176.3	165.2	169.400
144	197.9	183.2	180.3	175.8	171.408

the rate of reproduction at early stages of the subcultures was more than doubled by seeding with two rather than with one organism. It may be remembered that Peskett (1924, 1925) using isolation cultures failed to find increased growth-rate with increased number of cells present at any number level tested.

The experience with bacterial cultures parallels experience with protozoan and other animal cultures in another respect, in that the final yield is practically independent of the seeding as long as the seedings are smaller than the normal maximum yield of the medium.

MASS PROTECTION FOR BACTERIA

The evidence for mass protection for bacteria to be presented here is based almost entirely upon the studies of Churchman and

Kahn (1921) and of Burnet (1925). It is presented in some detail both because the results broaden the base of our knowledge of the physiological effects of numbers, extending this to the bacteria, and also because these results have been attained independently of similar work on different material. Verification studies are much more impressive when carried on in a different laboratory from that bringing out the original report; and when many different investigators working in widely separated laboratories and on radically differing materials reach similar conclusions independently and about the same time, the fundamental nature of the phenomenon under discussion becomes the more striking.

The bacterial studies in question are concerned with the behavior of certain types of bacteria in the presence of gentian violet. *Bacillus coli* grows equally well on both sides of a divided agar plate, one-half of which contains gentian violet, providing both sides are stroked with a heavy suspension of the bacteria. As the strokings are made with increasingly dilute suspensions, the colonies become less numerous on the gentian-violet side and finally disappear completely. In part the results may be due to differential susceptibility to the dye, shown by different *B. coli* individuals, for if a plain agar plate is inoculated from broth cultures of this bacterium which have been exposed to gentian violet, the stained organisms appear to grow as well as the controls, but if the experiment is repeated with increasingly dilute suspensions of *B. coli*, it will be seen that many of the bacteria have not survived the treatment with the stain.

Such results are what might be expected from the well-known relations between growth and size of the inoculum. In an ordinary bacterial culture, many organisms are known to be dead. Others, though living, are more readily affected by the somewhat unfavorable conditions found in the new medium to which they are transplanted, which makes it much more probable that heavy growth will occur if the inoculum consists of thousands or millions of organisms than if it contains merely hundreds. The fact just observed concerning the better growth of a large than of a small inoculum exposed to gentian violet becomes merely the starting-point of our interest.

The technique developed by Barber, which allowed the transfer of single cells to new media, gives a proper method for attacking this problem. This technique was employed by Churchman and Kahn. They took their transplants from a strain of *B. coli* which had been isolated from a single colony growing on gentian-violet agar and kept growing by frequent transplants for several weeks on media containing that dye.

Care was taken to avoid as much as possible the lag phenomenon. When strokes of a heavy suspension were made on gentian-violet agar, growth occurred with almost no inhibition; but transplants of single cells almost never grew, although only motile organisms were picked for transfer. Single transplants to controls lacking the dye gave as high as 85 per cent growth. "In the two cases in the whole series of cell transplants in which growth occurred, marked delay took place, a delay which was never observed for the controls. Moreover transplants of small groups of organisms (five to fifteen) did not grow, though transplants of thirty individuals grew regularly."

Churchman in another series of experiments, using *Bacillus subtilis*, which is definitely susceptible to gentian violet, tested the effects of repeated inoculations on the ability of this organism to grow in the presence of this dye. The susceptibility of *B. subtilis* is shown by the fact that its growth is prevented by dye dilutions of 1:1,000,000; inhibited by 1:2,000,000; and only becomes vigorous at dilutions of 1:3,000,000. "If, however, the gentian violet half of the plate is repeatedly and heavily inoculated on successive days at the same place, a fair growth—in some cases a rather vigorous growth—may finally be obtained." Such an experiment is shown in Figure 22. The figure shows the results of two initial strokes of a thick suspension of *B. subtilis* on this divided plate, the lower half of which is plain agar while the upper half is gentian-violet agar, at a strength of 1:100,000, which is ten times stronger than a dilution known to prevent growth.

After the first stroke, nothing further was done to the right-hand side, and the usual effect of the dye is clearly shown by the lack of growth on the gentian-violet part of the plate. On the left-hand stroke organisms were repeatedly introduced on the gentian-violet agar, and inspection of the figure shows that a fair growth resulted. This growth was not due to an acclimatization to the dye, because all experiments made looking directly toward such acclimatization have been unsuccessful. Further, attempts to reinoculate gentian-violet agar from this growth failed, at least insofar as the first smears were concerned.

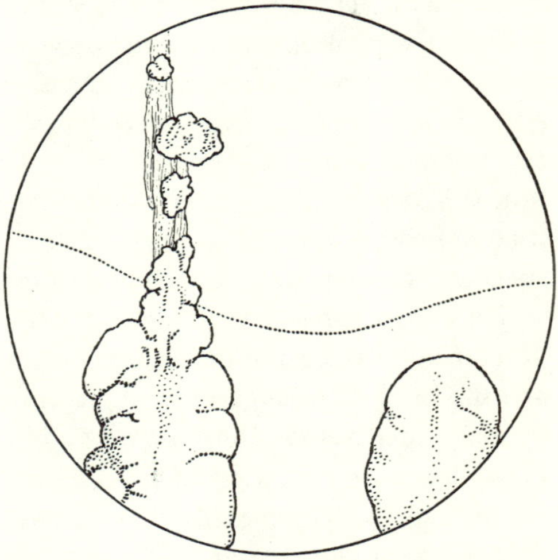

FIG. 22.—Showing the effect of reinoculations on a divided agar plate. The upper part contains gentian violet; the lower part is plain agar. Two initial strokes of *Bacillus subtilis* were made. On the right side, nothing further was done. On the left, reinoculations were made on the gentian-violet agar. (From Churchman.)

In discussing these results, Churchman suggests four possibilities: The original transplants, though not surviving, may live long enough to effect some change in the dye such that subsequent transplants can survive. Or, the result may be due to some protective barrier laid down by the dead bodies of the bacteria. Or, nutritional or growth-promoting substance may be provided by these same dead bodies. Or, finally, the results may be due to communal activity of the living bacteria, a possibility which will be examined later. We have already discussed possible causes of survival and growth in such cases in preceding chapters.

In experimenting to find which of these possibilities actually obtained, Churchman grew *B. subtilis* on medium containing only distilled water and agar, in order that the cells might contain a minimum of nutritive material. These bacteria were washed six times with distilled water and were killed. In order to test whether food material was present under these conditions, these dead bacteria were smeared over a Petri dish bottom which was then dried, moistened, and inoculated with *B. subtilis.* There was no growth when this was incubated.

A control tube of gentian-violet agar was inoculated with this organism. No growth occurred. Another similar tube was covered with a thin layer of the killed and washed organisms just described. After this layer had dried, an inoculation of living *B. subtilis* was made on top of it and growth occurred. Evidently the shield of dead, washed organisms, without sufficient nutritive value to support growth in themselves, served as some sort of a barrier between the poisonous dye and the living cells. Similar results were obtained if dead bodies of *Micrococcus aureus* or of *Bacillus coli* were used in place of dead *B. subtilis*, so that the protection is not species-specific. These results are suggested graphically in Figure 23.

These experiments with agar media indicate that there is a real difference between the behavior of a single cell and that of a group of cells in the presence of gentian violet. This was tested further by experiments upon the survival and growth of *B. coli* in a gentian-violet broth at a dilution of 1:100,000. The strain used came from one of the two colonies that appeared from single-cell transplants onto gentian-violet agar. Not only did these cells come from a strain known from several weeks culturing to be gentian-violet tolerant, but they all came from a single organism of this strain which had successfully produced a colony when isolated onto gentian-violet agar.

When single organisms of this strain were cultured into plain broth, 80 per cent growth was obtained. When 30 or more cells were cultured together in gentian-violet broth, almost 100 per cent growth was obtained. The first 140 single-cell inoculations into gentian-violet broth yielded no growth. In a final series of 8 inoculations, delayed growth was obtained in one tube in the second 24 hours. The chances are at least 147 in 148 that growth will not occur under these conditions. These results are to be compared with those showing that there are almost 100 chances in 100 of inocula of 30 such cells growing under similar conditions.

It may be that 30 cells succeed in growing because these 30 cells are able to produce some anti-dye substance in an amount sufficient to destroy the harmful effect of the dye when a single cell is unable to do so. Obviously, from the conditions of the experiment we are not dealing here with the usual chances that in a large inoculum there are more vigorous cells than in a small one. The probabilities just mentioned argue strongly against such a conclusion.

Churchman and Kahn performed a number of experiments to

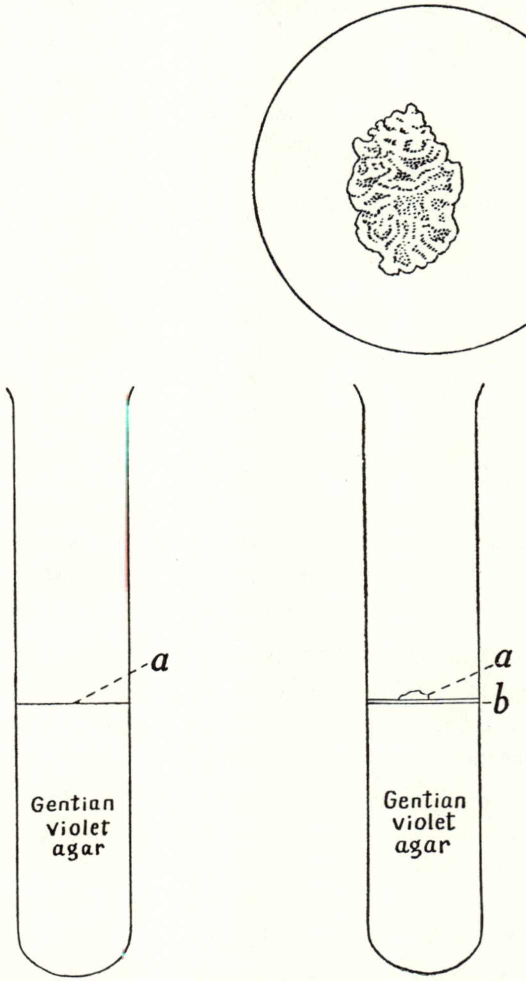

FIG. 23.—Showing results of implantation of *Bacillus subtilis* on gentian-violet agar without (left) and with (right) a thin layer of killed washed bacteria (*b*) below the implants. *C* is redrawn from published halftone of the mass *a* below it. (From Churchman.)

determine whether the observed facts were to be explained by the relation of the number of organisms to the amount of gentian violet to which they were transferred. This is the same sort of relationship which Allee and Schuett found effective with various organisms exposed to colloidal silver, and which Carpenter found in the case of fishes exposed to lead nitrate. In the language of the original report:

"Large inoculations of this gentian-negative strain grow in the presence of gentian violet without any apparent restraint; so, too, do inoculations of 30 cells; whereas single cells, under identical conditions, do not grow at all. This might be due to the fact that

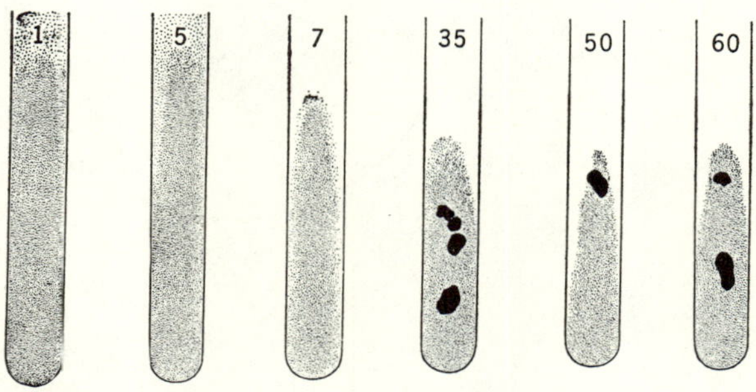

FIG. 24.—Showing effect of size of transplant on amount of growth of a gentian-tolerant strain of *Bacillus coli* on gentian-violet agar. Numbers in the test tubes give the number of cells transplanted. (Redrawn from photograph published by Churchman and Kahn.)

groups of cells, even small groups of 30 individuals, were able to make some change in the dye, gentian violet being assumed to offer a slightly unfavorable medium even for this gentian-negative strain, in spite of the absence of any apparent inhibition to the growth of groups of cells. Single cells might be unable to effect this change in dye in sufficient amount to allow growth to take place."

Acting on the assumption that this reasoning is correct and on the knowledge that the single-cell transplants had been made into 5 cc. of 1:100,000 gentian-violet broth in which 1 cell would not grow while 30 cells transplanted together would do so, transplants were made of 30 cells into 150 cc. of this broth. Growth occurred with a fair degree of constancy. Even when 30 cells were transplanted to a liter of such broth, growth took place in 60 per cent of the flasks. If the purely quantitative relations stated above held here, it would be necessary to transfer 6,000 cells into this volume of gentian-violet broth in order to obtain growth. Here 30 organisms perform not merely 30 times the work of one but at least 200 times that amount. This discrepancy between the work actually accomplished by the 30 individuals and that which they might be expected to accomplish on a quantitative basis is what is meant by the expression "the communal activity of bacteria." Such an expression implies interreactions not yet proven for bacteria. Here, as in many other places in the present summary, we come upon apparently well-demonstrated facts for which the physiological explanation is as yet lacking.

On single bacteria, the amount of gentian violet does have some effect. Thus *B. coli* isolated into more than 0.8 cc. of the broth used in these experiments did not grow. If isolated into 0.1 cc., 60 per cent of the isolations yielded growth. The corresponding ratio for 30 cells would be 3.0, yet the facts are that 60 per cent of successful inoculations occurred when 30 individuals were introduced into a liter of such broth, which is more than 333 times the expectation based on the performance of a single cell. Volume alone was not responsible for these results, as some of the early work of Drzewina and Bohn might indicate, for when single-cell inoculations of *B. coli* were made into a liter of plain broth, growth occurred in 75 per cent of the cases.

It seems clear, as Churchman and Kahn conclude, that 30 cells are able to accomplish much more than 30 times as much as a single cell, and that this excess of ability of the 30 is an expression of communal activity of bacteria, whatever that may be.[1]

[1] The results just recorded are mainly concerned with homotypic relations. Castellani (1926) has described somewhat similar relationships as holding for heterotypic colonies of yeasts and bacteria. Bakers' yeast, for example, is a mixture of one or more yeasts and one or more bacteria and produces fermentation over a wider range of carbohydrates than will the individual organisms used separately in pure cultures. The mixture *Bacillus typhosus* plus *B. morgani* produces gaseous fermentation in maltose, mannitol, and sorbite; *B. typhosus* alone produces acidity only, never gas; and *B. morgani* alone produces neither gas nor acidity from these substances. The opposite result may also be found. For example, *Monila tropicalis* ferments saccharose, producing gas; but when mixed with *Bacillus typhosus*, it loses this property. Buchanan and Fulmer (1930*a*) summarize many similar relations between bacteria and other organisms.

Burnet (1925, 1925*a*) has extended and verified certain aspects of the work just reported. He found that "a batch of nutrient agar plates which had been allowed to remain in the light for some time, while still capable of growing staphylococci when heavily inoculated, did not allow the growth of isolated organisms. If, however, a broth culture, diluted in saline so as to give discrete colonies, were spread on such a plate, and an adjacent area were inoculated heavily with staphylococci, growth of isolated colonies occurred within a range of approximately one centimeter from the region of heavy growth."

The inhibiting agent in the plates exposed to light was found to be hydrogen peroxide, whose effects are destroyed by substances produced by an active and heavy growth of staphylococci, so that in regions reached by the diffusion of such growth-products, colonies can develop from isolated organisms. The growth-facilitating substances are in part thermolabile and appear to have some of the properties of enzymes; in part they are thermostable, non-enzyme materials which Burnet calls "thermostable *X*." This substance will neutralize growth-inhibiting powers of hydrogen peroxide and will prevent its accumulation when plates are exposed to the action of light. There is a definite quantitative relation between the amount of thermostable *X* and the amount of peroxide that can be neutralized, and it seems probable that the interaction is a simple reduction of the peroxide, both substances being destroyed.

An inhibition of growth due to the presence of potassium cyanide can also be neutralized by these substances. Burnet regards the mechanism of bacterial mass resistance to KCN to be essentially the same as that which gives protection from peroxide introduced into cultures directly or by the action of light on culture plates. He thinks that the KCN is not destroyed by the thermostable-*X* substance as is the peroxide, but that it is inactivated more indirectly, thus: KCN acts to inhibit the normal mechanism for removing peroxide, which accumulates until it is present in toxic amounts. The diffusible substances produced by bacterial growth remove the peroxide as shown above, and by so doing render the KCN innocuous. Burnet gives evidence which indicates that the effect of certain dyes, of which acid fuchsin is an example, is either to increase the production of peroxide or to hinder the removal of normally produced amounts.

"A bacterial colony may be merely a fortuitous result of continued multiplication, but it seems too to furnish a means whereby numerous potentially nocuous influences are rendered less effective. An isolated organism may produce diffusible substances as readily as one within a colony, but it cannot retain them within its immediate neighborhood, and a high concentration of such substances can only occur when numerous adjacent organisms collaborate in their production. The most important peroxide destroying bodies have been shown to be diffusible and will therefore reach a high concentration only when a colony is formed. By their presence traces of peroxide in excess of the amount which can be dealt with by the oxidative mechanism of individual organisms are immediately destroyed and in this way the viability of cells damaged by small changes in the environment is retained. It may be considered as a primitive maintenance of constant internal environment by ensuring that the immediate toxic body in many cases of bacterial injury (hydrogen peroxide) shall not accumulate. In this respect the bacterial colony may almost be regarded as a metazoan individual."

This analogy between the bacterial colony and the metazoan individual does not mean that such colonies are on the evolutionary high road which leads to multicellular individuals, but it does give some insight into certain survival values which favored the development of metazoans from cells which had these and other potentialities. Similarly, the facts brought forward in the present comprehensive survey of the survival values of aggregations of organisms indicate that groups of animals having such values have the possibility of becoming what is usually known as "social animals." To be sure, they must possess other attributes. An aggregation implies that the grouped individuals have tolerance for the presence of other organisms in the same limited area, and that they have a reaction system which causes them to aggregate or to remain aggregated if passively collected. In addition certain other qualities are needed, particularly the ability to establish close group integration. The facts indicate that not all animals whose groups show survival values are to become more closely social, but that animals, whatever their endowments, could not have developed the social habit had the incipient social stages lacked the type of survival values which we have repeatedly demonstrated for different sorts of animals, and now also for bacteria. These values are a function, other conditions being equal, of the mass of animals in relation to the volume of their effective environment.

CHAPTER XVI

MASS PHYSIOLOGY OF SPERMATOZOA

The relation between numbers of spermatozoa present and their functional ability and longevity has long attracted attention, and a large literature has been built up about various aspects of the subject. The general facts seem well established, but there is much discussion concerning causal relations underlying the observed facts.

Spallanzani in 1785 made quantitative studies of the amount of seminal fluid necessary to fertilize amphibian eggs. He reports that, while the fluid is active in extreme dilutions, the percentage of fertilized eggs diminishes with such dilutions. Spallanzani did not understand the true nature of the spermatozoön, but his observations are essentially correct. Prevost and Dumas in 1824 found evidence that the spermatozoa are the essential elements of the seminal fluid and attempted to repeat Spallanzani's results. While they obtained indications supporting his conclusions as given above, their results were too variable to allow a precise statement of quantitative relations.

Gemmil (1900), using sea-urchin sperm, found that the length of the functional life of spermatozoa is directly related to the numbers present: the greater the concentration the longer the retention of ability to fertilize eggs. He records that the duration of vitality of sperm of *Echinus* varies from 3 to 72 hours according to the degree of dilution, those in the more dilute suspensions dying first. The decrease in vitality he thought to be due to exhaustion by increased movement and to the dilution of nutritive strength of spermatic fluid. He found similar relations concerning the spermatozoa of limpets (Gasteropoda) and of nemertine worms, except that the length of vitality varied with different sorts of sperm. As will soon appear, the work of recent investigators indicates that one cause of the lessened longevity in dilute sperm suspensions is to be found in the relatively greater activity shown under these conditions, as Gemmil thought; however, it is doubtful whether longevity is related to the dilution of the nutritive fluid, which Gemmil supposed to be the main factor involved.

It is unnecessary to follow here the details of our advance in knowledge concerning the mass relations of spermatozoa. The importance the subject came to occupy may be realized by the fact that in 1915 Glaser, as Schücking had done earlier (1903), questioned whether a single spermatozoön was capable of initiating development of a single egg, even though only one sperm nucleus is concerned in the biparental inheritance. To explain these relationships Glaser postulated a mass effect of spermatozoa, as well as an individual effect. Lillie (1919), reviewing his own earlier work on this question, states that he can demonstrate that mass effect is not necessary if the sperm suspensions have recently been made, and that "the appearance observed by Schücking and Glaser is found only with suspensions not perfectly fresh."

Lillie reports that fertilizing power of perfectly fresh sea-urchin sperm may extend to 1/90,000,000 of 1 per cent, although at a dilution of 1/10,000 per cent one can rarely find a spermatozoön in the jelly of the fertilized eggs. He states further, "If one determines by comparison the rate of loss of fertilizing power of sperm suspensions of different concentrations it is found that sperm suspensions of 1/240,000 per cent decline to zero in their fertilizing power in about six minutes, those of 1/30,000 per cent in about fifteen minutes, those of 1/300 per cent not until after more than two hours, while one per cent sperm may retain their fertilizing power for two or more days."[1]

The rate of loss of fertilizing power of fresh *Arbacia* sperm in

[1] In nature the sperm of the sea urchin, *Arbacia*, used in many of these studies, are shed from the genital pores into the sea-water, where they fertilize the eggs which have been similarly shed by the females. In making sperm suspensions for such work as recorded here, the sea urchin is cut around the peristome and placed aboral side down. The sperm is then collected in a clean, dry watch glass. Such sperm is designated "dry" sperm and may be diluted as desired by adding the appropriate amount of sea-water. Thus a 1 per cent suspension is made by adding 1 drop of sperm to 99 drops of sea-water.

relation to dilution is shown in Figure 25, taken from Lillie and Just, in which the ordinates give percentage of eggs fertilized and the abscissas give a geometrical series of dilutions of 1 per cent sperm in powers of 2.

Lillie holds that fertilizing power of the sperm vanishes before motility, and that the spermatozoa tend to lose their fertilizing ability in proportion to dilution, apparently because of the greater speed of diffusion from the spermatozoa of some substance or substances necessary for fertilization. Obviously, this diffusion gradi-

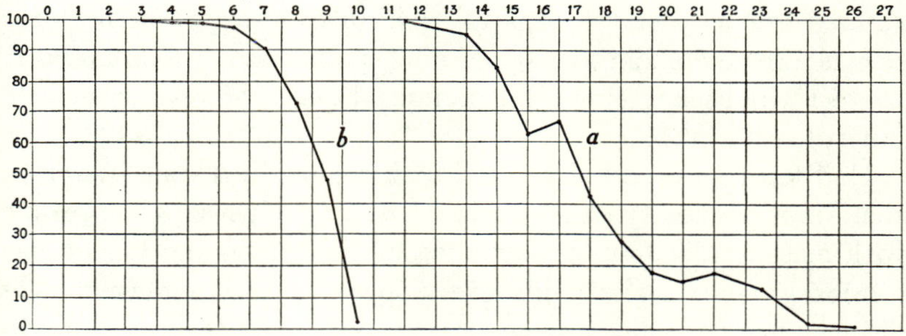

FIG. 25.—Logarithmic curves showing the fertilizing power of dilutions of *Arbacia* sperm. The vertical axis shows percentages of eggs fertilized; the horizontal axis gives dilutions of sperm in powers of 2. Graph *a*, perfectly fresh sperm; graph *b*, sperm suspensions 20 minutes old. (After Lillie and Just, 1924.)

ent would be steeper the less concentrated the sperm suspension; hence the length of functional life would be shorter.

Hinrichs (1926*a*), after producing further supporting evidence from the greater resistance shown by more concentrated sperm suspensions to ultra-violet radiation, sums up the situation as follows:

"Time and dilution are both known to be factors affecting the fertilizing power of *Arbacia* sperm (F. R. Lillie, 1915). Drzewina and Bohn (1923*b*) also showed dilution to be a factor in the susceptibility of sea urchin sperm to the combined action of neutral red and light. Motility and fertilizing power were lost more quickly in dilute than in concentrated sperm suspensions. Usually loss of motility is associated with loss of fertilizing power, but the two do not exactly parallel each other. Fertilizing power is not a function of motility alone, and declines more rapidly than does motility (F. R. Lillie, 1915; Lillie and Just, 1924). Sperm may be injured in such a way that its fertilizing power, as measured by the proportion of eggs fertilized and the normality of cleavage and development, is materially lessened while motility is not visibly impaired (Hinrichs, 1926; Lillie and Baskervill, 1922)."

That such is the case is evidenced by one experiment reported by Lillie (1915) and cited by Lillie and Just (1924) in which eggs were added to a $1/2^{10}$ sperm suspension that was near the point of complete loss of fertilizing ability. "The sperm were still active and entered the jelly of the eggs to such an extent that in ten eggs selected at random an average of nine spermatozoa was counted in contact with each egg in optical section; the eggs, however, remained unfertilized."

Had fertilization occurred, and with slightly less stale sperm, as there is a possibility that it might, we might perhaps be dealing with mass fertilization, as suggested by Schücking and Glaser. On a priori grounds, if the ability of the sperm to initiate development depends upon a material which leaches out before the loss of motility, then what one spermatozoön is unable to supply might be supplied by many. Barthelmy (1923, 1926), in discussing polyspermy, states that the egg reacts in proportion to the stimulus given by the sperm, so that when this is weak, as in the case of aged sperm, the egg reacts slowly and more sperm enter, producing polyspermy. Although there is evidence that the defect in polyspermy is in the egg rather than in the sperm, there is still a possibility that the phenomenon of physiological polyspermy may also be an instance of the same sort of mass fertilization behavior. Such polyspermy is characteristic of forms with large eggs, such as sharks, some Amphibia, reptiles, and birds. Insect eggs of several sorts are said to possess more than one micropyle as a structural adaptation for polyspermy (Henking, 1891). Lillie and Just state that among animals having small eggs, physiological polyspermy occurs only in the Bryozoa, where the sperm are united in bundles, and where the research worker describing the phenomenon (Bonnevie, 1907) thinks that polyspermy has definite value for the organism. Lillie and Just do not accept her interpretation on this point.

Among reviewers of the fertilization problem there is general agreement as to the importance of the analysis made by Cohn in 1918. After demonstrating that fertilizing power of sea-urchin sperm suspensions falls off more rapidly in dilute than in concentrated suspensions, Cohn undertook the measurement of the carbon-dioxide production in sperm suspensions of different concentrations by following the hydrogen-ion potential, with the results shown in Figure 26. The ordinates give the hydrogen-ion potential of the suspension

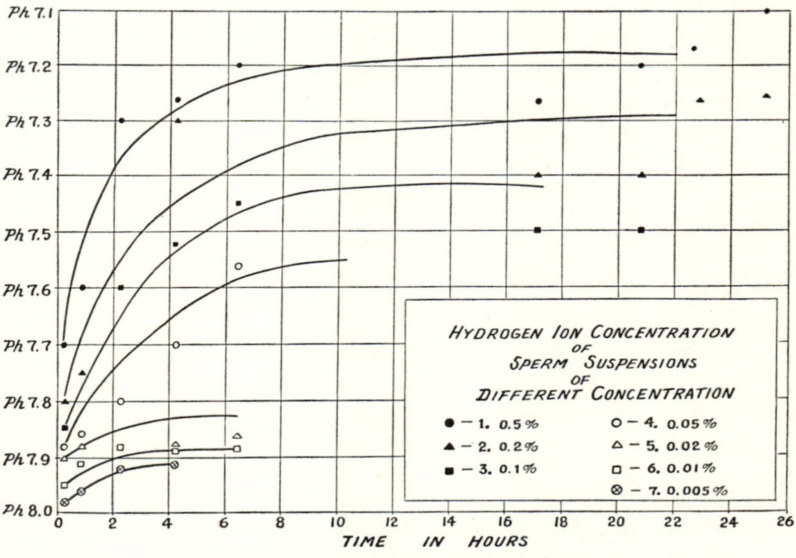

FIG. 26.—Diagram from Cohn (1918) showing the hydrogen-ion concentration of sperm suspensions of different concentrations after different time intervals.

in terms of pH; the abscissas, the time in hours. The concentration of each suspension is given in the accompanying legend. The increase in H-ion concentration is due to the carbon dioxide produced; and in turn, the rate of production of CO_2 is a function of the H-ion concentration of the suspension; and, as would be expected from these conditions, the rate of carbon-dioxide production of sperm suspensions decreases with lapse of time. The length of functional life as measured by the percentage of eggs fertilized by identical concentrations of sperm under identical experimental conditions is summarized in Figure 27.

An inspection of this figure will show that the spermatozoa lived longest in the most concentrated sperm suspensions. Further, Cohn undertook to study the total carbon-dioxide production of sperm suspensions at different dilutions, with the results shown in Table XXIX. These results show that in the less concentrated suspen-

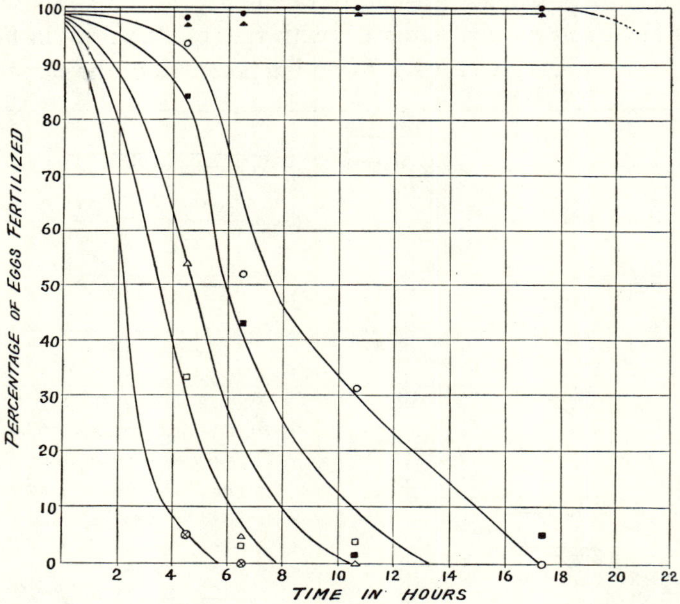

FIG. 27.—Showing the length of life of the different concentrations of sperm suspensions shown in the preceding figure as measured by the percentage of eggs fertilized (ordinates) after given lapses of time (abscissas). (From Cohn 1918.) The legend for concentrations is the same as in the preceding figure.

sions which more nearly approximate normal conditions, spermatozoa that live for longer periods of time produce no more carbon dioxide than do those living for only 4 hours. Roughly speaking, the total production of carbon dioxide approaches a constant; hence the total available energy of spermatozoa must be a constant, and the rate at which this energy is expended is a function of the activity and an inverse function of the length of life of the spermatozoa. These results argue in favor of Gemmil's assumption that length of life is associated with the amount of activity, and against his suggestion that the effect of dilution is related to a dilution of possible foodstuffs.

From such observations Cohn concludes that the basic factor concerned in the greater functional longevity of the sperm composing the more concentrated sperm suspensions is to be found in the depressing action of the self-produced carbon dioxide upon the activity of the relatively massed sperm. Similarly cyanides, dilute acids, depleted oxygen, or anything that will paralyze the sperm without killing them will prolong their ability to fertilize eggs. When dense sperm suspensions, either fresh or after a period of some time, are

TABLE XXIX

SHOWING THE TOTAL CARBON-DIOXIDE PRODUCTION OF DIFFERENT SPERM SUSPENSIONS

Sperm Concentration in per Cent	Calculation of Relative CO_2 Production per Unit Sperm Concentration	Approximate Life of Sperm in Hours
0.5	13	17+
0.2	18	17+
0.1	26	17+
0.05	37	10+
0.02	32	6+
0.01	45	6+
0.0005	60	4+

diluted with large quantities of normal sea-water, the H-ion concentration is markedly lowered, the activity of the sperm is much increased, and the fertilizing power of the sperm suspension is likewise increased. But sperm so treated will lose their ability to fertilize ripe eggs in sea-water long before spermatozoa that have been relatively inactive in more acid sperm suspensions. Here we have our fourth explanation of the unquestioned mass effect upon functional longevity of spermatozoa, and the end is not yet.

Cohn recognized that there is an H-ion aspect of these relations not necessarily identified with the CO_2 factor. Smith and Clowes (1924) present direct evidence that such is the case, based on experiments of CO_2-free sea-water. They also show that here, too, there is a definite extension of ability to fertilize eggs in higher H-ion concentrations, associated with increased density of the sperm suspension used, as shown in Table XXX.

The smallest quantity of sperm was so thin that it was unable to fertilize all eggs even in sea-water, and the largest was so thick that it was opalescent; but the block to development did not shift beyond 6.9 to 7.1. Increasing the quantity of sperm increases the number of eggs fertilized in acid solution, but the shift is not so great as expected if the failure to fertilize were attributable to impairment of sperm. The slight shift favors the belief that the block is due to an alteration of the properties of the egg. Unless the sperm are injured by toxic action of egg secretions, all eggs which are fertilized in these solutions develop normally, indicating that the fertilization reaction, when once initiated in the neighborhood of the block, is completed without impairment.

TABLE XXX

SHOWING THE EFFECT OF DIFFERENT H-ION CONCENTRATIONS IN CO_2-FREE SEA-WATER UPON FERTILIZING POWER AT DIFFERENT SPERM DILUTIONS
(Data from Smith and Clowes)

pH										Amount of Sperm in 25 cc. of pH Solution
6.6	6.7	6.8	6.9	7.0	7.1	7.2	7.3	7.4	8.15	
0	0	0	50	100	95	95	100	100	100	1 cc. 1/20
0	0	0	0	35	80	95	100	100	100	" 1/200
0	0	0	0	0	20	85	87	95	90	" 1/2,000
0	0	0	0	1	40	45	25	35	30	" 1/20,000

Before continuing with this search for further suggested explanations of this phenomenon, it is of interest to stop for a moment over the controversy concerning the mechanism of aggregation of spermatozoa in sperm suspensions. Lillie (1919) points out that with active spermatozoa in a suspension we should expect to find the individual spermatozoa colliding with each other and with the walls of their container, but that on the whole their distribution is approximately uniform throughout, much as is supposed to be the case with gas molecules. This arrangement is found in perfectly fresh suspensions, but usually lasts only a short time before various kinds of aggregations occur.

Figure 28, from Lillie, shows a photograph of aggregations of sperm of *Nereis* in sea-water. His account of the formation of these aggregations is as follows:

"If a drop of dry sperm from a mature *Nereis* is mixed in about 6 cc. of sea water in a Syracuse watch crystal it makes a uniformly milky suspension; in a few seconds clouds begin to appear, and in

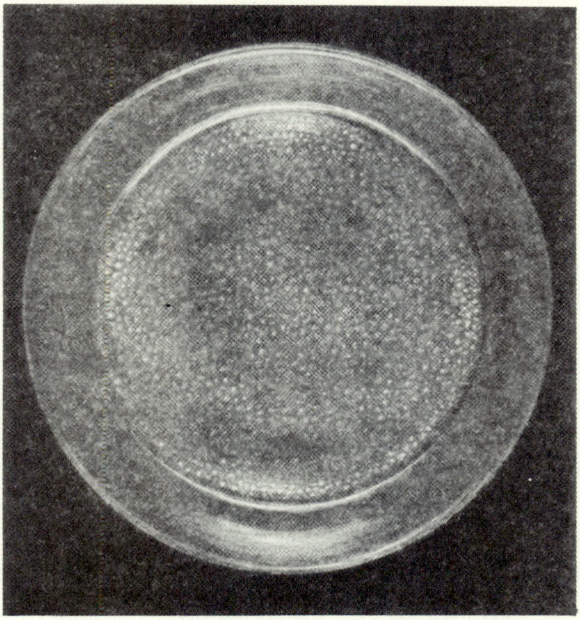

FIG. 28.—A sperm suspension of *Nereis* taken 90 seconds after mixing the suspension. (Drawing by Toda from photograph published by Lillie, 1919.

15 to 45 seconds these usually draw together into white solid masses uniformly spaced through the fluid. The intervening fluid becomes quite clear and the masses quickly settle on the bottom. The rate of formation of these masses and their number and size depend on the condition of the animal furnishing the sperm, temperature, 'freshness' of the sperm, reaction of the medium, etc. Sperm suspensions of most animals do not, however, exhibit such marked aggregations."

Lillie believes, as a result of well-planned experiments, that these aggregations form as a definite positive reaction of the individual spermatozoa to regions of self-produced higher concentrations of carbon dioxide; that is, they are tropistic reactions. Cohn and others believe that the aggregation is the result of a trap action whereby the spermatozoa, on invading a region of higher CO_2 concentration, have their activity reduced and tend to remain trapped. We are not, at the present time, interested in the relative merits of these two points of view. It is of interest, however, to note that, according to both, the aggregations form without any evidence of what we have called "secondary" or "social" reactions, but merely as a result of the primary or individual reactions to the environment, just as we have already seen that aggregations of animals frequently form. It is also of interest that here with the spermatozoa, as in the case of animal aggregations, far-reaching physiological effects may result from these groupings.

In the reaction of spermatozoa another aspect gives their behavior more direct significance. Lillie and Just summarize evidence which shows that among the other effects of egg secretions upon homotypic sperm there is, at least in the sea urchin, a marked tendency to aggregation in which "the spermatozoa are attracted to the drop of egg secretion and gather in or around it, depending on its concentration." Such reactions are not easily demonstrated, if they can be demonstrated at all, by Pfeffer's capillary tube method, but can be directly shown by introducing a drop of egg-water in a sperm suspension on a slide covered by a raised cover slip. Under these conditions, Lillie and Just report: "If a drop of egg water of *Arbacia*, with agglutinating substance removed, be injected into a sperm suspension of the same species, a ring of active spermatozoa forms around the drop, separated by a clear zone almost devoid of spermatozoa from the main suspension. If the clear zone be examined carefully, spermatozoa may be seen swimming directly across it from the general suspension to the drop of egg water for some minutes. The clear zone thus gives the range of some directive influence proceeding from the drop." This same phenomenon is exhibited clearly by *Nereis* sperm when a drop of 1 per cent CO_2 sea-water is similarly introduced, and is illustrated in Figure 29 and explained in the accompanying legend. The significance of the reaction to emanations from a homotypic egg in relation to fertilization is obvious, but does not constitute a part of our present problem.

Lillie (1921) studied the effect of copper chloride upon fertilization in connection with his analysis of the fertilization reactions. Briefly, he found that fertilization was markedly inhibited at as low a concentration as 1 part of copper chloride in 2,500,000 parts of sea-water, using a normal sperm suspension. Most of his experi-

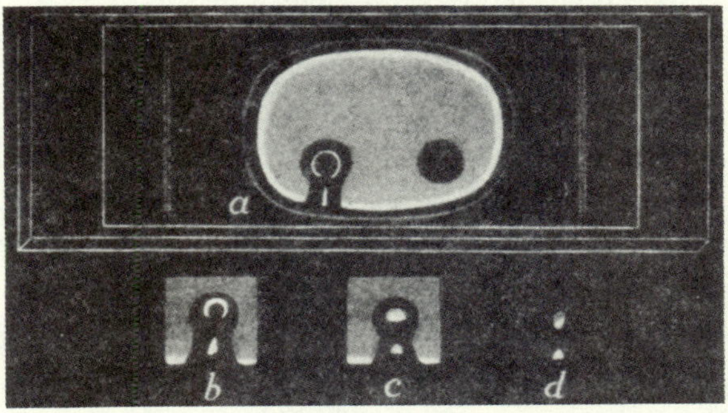

FIG. 29.—Reaction of a sperm suspension of *Nereis* to a drop of 1 per cent CO_2 sea-water (natural size). The preparation (*a*) is mounted on a slide beneath a raised cover slip. *a*, Showing the form of the reaction after 15 seconds; *b*, after 75 seconds; *c*, after 105 seconds; and *d*, after 195 seconds. In *d* the general suspension has aggregated. The drop to the right in *a* is a control drop of sea-water. Note that in *a* the spermatozoa also withdraw from the margin of the preparation, thus increasing the CO_2 tension. (Figure and legend from Lillie, 1919.)

ments were done at a copper concentration of 1/500,000 copper chloride in sea-water, a concentration at which no eggs fertilize at normal sperm concentrations. If much higher concentrations of sperm are used, small percentages of fertilization are obtained, the number varying somewhat in different similar experiments. Lillie comments: "There is thus a certain virtue in mass action of the sperm in the presence of this inhibitor of fertilization; this is somewhat difficult to understand, because only one spermatozoön penetrates normally." He suggests the possibility that an excess of sperm protects the eggs to a certain extent by combining with the copper and thus reducing the amount acting directly on the eggs, a suggestion he and Just repeat (1924).

Drzewina and Bohn, in their studies upon the effects of dilution of sperm upon their physiology, record (1923) another interesting phenomenon. As we have seen before, dilution of sperm in sea-water hastens their loss of activity, and the greater the interval after dilution the greater is the loss of activity. Likewise, the greater the dilution the more rapid is the loss of activity. Drzewina and Bohn found that sperm immediately after dilution (1/5,000–1/100,000) are very active, but are reduced in vitality so that, while fertilization is effected, abnormal development and cytolysis result. If an interval is allowed between dilution and fertilization, normal development is obtained. It is believed that the sperm, when diluted, undergo a disturbance of equilibrium, exhibiting a "differential sensibility." There is little, if any, change in fertilizing power, but a considerable change in cytolyzing power. These phenomena are exhibited only within certain limits of dilution and vary with the season and locality.

Treatment of sperm of sea urchins, *Echinus microtuberculatus* or *Strongylocentrotus lividus*, with KCl or distilled water shortens the duration of fertilizing power in proportion to duration of treatment and concentration of the sperm, and also in proportion to the light intensity; the weaker the dilution the more rapid the loss. One or 2 drops of normal KCl in distilled water, mixed with 1 cc. of sea-water is a rapidly lethal dose for Infusoria, *Planaria*, etc. Sperm diluted to 1/100 survive for 24 hours, but dilutions of 1/1,000 and 1/5,000 succumb much more rapidly. Certain of these relations are illustrated in Table XXXI

Further, Drzewina and Bohn observed that in KCl, in fresh water, and in the controls, within certain limits of duration of treatment and of dosage of injurious substance, the nature of the liquid has little effect, but that the concentration of the sperm is the decisive factor in duration of fertilizing power. A given concentration in any of the foregoing media will, after a certain time, give about the same percentage of fertilization. This condition is compared to a colloidal suspension in that the nature of the solvent does not make as much difference as the number of particles per unit volume.

They pointed out that the spermatozoa behave as if a substance were emitted which, when the sperm are numerous enough or when the volume of liquid is small enough, is sufficient to protect the group. This action is regarded as being similar to that obtaining when *Procerodes* are isolated into fresh water in which other organisms have lived and died. This is Drzewina and Bohn's hypothesis of the production of an autoprotective substance applied here to

TABLE XXXI

EFFECT OF DILUTION OF SPERM SUSPENSION IN THREE DIFFERENT MEDIA
(Data from Drzewina and Bohn)

	4 Hours	8 Hours
1/1,000 sperm in sea-water	No effect	25% fertilization
1/1,000 sperm in sea-water plus 2 drops of KCl per cc.*	No effect	50% fertilization
1/1,000 in 2/3 sea-water plus 1/3 distilled water	No effect	10% fertilization

	20 Minutes	1 Hour	4 Hours
1/5,000 sperm in sea-water	No effect	Almost 100%	$<1\%$
1/5,000 sperm in KCl	No effect	90%	0%
1/5,000 sperm in 2/3 sea-water plus 1/3 distilled water	No effect	70%	0%

* It is noted here that in weak doses and certain concentrations of sperm KCl increases survival of sperm.

spermatozoa, and adds the fifth explanation of the greater longevity of relatively heavy sperm suspensions.

The same workers (1926*a*) reinvestigated the effect of carbon dioxide upon the longevity of spermatozoa. They exposed sperm suspensions of different dilutions, from 100 to 5,000 times diluted, to sea-water kept saturated with CO_2, with results which may be summarized as follows:

Diluted 1/100............No paralysis; sperm immediately gather around the eggs, but no fertilization membrane is formed. Segmentation is irregular, and larvae cytolyze easily.

More dilute..............Paralyzed ½ hour, but then recover and fertilize with normal membranes and development.

Apparently the sperm most severely affected give the better results in fertilization. The explanation of this paradoxical situation is that CO_2 determines a crisis. In the former case, the fertilization occurred during the crisis with abnormal results; in the latter case, fertilization took place after the crisis was over, and the results were normal.

By a longer exposure of 20 to 60 minutes the contrast between concentrated and dilute suspensions is reversed. The concentrated recover gradually and in proportion give normal development, while the dilute sperm suspensions lose fertilizing power more and more. The same results are obtained with one-half the dose of CO_2. It is scarcely possible even to suggest more rapid exhaustion of toxic substance as the cause for the protective effect of numbers in this case, since the CO_2 is kept constant. Obviously Cohn's explanation that the more concentrated sperm suspensions survive the longer because of the greater CO_2 tension which, by inhibiting activity, promotes functional longevity, does not represent the whole story of the protective action of the mass of spermatozoa, however true it may be under many conditions.

BIOELECTRICAL EXPLANATION OF MASS PROTECTION

Again Drzewina and Bohn (1926) call attention to their experience that the more concentrated suspensions of sperm resist a raised temperature better than the more dilute (see Young, 1929, on mammalian sperm). Also, with low temperature (1° C.), if one compares the effect on 1/100 or on 1/1,000 sperm, the cold activates the first and inhibits the second, as if there were a change in the sign of the reaction with the change in mass. Drzewina and Bohn report similar effects from darkening of sea-urchin sperm. They are particularly interested in this protection furnished by the more concentrated masses of sperm, not only because of the similarities to their observations on mass relations with various animals previously cited, but the more so because they believe that the spermatozoa are the bearers of electrical charges, which are not only associated with the fertilization phenomena but are also concerned with the problem of mass physiology of spermatozoa with which we are dealing here.

Here we have a suggested sixth explanation of the greater longevity of the more concentrated sperm suspensions. Some of the further evidence at hand concerning the electrical nature of this relationship, taken from the summary by Drzewina and Bohn (1928) of their work, will be given.

Sperm of sea urchins lose fertilizing power rapidly if a medium concentration be placed in a silver dish. The greater the concentration the longer they retain this power. Diluted 100 times they fertilize eggs up to an hour's exposure; diluted 10,000 times, up to 20 minutes; while at a dilution of 100,000, fertilizing power is lost immediately. The effect is in part a function of the mass of the silver acting: a thin silver foil is completely inactivated after three consecutive exposures. It is reactivated if placed over a block of silver but not if placed over glass. The effect on sperm is paralleled by the action on the flatworm *Convoluta*.

Further, hydroquinone is active when very dilute, and there is good reason for belief that its activity is, at least in part, electrical in nature. If so, the effect upon suspensions of sperm should be illuminating. It is found that if sperm are diluted 100 times and exposed to hydroquinone diluted 1,000,000 times, and then given a chance within 30 seconds to fertilize normal eggs, normal fertilization membranes are formed. This is not the case when sperm suspensions diluted 1,000 times are given the same treatment; and if the hydroquinone, rather than the sperm suspension, is made more dilute, it also loses its effect.

Convoluta placed in a dish lined with stearin survive only half an hour, although if placed in a similar dish lined with paraffin they live normally and even longer than in a glass dish. The paraffin appears to exert a protective effect for the *Convoluta*, while the stearin has a destructive action. When exposed to silver in a glass dish, all *Convoluta* die, while those exposed similarly in a paraffined dish are still alive. A paraffined dish favors survival of sea-urchin eggs but is harmful to sperm. Stearin, on the other hand, is harmful to eggs.

Both paraffin and stearin are more chemically inert than glass and are less soluble in water; yet it is sufficient merely to line the containers with these substances in order to change the survival time. Drzewina and Bohn suggest that in such cases we are dealing with an electrical phenomenon which they designate as "catalysis by contact." They suggest that animal groups or masses of spermatozoa, by the mere fact of their presence in numbers in a limited space, influence each other, causing the individuals to become more or less sensitive, just as they are activated or disactivated above paraffin or stearin; somewhat as inorganic chemicals are more or less sensitive according as they are placed over paraffin or stearin.

From an entirely different approach Gray (1915) had suggested that at least a part of the phenomena connected with the physiology of spermatozoa is electrical in nature. We have stated above that the spermatozoa of sea urchins become quiet in acid of proper concentrations and are again activated by making this sufficiently alkaline. He thinks that the movements of the spermatozoa are dependent upon the electrical properties of the cell and of the surrounding medium, and cites two lines of supporting evidence.

If sperm are suspended in a neutral isotonic cane-sugar solution, their activity ceases, but is recovered if a trace of alkali is added. If an electric current of appropriate strength is passed through the neutral solution just mentioned, the sperm tend to collect rapidly about the positive pole. Those remaining around the negative pole become very active. This activity is probably associated with the fact that the region becomes definitely alkaline. If the neutral solution is made slightly acid, neither the migration to the positive pole nor the activation about the negative one takes place. In such a solution the action of the electric current causes the spermatozoa to collect in a well-scattered netlike aggregation. The activity shown can readily be interpreted on the assumption that the active spermatozoa carry a negative surface charge which is lost when free hydrogen ions are present.

Further (Gray, 1920), if a drop or two of a dilute solution of cerous chloride is added to a weak suspension of *Arbacia* sperm, the sperm become highly active and aggregate rapidly. The aggregated sperm remain active for some time and then become motionless. These results are attributed to the action of the trivalent cations present.

Their effect can be neutralized by treatment with sodium citrate. Gray suggests that these trivalent ions have this effect upon the spermatozoa because of the electric charges carried by the former.

Gray returned to a consideration of the whole problem of sperm-suspension dilution in relation to functional longevity in 1928. He repeated the well-known observations that as long as seminal fluid of a ripe sea urchin remains undiluted, little or no movement occurs on the part of the sperm. If a small drop of this fluid comes into contact with sea-water, the cells at the surface at once become intensely active, and eventually the whole lot of spermatozoa exhibits lively movement. Gray suggests four possible causes: (1) There may be an inhibiting substance in testicular fluid. (2) Sea-water may contain some element absent from testicular fluid necessary for movement. (3) The viscous resistance of testicular fluid may be too high to allow movement. (4) Each spermatozoön may exert some form of inhibition on the movement of its neighbors.

If undiluted sperm be centrifuged, they can be separated from the surrounding testicular plasma. If a drop of undiluted sperm be added to this plasma so that it becomes diluted thereby, activity results as in sea-water. Appropriate checks show that the results are not due to changes in CO_2 or O_2 tension due to centrifuging. Gray thinks, therefore, that the activating effect of dilution in sea-water is due to mechanical dilution whereby each spermatozoön is given more space, allowing free movement. The spermatozoa do not move when closely packed; but what is more striking, they make no effort to do so; this can be demonstrated by comparing O_2 consumption of undiluted and diluted sperm. As movement becomes retarded in the diluted sperm due to increased CO_2 tension, artificially produced, or to the passing of time, the rate of O_2 consumption approaches that of the originally undiluted sperm. In this connection it would be interesting to know if the same effect can be obtained by diluting the sperm and then packing it together again in sea-water with a low CO_2 content.

Except in very dilute suspensions the respiratory level at the beginning of active life is not simply proportional to the number of spermatozoa present, but also depends on the degree of dilution of the original testicular fluid. The greater the dilution the greater is the initial activity of individual sperm; but as dilution increases, the effect on activity is less marked. In very dilute suspensions the activity of a spermatozoön is more or less independent of dilution, but in stronger suspensions the activity is much affected by further dilution. That the reduced activity in strong suspensions is not due to lack of O_2 is shown by the fact that in one of Gray's standard 6 cc. of a suspension containing 25 mg. nitrogen[1] equivalents of sperm, there is actually less O_2 consumption than is shown by the same volume of a suspension containing 5 mg. nitrogen equivalent; and also by the fact that strong aëration failed to increase activity.

Gray's interpretation is that inactivity of sperm in concentrated suspensions is due to lack of free space. Over a considerable range of dilutions the specific activity of a spermatozoön is a linear function of the cube root of the volume of sea-water. The total initial activity is proportional to the number of sperm and to the average free space for each cell. The initial degree of activity is thus lower in the less diluted suspensions; but, interestingly enough, the total energy expended during a period of at least 2.5 hours is also distinctly less. The relation between density of spermatozoa and rate of oxygen consumption may be accurately expressed by the same formula used by Pearl to show the effect of population density of the rate of reproduction in *Drosophila* (Pearl, 1925).

Gray states, without supporting reference or evidence, that mutual inhibition of activity of unicellular organisms is known among the Protozoa, and particularly with Paramecia. The decreased activity with massed spermatozoa may be the result of inhibitions due to continual collisions, or it may be comparable to the effect of thigmotactic reaction to foreign bodies in Paramecia. The behavior of spermatozoa almost suggests a voluntary phenomenon in which contractile effort is proportional to free space in which the organism can move. Until more is known of the mutual effect of one cell upon its neighbors, the phenomenon which Gray calls "allelostasis," that is, the mutual depressing effect of one cell upon another, must

[1] Total nitrogen present is used as a measure of sperm concentration. Oxygen consumption is used as a measure of activity of sperm in suspension.

remain obscure. This last explanation of the effect of dilution upon spermatozoan physiology is fundamentally the same as Gemmil's first assumption.

It is a well-known fact that the fertilizing power of sperm falls more rapidly when sperm are kept in dilute suspension than when they are concentrated. The usual explanation is that in stronger suspensions activity is inhibited by CO_2 given off by the sperm, and so the cells conserve their energy (Cohn, 1918). Gray's data show that this explanation is only partially correct at best and may be quite erroneous. In his experiments the CO_2 product was continuously removed, and the respiration per unit of sperm suspension in the concentrated mass at no time equals that in the diluted suspension.

If the conditions of a dilute sperm suspension are such that the CO_2 generated is allowed to accumulate, the rate at which energy is expended will decrease, which will tend to increase the length of life during which the cells are motile when the CO_2 is removed by subsequent dilution. Such conditions appear to have existed in some of Cohn's experiments where the suspensions were very dilute and the full initial activity was mechanically possible. In such suspensions it should be possible to show that the rate of respiration in relation to the percentage of sperm was the same in all cases.

Gray's conclusions are:

1. Relative inactivity of undiluted *Echinus* sperm is not due to physical and chemical constitution of their natural medium in the testis, since the cells are intensely active in this medium when the majority of the spermatozoa are removed by centrifuging.

2. Total activity of any suspension, as measured by its O_2 demand, is proportional to the number of sperm present and to the average space in which each cell is free to move. Inactivity in the testis appears to be due to mechanical overcrowding, each cell appearing to exercise a restraining, or allelostatic, effect on the activity of its neighbors.

3. Total energy expended during life of sperm, as well as the level of activity exhibited immediately after activation, depends on the degree of dilution of the suspension examined.

4. The relatively long life of concentrated suspensions is not due to the narcotic effect of accumulated CO_2 but is the result of an incomplete state of activation on the part of each spermatozoön.

MASS PHYSIOLOGY OF MAMMALIAN SPERM

The experiments reported so far have dealt with spermatozoan studies in those sea-dwelling animals where the sperm are shed freely into the sea-water and where fertilization occurs in this medium. There is a small amount of data at hand concerning the physiology of mammalian sperm where fertilization takes place within the body of the female after introduction by artificial or natural means. It has already been indicated that in regard to resistance to high temperatures the relation between volume of sperm and degree of resistance is similar in the two types. The work of Lloyd-Jones and Hayes (1918) and of Walton (1927) allow these comparisons to be carried further. The latter performed experiments upon the physiology of the spermatozoön, using insemination as a test for fertility. One of the variables found to enter is that of the mass of sperm present per given volume of medium.

Walton, in his work, obtained spermatozoa from a buck rabbit killed by a blow behind the ears. The abdomen was opened immediately, the testes removed, and the *cauda epididymi* were opened in about 5 cc. of 0.15 NaCl solution. This gave a dense suspension of spermatozoa. After examination under the microscope to check for motility, these were diluted in ascending powers of 10.

Doe rabbits were allowed to copulate with vasectomized males to insure ovulation and were then artificially inseminated immediately. Directly thereafter the densities of the sperm solutions used were determined with the aid of a haemocytometer for those between 10^5 and 10^7, and the others were estimated from the dilution in comparison with those counted. The results from 130 experiments performed are given in Table XXXII.

Walton recalculated the data obtained by Lloyd-Jones and Hayes from studies on the effects of insemination after excessive sexual activity of male rabbits. The results are shown in Table XXXIII.

Even in these internal impregnations a large number of spermatozoa is apparently necessary to insure fertility, and there is a significant decrease in percentage of fertile matings as the artificial or natural sperm suspensions become less dense. In the artificial inseminations just summarized a mark of decline came when the number of spermatozoa was less than 10^6 per cc., and complete sterility occurred when this value fell below about 10^4 per cc. In the series with overworked males the point of complete sterility was not

TABLE XXXII

NUMBER OF SPERM PER 3 CC. PLACED IN VAGINA OF DOE

	10^3–10^4	10^4–10^5	10^5–10^6	10^6–10^7	10^7–10^8	10^8–10^9
Total inseminated	14	19	27	28	30	12
Fertile	0	1	6	14	19	7
Percentage	0	5.3	22.2	50	63.3	58.4

TABLE XXXIII

Copulation	1	5	10	15	20
Volume, cc.	0.34	0.21	0.22	0.13	0.1
Density per cc.	104.5×10^6	39.5×10^6	14.7×10^6	4.2×10^6	3.5×10^6
Total number of sperm ejaculated	35.6×10^6	8.4×10^6	3.2×10^6	0.6×10^6	0.3×10^6
Percentage of fertility	72.09	61.1	41.37	41.37	35.55

reached. It is of interest that the size of the litters is reduced after the first, fifth, and fifteenth service.

In the latter series there are complicating factors other than density of the sperm population. Thus, there is not only a decrease in the density but also in the percentage of functionally mature sperm, a decrease in progressive motility, and a decrease in the duration of the motility of the sperm. That the latter effects are not due to mere dilution is shown by the fact that similar sperm diluted with 10 times its volume of isotonic solution showed greater duration of longevity than did sperm in the natural medium at the density existing when ejaculated. The probable cause of this phenomenon is that in the natural semen the by-products of metabolism and of developing bacteria more quickly reach a harmful or fatal concentration for the sperm than in the case of artificial dilutions.

In Walton's cases, where all dilutions were under controlled artificial conditions, three separate factors, each influenced by quantitative considerations, may influence the result. (1) The probability of any one spermatozoön reaching the fertilizable ovum within the time that the ovum is fertilizable is small; and the probability of such success is greater, within limits, the greater the number of sperm introduced. (2) Spermatozoa are variable and are not all equally capable of fertilizing, so that, again, the probability of fertilization is greater, within limits, the greater the number of sperm present. Finally (3), the toxicity of the medium may well act differentially on sperm suspensions under natural conditions, just as it has been shown to act in laboratory experimentation. This assumes that there is a general toxicity to sperm in the female genital tract which would act more vigorously upon the more dilute sperm suspensions.

In summary, we have found uniform agreement concerning the greater functional longevity of the more dense sperm suspensions, but a not unexpected lack of uniformity in the explanations advanced for this phenomenon. Recapitulating the latter, we find the more rapid exhaustion in dilute suspensions attributed to—

1. More rapid movement, exhausting sperm energy (Gemmil).
2. Dilution of nutritive spermatic fluid (Gemmil).
3. Greater diffusion of substance from spermatozoa necessary for fertilization (F. R. Lillie).
4. Effect of self-produced CO_2 (Cohn).
5. Decreased production of an autoprotective secretion (Drzewina and Bohn).
6. Unfavorable electric relations (Drzewina and Bohn).
7. Allelostasis (Gray).

Of these, evidence has been produced that throws grave doubt upon the second. Cohn's observation that the total amount of CO_2 per unit is practically the same, regardless of dilution, is strong negative evidence; and the supporting evidence produced by Gemmil as a result of treating sea-urchin sperm with nutritive solution has been shown by Cohn to be what one would expect from the pH changes known to accompany this treatment. The third explanation seems to be well established for a number of animals and is supported by various observers. If this be sound, the movement hypothesis—the first and in a modified form the seventh on our list—is not closely pertinent in the problem of functional longevity, although still of value in the matter of total longevity. There seems to be no doubt but that self-produced CO_2 slows down metabolic processes of crowded spermatozoa, and so delays their final loss of fertilizing power; but following the experience of Drzewina and Bohn and of Gray, with controlled CO_2 tensions, it is no longer possible to believe that this is the whole story. The autoprotective secretion hypothesis is not strongly supported by direct evidence. A final decision concerning the applicability of electrical phenomena, as advanced in our sixth hypothesis, must await the further development of biophysics in this field.

Finally, we may pause for a moment to consider the justification for bringing the mass relations of spermatozoa into a discussion of the physiology of numbers as a foundation for the formulation of a consistent view of general sociology. If the phenomena with which we have been dealing have general rather than special significance, we shall expect them to appear wherever there are collections of living material, whether these are gametes or zygotes or the products of zygotes. The advanced state of investigation into the physiology of the spermatozoa allows us to make this an important test case of the general application of the relationships already found to hold rather generally throughout the animal kingdom. The fact that they also operate here strengthens by that much our belief in their general applicability. A similar inquiry into the same set of relations in the culturing of bacteria and in tissue culture has shown that in those fields also there are definite protective values in optimum mass relations, and that, as in many animal relationships, the optimum usually does not coincide with the minimum population either of spermatozoa, of bacteria, or of cells in tissue culture.

GENERAL EFFECTS OF AGGREGATIONS

CHAPTER XVII

INFLUENCE OF CROWDING UPON SEX DETERMINATION

We have just seen that when gametes are shed free into the surrounding sea-water, their period of life is distinctly limited. If two gametes of opposite sex are unable to meet during this fertilizable period, death results for the spermatozoa, probably from starvation, and for the egg, perhaps from suffocation (Child, 1915). This means that animals of different sexes must be relatively close together in order that there may be a successful union of the shed gametes. Grave and Downing (1928) give us a chance to estimate some of the space requirements for the successful operation of this system in sea-water. They report that the most vigorous sperm of the sea urchin, *Arbacia*, and of the mollusk, *Cumingia*, can travel 30 cm. by their own effort in still water. *Hydroides* sperm are less active. Spermatozoa of the first two animals in suspensions of 1/2,000 to 1/10,000 per cent survive from 3 to 12 hours. Spermatozoa under natural conditions in sea-water may survive and may fertilize eggs for this period, but many die after 3 hours and the majority succumb after 7 hours. Where currents are present, these survival periods would allow of much greater distribution than 30 cm.; but the fact remains that even in the most favorable sea-water, animals must be relatively closely aggregated for fertilization to be successful. In fresh water the life of the shed gametes is quite short. After 10 minutes (Reighard, 1893) the eggs of the wall-eyed pike lose the power to be fertilized. The same observer has stated in a lecture at Woods Hole that the sperm of certain fishes loses fertilizing power within a half-minute after shedding. With animals requiring internal impregnation, the necessity for close co-operation of at least two individuals is obvious. These considerations must be fundamental for the long-recognized breeding aggregations of animals, particularly of animals shedding their gametes into the surrounding water during the breeding season.

The aggregation phenomenon may be of still greater importance in sex biology than the foregoing illustrations would indicate. The following considerations strongly suggest that it may have been an essential element in the evolution of sex itself. Presumably, this evolution started with a time when all gametes of any one species were similar. Under these conditions a first step toward union of two isogametes could be supplied by the greater well-being fostered by the presence of more than one gamete within a limited area, such as we have seen holds, under certain conditions, with the Protozoa and bacteria. From the survival value thus present before actual union took place, we can find a logical beginning for the action of a selection, which would in time, and with present known values, result in the establishment of the sexual phenomena as they appear today.

These fields, important as they are, have not yet been explored sufficiently to allow more than this suggestion of their fundamental significance. In the matter of sex determination, however, there is a mass of evidence concerning the importance of the close association of animals which merits presentation.

THE EFFECT OF CROWDING ON SEX IN THE MONSTRILLID COPEPODS

The copepod crustaceans of the aberrant family Monstrillidae are marine and free-living as adults, but in their larval stages certain of them have been shown to be parasitic in the blood-vessels of marine annelid worms. Malaquin (1901) has described in detail the life-history of *Haemocera danae*, which passes its parasitic stage in a serpulid worm. The adult males and females are highly dimorphic. Malaquin reports that when a single parasite is found in the host worm it may be either a male or a female, but that when two or more are present they develop into males, with the exception of very rare cases; in two cases only, out of some thousands examined, were two females found in the same host. The crustaceans gain entrance to their host in the nauplius stage, and it is extremely unlikely that the observed sex distribution is the result of a differential penetration of predetermined sexes. Since, when a single parasite is present, either sex may develop, the chances that the first parasite to enter may become male or female are 50 to 50. It is also highly improbable that a second or third nauplius enters only those worms which have been previously parasitized by a copepod predetermined for maleness. The possibility of differential mortality, however, is not so easily dismissed.

Malaquin interprets the observed results as being due to crowding, and suggests that the effective agent is the decrease either in amount of food or in available space. With the higher numbers of parasites per host he finds a marked decrease in size of the parasites and, accompanying this, partial or nearly complete suppression of the testes, which he regards as an example of nutritive castration. The statistical facts concerning the tendency toward maleness of crowded monstrillid parasites appear plain, but the physiological explanation is not so clear; and until necessity compels us to change, we must keep in mind the possibility that the results are due to differential mortality. In this regard, at least, the effects of crowding on sex ratios will be more clear in cases to be reported immediately.

SEX DETERMINATION IN BONELLIA

Bonellia is an aberrant annelid worm with strong sexual dimorphism. In the female, the proboscis is long and extensible and is bifurcated at the anterior end. The enlarged body contains a well-developed alimentary canal, a pair of nephridia, and a single anterior nephridium which is enlarged and serves as a uterus. The male is a small turbellarian-like worm, about 1 mm. in length, ciliated, lacking a proboscis, and with a reduced alimentary canal which lacks both mouth and anus. These and other details are well illustrated in Parker and Haswell's textbook of zoölogy. Although the adult males differ so much from the females, their structures can be shown to be homologous. A series of studies, chiefly by Baltzer, has presented us with an analysis of the factors leading to this dimorphism, which he conveniently summarizes (1928).

According to this account, the male living within the uterus of the female fertilizes the eggs there. These are laid, and in about 4 days develop into free-swimming larvae about 1 mm. long. These larvae are at first sexually indifferent. About 10–12 days after fertilization the first signs of transformation into females may appear, providing the larvae remain out of contact with the proboscides of older females, although the larvae can remain indifferent from 2 to 4, or even 5 weeks, and then develop into females.

When they develop into females, the larvae first lose their ciliary bands, the posterior part of the body enlarges, due to the development of the coelom, and the peristaltic contractions, characteristic of mature females, also begin. The alimentary tract differentiates, setae and anal vesicles are formed, and the animal sinks to the bottom, where it lives from now on. It normally depends on yolk for its food for about another week, before it begins active feeding on mud and detritus.

Male development is almost always initiated by the effect of a special metagametic factor or factor complex present in the proboscis of the female. If the indifferent larvae settle upon the proboscis lappets of a female, male development starts. It is not necessary that the female should be adult, for immature females have the same effect. This sessile period lasts some 3 days. During this period the typical male pigmentation develops, together with the shortening of the forward end, which is also a characteristic male feature. In the later part of the sessile period spermatogenesis is initiated.

After 3 or 4 days on the proboscis lappets the developing male creeps into the foregut of the female, where development as a male is completed; and after 16 or 18 days from the beginning of male differentiation the adult male makes its way into the uterus of the female by way of its external opening. During the time spent on the proboscis the developing male takes no food, living on its stored yolk. The food relations of the male during its life in the gut and later in the uterus, though presumably of a parasitic nature, are not exactly known.

There is good evidence, both from vital staining and from experimental removal of the sessile males at different periods after attachment to the proboscis lappets of the female, that the attached larvae receive material from the female which induces male development. If the larvae are removed after a very short time on the proboscis and then are carried along as free-swimming larvae, there is less tendency toward male development than is shown by other larvae left longer on the proboscis. However, the sex determination in the attached larvae is very rapid, and in a few hours can progress far enough to produce intersexes. In one experiment 16 larvae left on the proboscis for between $3\frac{1}{2}$ to 8 hours after attachment showed no complete females, 2 slightly female intersexes, 2 male hermaphrodites, and 12 complete males.

Experiments with stains show that male characteristics begin to appear at the anterior end and gradually spread through the body (Baltzer, 1928*a*). Larvae which have been left for several hours on proboscides dyed blue and then transferred for 2 or 3 days to proboscides colored with a red stain, show the blue limited to the anterior part and the red distributed over the whole body. So with the male-determining substance, a smaller quantity gives a narrower action radius while a longer attachment with more absorption acts over the whole body.

When larvae are experimentally reared in vessels lacking females, and when those starting to change in the female direction are removed, it is found that the great majority of the animals become females but that those transforming near the end show intersexuality, including even spermatogenesis. Under similar conditions Herbst (1928) found male development to be brought about by the use of a weak solution of HCl; the best results were obtained with an N/400 solution. In these experiments not all larvae transformed into males. This observation of Herbst may account for the occasional transformation in the male direction in Baltzer's experiments just recorded.

When larvae are reared in the presence of older females, or in the presence of female proboscides, the majority of the larvae settle upon the proboscides and develop into males. Similarly, extracts of proboscides or of intestinal issue of females placed in sea-water with the indifferent larvae bring about male differentiation. Extracts from muscle tissue made in the same manner did not have this effect. Extract of intestinal tissue gave better results than did that from proboscides, but in both instances there was slower initiation and the final results were less complete than when attachment to living proboscides was allowed.

The situation, so far as *Bonellia viridis* is concerned, has been summarized by Baltzer as essentially follows:

Sex determination is partly predetermined, partly epigenetic. Both sex tendencies are probably predetermined in the fertilized egg in varying degree, with the male tendency predominant. The fertilized egg, as well as the indifferent larvae which is not yet sexually differentiated in its organization, is most probably hermaphroditic. Only in their further development do they become wholly male or female. In this development the evolution of the organization in males or in females proceeds in an unlike manner, corresponding to the strong sexual dimorphism of the species.

With few exceptions the development of males is possible only when the indifferent free-swimming larva finds an opportunity to live parasitically on the proboscis of an adult female. During this period there is an absorption of sex-determining substance from the host by the larva. If this opportunity for parasitism exists, all the larvae usually become males. If it is lacking and the larvae are forced to live freely, females arise almost exclusively, though there are occasional hermaphrodites. Females appear only after a longer indifferent period during which development is approximately at a standstill; during this time the female tendency gradually gains in intensity and finally assumes ascendancy over the male tendency. The females that arise late are protandric hermaphrodites and contain sperm in the coelom, but in further development become typical females.

If opportunity for parasitic development is given to swarming indifferent larvae, and if the parasitism is prematurely interrupted by separating the larvae from the proboscis of the host, then, after further culture in the free state, hermaphrodites arise, together with a few males and females. The duration of the parasitic period determines whether hermaphrodites actually bisexual are obtained, or gynandromorphs in which only the secondary sexual characters are combined. In *Bonellia* these secondary sexual characters include most of the organizational characters.

The method of effecting male determination in the case of the *Bonellia* larvae appears not unlike that which effects the transformation of a zygotically determined female calf embryo which is in blood-stream connection with a twin of the opposite sex. The well-known work of F. R. Lillie (1917) has shown that under these conditions a transformation in the male direction takes place, due to the effect of a hormone produced by the male upon the organs of the developing female.

SEX IN CREPIDULA

In the gasteropod mollusk, *Crepidula plana*, Gould (1917) has shown that the sexual life of adults of this marine snail may be divided into (*a*) the male phase, (*b*) the transitional phase, and (*c*) the female phase. In other words, these animals are protandric hermaphrodites with the opposed sexual phases completely separated. The development of the male condition does not always take place at the same stage with respect either to age or to size of the individual, and at times may be omitted entirely.

Male development occurs, if at all, during the early life of the adult. The growth of the animal during this period is highly variable, depending, among other factors, on the amount of movement, the extent of available space, and the season of the year. Clearly distinguishable primordial male and female germ cells are both to be found in the gonads of these animals from the postlarval stage up to the time of complete female development, and can be distinguished from each other under the microscope. During the period of transformation from the male to the female phases, the testis degenerates; and finally all the primordial male germ cells disappear as the gonad becomes reduced in size. With the assumption of the female phase the gonad again enlarges.

The same duct serves for the passage of the sperm in the earlier phase and the eggs in the later, but it undergoes marked changes in the transition. Secondary male characteristics, such as the penis, the sperm groove, and the seminal vesicles, are present only when the testis is developed; they appear when the testis appears and disappear when it degenerates.

The variability of the male phase is summarized by Gould (1917*a*) as follows: "A number of specimens of the same size and apparently of the same age, taken at the same time of year, may show widely different sexual states. One may be a fully developed male; one may exhibit evidence of having been a male, though the male characters are being lost; and one may furnish no suggestion that any male characters have ever developed."

An analysis of these conditions shows that the development of the male phase is dependent on the nearby presence of a larger individual of the same species, which is usually, though not necessarily, a female. The greater the difference in size between the smaller and the larger animal, the more certain and complete is the development of the smaller in the male direction. A greater stimulus is necessary to complete male development than to initiate it.

In nature these snails develop from free-swimming veliger larvae which settle within the gasteropod shells occupied by hermit crabs. If the snail which has formed the shell has recently died, the *Crepidula* population will be small; but in the shells long occupied by the crabs, a large number of *Crepidula* may collect. Those located near the outer margin will show little sign of movement. Their own shells will have grown to fit the irregularities of the substratum. Usually these are large females. Crowded about them, occupying vacant patches, and back in the deeper recesses of the shell, the *Crepidula* are smaller; and, since their shells are not intimately fitted to their surroundings, it may be assumed that they move about.

If a larval *Crepidula plana* settles into a shell where no larger members of its species are present, the male phase is normally omitted. In cases where Gould found neuters and returned them to hermit crab shells free from large individuals, he recovered at the end of 34 days 24 specimens, which, when sectioned, showed 3 with adult testis, 2 with spermatids, 1 with spermatocytes, 11 with spermatogonia only (that is, no further development toward spermatozoa), 5 sexually inactive, and 2 with oögonia. Of 2 specimens found to be neuters, isolated on hermit crab shells and sectioned after 67 days, 1 had early oöcytes and the other had ova with yolk.

Of 4 others tested after 75 days, 3 were definitely females, while the other, the smallest of the lot, which had been closely associated with a larger specimen, was a fully developed male.

In contrast with the preceding experience, when known neuters were transferred to a colony containing larger individuals, there was a definite development in the male direction regardless of the sex of the large individuals. In one experiment *Crepidula* examined and found to be in a neutral stage were transferred to the vicinity of large females. At the end of 34 days the specimens were sectioned for microscopic examination. Of 27 individuals so treated, 18 had adult testis, 6 were in the spermatid stage, 2 had spermatocytes, 1 showed spermatogonia only, while none of the entire lot remained in the sexually indifferent stage and none had developed in the female direction.

When small males were placed near larger males in an attempt to find the effect of such proximity upon the continuation of the male phase, it was found that fewer small males underwent degeneration of the sexual organs than when they were completely separated from larger animals. More showed degeneration than if they were in the presence of large females. Even a large animal with a degenerate testis will give a stimulus toward male development to a smaller one near-by. An immature female, formerly a male, will also have the same effect. After degeneration of the testis and the accompanying secondary organs, these may be regenerated if the individual comes under the influence of a larger animal; but the largest animal in an experimentally arranged colony never shows regeneration of the testis following degeneration. There is no evidence that a smaller animal can affect a larger one or that a group of smaller ones can have this effect, although the last point does not seem to have been specifically investigated by Gould.

A mature or nearly mature female does not undergo degeneration of sexual organs however placed; but if a partially developed female is placed near a large, mature female, the female sex organs degenerate and the male organs develop.

The degeneration of the testis in the absence of a large female is not due to the lack of opportunity for copulation, for degeneration does not take place when a small male is near a larger one. Though males are normally more motile than females, movement is not necessary, as is shown by those males the conformation of whose shells proves that they have not recently moved about. Neuters will develop into males without showing movement. Male development here is not a matter of food, for neuters will develop into males regardless of the richness or scantiness of the food supply, provided only they are near large animals. Starving or feeding mature males does not affect their state of sexual development if they are near large females.

Preliminary experiments designed to test for the presence of a secretion from the larger animal which would affect the development of the sexual organs of the smaller have been run without positive results. There is no specific ovarian secretion concerned, since the presence of a large male has somewhat the same effect as that of a female. Males in one finger bowl were covered with water from another finger bowl which contained a number of females. The water was changed daily and was replaced with water in which the females had stood. Degeneration of the male organs occurred. Similar results were obtained when water from a dish containing 20–25 large females was led into a finger bowl containing 20–25 males. The experiment ran for a month, and during this time the males were kept separated as carefully as possible. Again the degeneration of the sexual organs was shown by the condition of the penes. A similar experiment in which water from the mature females flowed over a number of small neuters did not reveal any tendency for penis development. Neither was there any initiation of male development from adding extract of crushed adult females twice daily to finger bowls containing neuters.

The stimulus to sex development or recession does not depend on the presence of the hermit crab with which these *Crepidula* are normally associated in nature, since the whole gamut of sex can be run under the artificial conditions of the laboratory, with glassware in place of shells, and with hermit crabs entirely absent.

Gould recognizes that his experiments concerning the causal relations involved in sex determination in *Crepidula* are inconclusive.

They do not make clear whether the transformations affected by the presence of the large individual can take place in the absence of physical contact. They are not conclusive concerning the possible presence of some chemical product or the effect of such a product transmitted through the sea-water; and the biophysical possibilities remain to date unexplored. All we know is that in some manner the presence or absence of larger individuals affects the state of development of the primary and secondary sex organs of associated smaller individuals, which is exciting enough information to deserve greater experimental attention than has been so far given to it.

SEX IN NEMATODE PARASITES

While working on grasshoppers to determine the fatal dose of the eggs of one of their parasites, the nematode hairworm, *Mermis subnigrescens*, Cobb, Steiner, and Christie (1927) discovered that the sex of the resulting adult parasite was male or female according to the size of the dose. The parasite is an important factor in the control of grasshoppers, and the experiments were primarily concerned with this aspect.

The parasite *M. subnigrescens* is found quite commonly in a number of varieties of common grasshoppers: for example, the red-legged grasshopper *Melanoplus femurrubrum* and its relatives. The grasshoppers become parasitized by swallowing eggs of the parasite that have been deposited on their food plants; these eggs contain well-developed hairworms.

Cobb, Steiner, and Christie (1927) state that their researches have shown that "when immature females of the hairworm are kept in solitary confinement before they can have copulated with males they produce viable eggs." In thousands of observations made by these workers the average number of hairworms per infested grasshopper in nature has been from 1 to 3, and these are always females.

Usually overinfestation of a grasshopper by hairworms resulted in the premature death of the grasshopper as well as of the parasites, though on rare occasions highly infested grasshoppers have been found in nature containing more than a hundred hairworms. In all such extreme cases the parasites were small and young. These observations led to experiments designed to determine the dose of hairworm eggs that would be fatal to the host; for the present purpose it may be said that for young grasshoppers in the second instar it averages well under 50 eggs. These later experiments led to the discovery that when a slightly sublethal dose of eggs is given, the resulting parasites are all males. In one case where 20 hairworm eggs were fed to a grasshopper first freed from this parasite, 19 parasites were recovered, all males. Many tests and observations were made, and all led to the same conclusion: Feeding a very few eggs resulted in female parasites; feeding a large number, but not quite enough to kill, resulted in male parasites.

Similar experiments in treating the larvae of a midge (*Chironomus*) with the hairworm *Pseudomermis zykoffi* gave similar results. High parasitism resulted in the parasites being all males. Similar observations were made on the hairworm *Agamermis paradecaudata* infesting the tea bug *Helopeltis theivora*. No artificial infestation was made in this case; the observations were confined to the parasite as it existed in nature. Here, again, when there were a large number of parasites all of them were males.

Observations on the mirmithid *Allomermis* sp. infesting the common dooryard ant *Lasius niger americanus* showed that when parasitism is high the parasites are all or nearly all males, and that when it is low the parasites are all females.

The phenomenon is the same in hairworms of four different genera, found in host insects of four different orders. Between the extremes of parasitism in all these cases there is a gradient with mixtures of males and females, the proportion of the males increasing with the severity of the infestation.

In the case of the hairworm *Mermis subnigrescens*, Cobb and his associates think (Christie, 1929) that by appropriate experiments they have excluded other possible explanations of these results; that they were not brought about by using the eggs from an individual hairworm, nor from a definite part of the uterus having a uniform sex-potentiality, nor by death or other selective elimination. The most convincing data at hand on these points are those collected by Christie (1929). Table XXXIV shows the results from one set of experiments in which he fed a known number of *M. subnigrescens* eggs to the common New England grasshopper, *Melanus femur-rubrum.*

In four cases of relatively heavy infestation a total of 100 eggs were fed, from which 86 parasites were reared, all of which were males. In the relatively light infestations, 102 eggs were fed, usually 5 to each grasshopper; of these, 73 parasites were reared, of which 6, or 8 per cent, were males.

TABLE XXXIV

SHOWING THE RESULTS OF FEEDING GRASSHOPPERS WITH KNOWN NUMBERS OF MERMITHIDAE EGGS UPON THE SEX OF THE DEVELOPED NEMATODES

(Data from Christie)

Number of Eggs Fed	Number of Parasites Obtained	Males	Females
30	25	25	0
30	24	24	0
20	19	19	0
20	18	18	0
5	2	0	2
5	2	0	2
5	5	0	5
5	4	0	4
5	3	0	3
5	3	0	3
5	3	0	3
5	5	1	4
5	4	1	3
5	1	0	1
5	5	1	4
5	4	1	3
4*	4	0	4
4*	4	0	4
4*	3	0	3
5	5	1	4
5	4	0	4
5	4	1	3
5	4	0	4
5	2	0	2
5	2	0	2

* *Camnula pellucida* used as host insect.

Caullery and Comas (1928), stimulated by the early reports of the work just reviewed, examined the distribution of sexes in the nematode worm, *Paramermis contorta*, which is a parasite upon a chironomid larva. A part of their results are summarized in Table XXXV. They interpret their findings and the findings of Cobb, Steiner, and Christie (1927) as meaning that in these parasitic nematodes sex depends largely on the number of parasites simultaneously present, and on the resulting nutritive values of the host. When a single parasite is present, some genotypic males appear to have been inverted into phenotypic females. When two are present, the normal genotypic relation is not sensibly altered. With higher

TABLE XXXV

Showing the Number of Nematode Parasites per Chironomid Larva in Relation to Sex Distribution

(Data from Caullery and Comas)

No account is given here of the intersexes also reported.

Number of Parasites per Larva	Number of Cases Observed	Females	Males	Ratio of Females to Males
1	272	255	17	15.0
2	173	180	166	1.14
3	43	47	82	0.57
4	16	23	41	0.56
5	6	5	25	0.2
6	3	3	15	0.2
7	2	3	11	0.27
9	1	1	8	0.125
10	3	4	26	0.154
11	1	2	9	0.22
17	1	2	15	0.133

numbers in the nematode population of individual bloodworm hosts, genotypic females apparently were transformed into phenotypic males. The production of females is not quantitative at any level in Caullery and Comas' work, but it may be summarized as showing that the more parasites present in a given host, the fewer females in the population.

There seems no escape from the conclusion that we are dealing here with cases where the numbers of individuals of the same species in the immediate environment is a sex-determining factor for all those present. Here, even less than in the preceding similar case of *Crepidula*, has analytical work been done. We do not even know, for example, whether nematodes of different species present in the same host will have the same effect as if all were of the same species; nor do we know whether the effect of malnutrition or other adverse environmental influences upon the host will affect the sex ratio of the parasites. Christie has noted that with heavy infestations there is a reduction in size of individuals present, whether males or females, but that there is no tendency toward a differential suppression of the gonads. He finds that there is a tendency for the parasites in a heavily infested host to complete their parasitic development more rapidly than those in a lightly infested animal. While this applies to both sexes, it is more apparent with the females.

SEX IN CLADOCERA

The sex situation in the cladoceran Crustacea differs from the instances previously discussed in that with these animals the usual method of reproduction is parthenogenetic: the females produce eggs that develop without fertilization to form other females. In nature, after a period of purely parthenogenetic reproduction of this sort, there may occur an outbreak of bisexuality in which parthenogenetic eggs develop into males,[1] and at the same time, or, more usually, slightly later, eggs are produced which require fertilization before development. The eggs so produced are resistant to many adverse conditions, such as drying or freezing. With favorable conditions some of them, after a dormant period, produce parthenogenetic females, and the usual type of reproduction begins again. In the other cases cited, the effect of crowding upon sex dealt with the effect on the expected sex ratio; here we are interested in its effect on the transfer from the production of parthenogenetically produced, self-sufficient females to the production of parthenogenetically produced males.

[1] It is interesting to note that Banta and Wood (1928) report genetic evidence that males of the cladoceran, *Daphnia longispina*, are diploid, and that E. Allen (1928) finds in cytological studies of *Moina macrocopa:* "After the egg is laid, the first division occurs in the parthenogenetic egg without reduction in the number of chromosomes. In the sexual egg, the first maturation division results in the haploid number, which is eleven. The diploid number is twenty-two in both types of egg. In the eggs of crowded mothers, which should produce a high percentage of males, no evidence has yet been obtained indicating that the male number is haploid. Several such crowded mothers have been studied."

Many workers have united in believing that the change from parthenogenetic to sexual reproduction is due to internal factors, perhaps to an innate sexual cycle. The problem has attracted much attention because of the frequent coincidence between the appearance of sexual forms, the production of the resistant sexual egg, and the beginning of periods of special environmental stress. The whole relationship appears to be highly adaptive in that the resistant egg allows the species to survive periods of drought or of low temperature which would otherwise be fatal to it. The phenomenon is the more impressive when we find that if a given pond contains more than one cladoceran species, all show development of sexual reproduction at approximately the same time.

In the laboratory, Banta has shown that one cladoceran, *Moina macrocopa*, can be reared for at least 780 asexual generations with complete vigor, and that another, *Daphnia pulex*, can be grown similarly for at least 767 generations without the appearance of sexual forms, providing cultural conditions remain favorable. Thus it appears that the hypothesis of an innate sexual cycle will not hold as a step toward a universal solution of the problem.

Grosvenor and Smith (1913) suggested, as a result of their experience with another species of *Moina*, that the production of males is initiated by the accumulation of waste products due to the crowded condition of the mothers. Langhans (1909) had shown that accumulations of excretory products reduce growth and reproduction and hence presumably favor the production of sexual forms. Papanicolau (1910) confirmed an earlier report of Langhans that repeated transfers of the *Daphnia* to fresh water postpones the appearance of sexual forms indefinitely; but as a result of further work (1910*a*) he concluded that crowding serves only to decrease numbers. McClendon (1910) also observed a hastening of sexual reproduction which he interpreted as being due to excretions. Smith (1915) believed the effect of crowding is due to the excretions which accumulate, rather than to deficiency of food. Hartmann (1919) considered the accumulation of excretions a probable cause of the outbreak of gamic reproduction.

Banta and Brown, reporting in a series of papers beginning in

1923 and not yet finished, have found that crowding of the mothers in 2 species of *Daphnia*, 3 of *Simocephalus*, 3 of *Moina*, and 1 of *Ceriodaphnia* causes a large number of males to appear. Agar has recently informed me personally that he has confirmed these results "to the hilt," although his published work (1914) showed that males will appear without crowding and with frequent changes of medium. Shelley (1929), working with another species of *Daphnia* (*D. magna*) in this laboratory, has obtained results similar to those reported by Banta and Brown. There can be no doubt of the fact.

The later work of Banta and Brown gives an opportunity to show the extent of the changes produced in the sex of the offspring. In all they have performed some 2,900 experiments dealing with the effect of crowding on male production in cladocerans. One lot of 109 uncrowded mothers which produced 1,954 young, averaged only 0.3 per cent males; 5 males were produced by one mother, and a single male by another. On the other hand, 33 moderately crowded culture bottles which contained from 7 to 14 females, usually 10 per culture bottle, gave an average male production of 41.6 per cent out of 3,638 young whose sex was determined. Six of these bottles failed to produce males; 2 produced 100 per cent males. In heavily crowded bottles, with 15 to 24 mothers to the bottle, from 2,240 young whose sex was determined, 62 per cent were males. One bottle failed to produce males, and 4 produced 100 per cent males. They regard these as typical results. It will be noted that the effects of crowding are not uniform and that quantitative male production does not occur in all the bottles of a densely crowded series.

We find that Banta and Brown conclude from their experiments, as Grosvenor and Smith did earlier, that the chief factor in stimulating male production in *Moina* is the accumulation of excretory products. Further, they present evidence which shows that this effect of crowding can be produced by the presence of other genera of Cladocera or by other aquatic animals, such as *Planaria*, *Asellus*, *Physa*, insect larvae, small fish, or frog tadpoles; monotypic excretory products are unnecessary. Here, in the heterotypically produced adverse conditions, there is a lack of species specificity of effects, just as we have already seen a lack of species specificity in group protection from toxic solutions or hypotonic sea-water. These excretory products effective in male production are apparently not volatile, but they are unstable compounds readily made non-effective by a variety of treatments. Standing for a short time, or treatment with NaOH or H_2SO_4, or a greater dilution of the medium, is effective in causing a lack of potency for male production.

The factors causing production of sexual eggs are apparently different from those causing males to appear. Sexual eggs are produced only rarely in the crowded bottles under the conditions used to produce males. In *Moina macrocopa* they are produced by rearing females in clear pond water, or in very dilute food, or in old culture media. Banta and Brown believe their experience shows that scarcity of food controls the production of the sexual egg, as the accumulation of excretory products as a result of crowding controls the appearance of males.

Further studies have indicated that the time at which the sex of the forthcoming young of this cladoceran is determined lies about 4 hours before the parthenogenetic eggs are laid, when the temperature stands at 20° C. (Banta and Brown, 1929*b*). Observation that there was some relationship between the time of the release of a female's first brood and the sex of the young led to definitive experiments concerning the relationship between the rate of the mother's development and the sex of her offspring. The evidence from these experiments is summarized in part in Figure 30, which indicates that the production of males is closely associated with a reduction in the rate of development of the females producing them. Apparently this is due to the accumulation of excretory products in the crowded bottles, since the reduction in time of production of the first clutch of young is proportional to the degree of crowding, and, further, the percentage of males produced is proportional to both.

Tests were made (Banta and Brown, 1929*d*) to find whether experimentally changing the rate of development would affect the sex ratios of the produced young. Treatment with small dosages of ethyl alcohol and with filtrates from dried adrenal cortex, thyroid, thymus, and muscle tissue serves to increase the rate of development of *M. macrocopa* females as measured by the time before the production of the first (parthenogenetic) brood. When tested under crowded conditions, the retarding effect normally accompanying

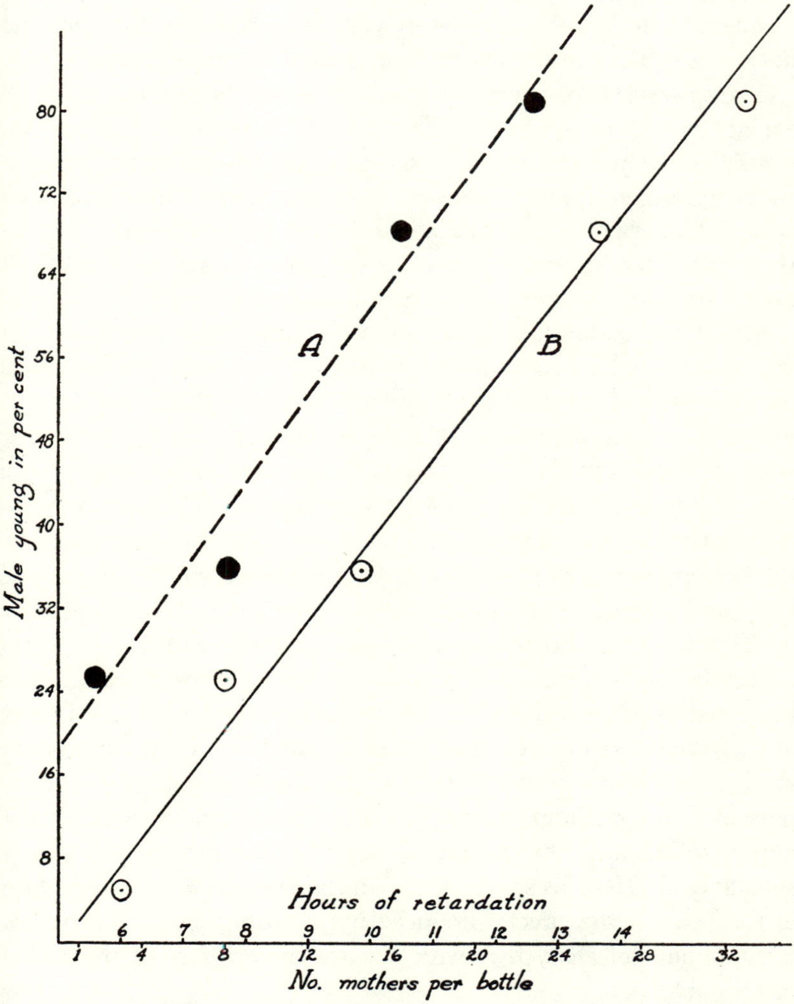

FIG. 30.—Diagram, from Banta and Brown 1929*b*, showing the relation (*A*) between the amount of retardation of the production of young and the percentage of males produced in *Moina macrocopa*, and (*B*) between the number of mothers per bottle and the percentage of males produced.

crowding was largely counteracted by the foregoing treatment, and the percentage of males expected from crowded cultures was materially reduced. Considering the wide range of tissue extracts used, the authors appear to be correct in regarding the observed results as due to some generalized factor, such as a change in the bacterial flora, rather than to the action of a specific tissue product.

The converse experiment, in which the general metabolic rate was experimentally depressed by the use of chloretone or of potassium cyanide in proper concentrations, caused the *Moina* tested to reach the reproductive age more slowly. The females so retarded produced a much higher percentage of males among their first-broods than did their untreated sisters, which were subjected to similar conditions except for the treatment with depressing agents.

Stuart and Banta (1929, 1931) report controlled bacterial studies which indicate that sister *Moina* females cultured in equal numbers have the sex of their offspring determined in part at least by the numbers of bacteria present. "These females in the highly bacterized bottle produce only female young. Between these two extremes there is a very uniform gradient in male percentage." These authors interpret their results as indicating that quantity of food appears to be the determining factor, which causes speculation as to whether the previous analysis of the situation by Banta and Brown is final or whether their findings may not be bound up with as yet unexplored differences in bacterial flora accompanying crowding. The latter is indicated by the conclusion of Stuart and Banta (1931) that: "Sex of the young produced by experimentally crowded *Moina macrocopa* mothers can be controlled by the amount of available food (bacteria) present in the medium. The bacteria present in the medium do not appear to influence the control of sex by the adsorption of excretory substances." Here, as so frequently in this survey, we can be certain of the facts of the effects produced by crowding, but, even with a large amount of analytical work, we are not yet sure of the causal factors involved.

There is no question but that sex is one of the most nearly fundamental properties of animals. On this account its determination has aroused great interest, and an enormous literature on the subject has resulted. Much evidence has accumulated of recent years that sex determination at or before fertilization is not the hard and fast fact it was regarded to be in the first blush of enthusiasm over the discovery of the sex chromosomes. The work of Lillie, Goldschmidt, Crew, and Domm has shown a much greater fluidity than was at first anticipated in the light of the chromosome theory. The results of metabolic effects produced by crowding, reported here for monstrillid copepods and Cladocera among the Crustacea; for *Bonellia*, the aberrant annelid worm; *Crepidula*, the gasteropod mollusk, and certain of the nematode worms, are not out of harmony with other instances of epigametic sex determination produced by other means. From our present knowledge it appears that sex determination, at least for the majority of animals, is normally associated with the chromosome mechanism, but that chromosome determinations may be overruled by other factors, among them the effects produced by crowding. In certain cases these effects appear to be produced by the transfer of material from one individual to another, as in *Bonellia;* in others, as Cladocera, by the effects of crowding upon animals that are not necessarily in physical contact with each other. From our point of view the important thing is that they occur at all and can be controlled, in certain animals of widely distributed taxonomic position, by the degree or kind of aggregation obtaining.

When we examine the possible survival value of the effect of crowding upon sex, we find a mixed situation. In the monstrillid copepods crowding has negative survival value, in that a normal sex ratio among uncrowded animals gives way to an overproduction of males among crowded individuals. With the parasitic nematodes the high proportion of males accompanying heavy infestations would appear to be as definitely harmful for the race as their absence in cases of low infestation. Here is evidence of an optimum population, where males and females are in fairly equal ratio, well above the minimum population, as we have seen in other relationships. It may be suggested that there is a more remote survival value in the overproduction of males, in that the neighborhood is less likely to be overstocked, to the destruction of all.

With *Bonellia* and *Crepidula* the tendency for an isolated individual to become a female and for those appearing later in the immediate vicinity to become males has survival value. Baltzer considered this point in 1914. He reports that *Bonellia* is not an abundant animal, and occurs in numbers in only a few places in the Bay of Naples. If one supposes that the males remained parasitic but, unlike their present condition, were unalterably predetermined for maleness, then a male larva would be useful only if it found an adult female. All males which did not by chance come into the neighborhood of an adult female must die without issue. This would include a large number of larvae, not only because the species is relatively rare, but because the larvae are positively phototactic at the beginning of their free period and frequently swarm away from their egg mass, even though they are able at this early age to attach to the proboscis of a nearby female. (It appears that these larvae may not attach to the proboscis of their own mother.) By means of their special method of sex determination this potential loss of large numbers of males is avoided, since if they reach a suitable environment lacking females they develop directly into females and are able then to effect the transformation in the male direction of the next larvae to arrive in that vicinity. The survival values of the *Crepidula* sex situation are similar.

Essentially the same situation is found in Cladocera. With these animals the tendency to change, as the culture becomes crowded, from parthenogenetic eggs rapidly produced but not resistant, to the sexual eggs which can withstand adverse conditions, has survival value. In nature such crowding usually precedes the drying-up or freezing of small bodies of water in which the animals have been living; and either usually follows a long reproduction period which has given time for the increase of the cladoceran population to effect a definite change in the environment. Under these conditions the production of resistant eggs has definite value.

It appears, therefore, that in 4 out of these 5 cases, in which crowding is known to affect the sex ratio, the result has survival value.

CHAPTER XVIII

MORPHOLOGICAL EFFECTS OF CROWDING

In addition to its effect upon primary sexual characters through sex determination, which has been discussed in the preceding chapter, crowding may produce morphological changes in secondary sexual characters especially noticeable in animals such as *Bonellia*, which have a strong sexual dimorphism. Crowding also exerts decided influences upon structures, entirely apart from its effect upon sex. Casual observation shows that a tree grown in the open country has a different growth form from that shown by the same sort of tree grown in a forest. Similarly, when sessile marine animals, such as barnacles, ascidians, corals, sea anemones and sea mussels, grow in closely packed masses upon a rock or wharf piling, their growth form differs from that shown if they grow separately. With these sessile animals the changes associated with crowding appear to be due largely to the limitations imposed by the physical contacts established under conditions of close aggregation.

The effect of crowding upon physical form is not limited to sessile organisms. There is the well-known case of the free-swimming pelagic *tunicate*, *Salpa*, with its alternation of the asexual solitary, casklike form with the quite differently shaped aggregated form of the members of a *Salpa* chain. These latter are sexual individuals that, apparently because of crowding, have lost their regular barrel-like appearance and are rather rounded, ovoid, or fusiform.

With non-sessile animals which lack the physical connections of a *Salpa* chain, even though the crowding be less dense, there are cases of marked morphological changes other than mere decrease in size. Whitefield (1882) observed that *Lymnaea*, a snail, when grown in a small volume of water for three successive generations, had not only become much reduced in size but had suffered other physical changes as well: the male organs did not develop, and the liver was also much smaller in proportion to the other organs, as compared with that of related snails grown under uncrowded conditions. Whitefield says that the shell proportions were so changed that experienced conchologists recognized the dwarfed forms as sufficiently distinct to be placed in a separate species.

Differential morphological effects associated with partial starvation are well known. Child (1915) has shown that starving adult planarian worms will cause them to return to a juvenile condition. The culture water in which *Planaria* have been crowded definitely depresses normal head formation in regenerating planarians (Child, 1911). In analyzing this effect, Child reports (1911*a*): "The presence of metabolic products of *Planaria* in the water undoubtedly decreases the rate of metabolism, and the effect on regulatory morphogenesis is similar to that of starvation or low temperature, though it may be greater and in extreme cases it approaches that obtained with anesthetics."

Vernon (1903) has reported differential inhibition in echinoderm larvae due to crowding. He attributes this to food deficiency. Warren (1900) reports modifications of the spine of *Daphnia* brought about by crowding. Other such cases could be cited, but it will be more profitable to examine in some detail the evidence that crowding affects (*a*) the appearance of winged forms in aphids, (*b*) coloration and changes in bodily proportions of certain grasshoppers, and (*c*) the suppression of certain bristles and eye facets in some races of *Drosophila*.

WING PRODUCTION IN APHIDS

Aphids show two changes in the course of their normal seasonal cycle, which may or may not be associated. Their annual history, briefly, runs as follows: In the autumn sexual forms appear, and an overwintering fertilized egg is formed which hatches out the following spring into a wingless female capable of producing young parthenogenetically. These, too, are usually wingless females capable of parthenogenetic reproduction. After a time winged forms appear which migrate to new host plants where they give rise to wingless females with parthenogenetic powers. The production of winged forms is thus a process distinct from the production of the sexual forms. In the case of the latter in most, but not all, aphids the sexual female is wingless but the male is winged. The problem with which we are especially concerned here is the relation of crowding to the production of the winged (alate) parthenogenetic forms from apterous mothers.

Grassi (1907; *vide* Shull, 1929) first reported that crowding of aphids of the genus *Phylloxera* on their food plant was followed by wing production. Davidson (1914), in speaking of his own work with the aphid *Schizoneura* and other species, attributes the appearance of winged forms to some change in the constitution of the cell-sap products of the overpopulated plant. He summarizes his work by saying: "Throughout my experiments it was observed that when the plant had finished its active growth, or became heavily infested with aphids, the changes resulting either in the quality or the quantity of the cell sap (or both) seemed to induce the production of winged forms."

Wadley (1923), using *Rophalosiphum*, reports in detail experiments which show that when aphids were allowed to multiply for two or three generations and to overcrowd the plants in their experimental cage, a high percentage of alate aphids were invariably produced. Even when apterous aphids were removed from these crowded plants and placed on new and flourishing host plants, their progeny, reared with abundance of food, gave from 30 to 86 per cent alate individuals, while the control gave from 7 to 20 per cent. Wadley attributes these results to the effect of limited nutrition, since he was able to obtain similar data from starvation without crowding. In both instances, alate forms tend to give rise to apterous forms, as they do normally under optimum conditions.

Ackerman (1926) confirmed the results obtained by Wadley, using the same species of aphid. In one experiment with 776 aphids reared under crowded conditions, 34 per cent were winged; while with no overcrowding, all of 280 aphids were wingless. Ackerman also found that partial starvation tended to produce a high percentage of winged forms in the offspring.

Reinhard (1927), using another species of *Aphis*, has undertaken the most extensive set of experiments to date dealing with the effect of crowding upon wing production in aphids. He first tested the effect of parentage upon wing production to see if there is an innate cycle which causes winged forms to appear at certain intervals. During a period of 12 months of continuous work, 59 complete generations were reared without the appearance of winged forms, except in 3 generations when the aphids became crowded. These results are supported by other data and indicate that with this species of aphid there is a normal tendency to be wingless and that the production of wings depends upon environmental influences rather than on an innate cycle, as many have thought (Shull, 1929). Reinhard's experience with starvation supported that of previous workers. He found that starvation of apterous parents increased the number of alate individuals, while starvation of alate parents did not affect the normal tendency of such animals to produce wingless forms.

Reinhard also found that with the species of aphid which he observed, the appearance of its winged phase was not determined by the temperatures to which the animals were exposed, nor by humidity. Other workers (*vide* Shull, 1929) have found that temperature, and light relations as well, do affect the appearance of winged aphids of other species. Having cleared the way by these preliminary experiments, Reinhard turned to more intensive experimentation upon the effect of crowding upon wing production.

In his long experiments, he found no winged forms except when the plants were allowed to become crowded. Subcultures from this experiment were made at different generations. The aphids were isolated on uninfested plants, and they and their progeny were undisturbed until a crowded condition resulted. In each case these subcultures yielded alate aphids, even though the uncrowded main experiment did not. The results of these experiments are summarized in Table XXXVI.

In another series of experiments wingless aphids of unknown parentage were placed on uninfested plants and allowed to develop crowded conditions. In 14 tests which are reported by Reinhard in detail the history of 438 aphids is given. Of these, 221, or slightly more than 50 per cent were alate; 5 winged aphids, or about 2 per cent, appeared among the 243 individuals reared under conditions similar, except that there was practically no crowding. Normally, as we have stated, the alate aphids give rise to wingless forms. Reinhard found that crowding these alate individuals caused an increase in the percentage of winged offspring in the next generation. In the light of these experiments, Reinhard concluded that "crowding is a potent, if not the dominant, factor in controlling wing development in *Aphis gossypii*." While this conclusion appears entirely reasonable, it must be remembered that the factors involved in bringing about this result are not fully revealed. Observers agree that starvation is also an effective agent in producing winged aphids, and anyone who has undertaken to rear aphids under crowded conditions will appreciate the difficulty of separating these two factors with certainty. One of Reinhard's observations indicates that crowding is the more effective agent of the two. Starvation did not cause winged forms to increase the number of winged progeny in the next generation, while crowding did do so. Other possible factors to be tested include the physical effect of the presence of

TABLE XXXVI

DATA ON THE EFFECT OF CROWDING ON WING DEVELOPMENT IN APHIDS. THE CHECK PROGENY WERE SISTERS TO THOSE CROWDED, AND OF THE SAME GENERATIONS

(From Reinhard)

Generation	Parentage	Date Isolated	Number Isolated	First Young Observed Developing Wings	Winged Forms after Progeny Well Crowded	Check Progeny Not Crowded	
						Apterous	Alate
		1925		1926			
8	Apterous	Dec. 9	9	Jan. 9	Many	25	0
10	Apterous	Dec. 20	17	Jan. 15	Many	19	0
		1926					
13	Apterous	Jan. 9	11	Jan. 20	Few	32	4
18	Apterous	Feb. 10	20	Feb. 16	Many	75	6
19	Apterous	Feb. 16	65	Feb. 24	Predominate	16	0
29	Apterous	April 17	19	April 23	Few	11	0
47	Apterous	July 25	13	Aug. 9	Few	16	0
48	Apterous	July 28	10	Aug. 18	Many	17	0
55	Apterous	Sept. 15	9	Sept. 23	Many	11	0
59	Apterous	Oct. 3	9	Oct. 19	Few	14	0

other aphids and the possibility of aphid secretions being transferred from one to another through the plant. Merely stating these alternative hypotheses helps to emphasize the probability that changed or decreased nutrition is the dominating factor in the situation.

This recital of references to observations by a number of workers in widely separated regions, and usually upon distinct species, indicates that the effect of crowding upon wing production must be a widespread phenomenon among aphids. The racial importance of this phenomenon is apparent when one remembers the added migratory power thus conferred upon members of a crowded colony living upon a host plant which may soon become exhausted from the feeding activities of its aphid population.

CROWDING AND THE PHASE THEORY OF LOCUSTS

In the following discussion it will be convenient to follow Uvarov (1928) in applying the name "locust" to gregarious members of the short-horned grasshoppers, family Acrididae, which migrate in swarms, while "grasshoppers," in a restricted sense, will be used in speaking of the non-gregarious, non-migrating members of the same group. The gregarious collections of adult locusts will be spoken of as "swarms," and the similar collections of immature hoppers (nymphs) will be called "bands."

In order to have clearly in mind the morphological relations of locusts and grasshoppers, it will be necessary to pass hurriedly in review some of the available knowledge concerning the behavior of these animals. The egg pods of the gregarious locusts are deposited close together so that when the young hatch and emerge from the ground they are immediately in close contact with each other. After their intermediate molt these recently emerged hoppers soon form primary bands, due largely to the reaction of the young animals to light and heat. On warm sunshiny days these gather in exposed sunny places; on cool wintry days the bands collect in sheltered spots (La Baume, 1918). In addition to these place aggregations there is said to be a distinct gregarious tendency which has not yet been analyzed to see whether it is a behavior unit or whether it may be split into more elementary reaction complexes.

The alternation of behavior, night and day, appears to be related to that on cool days and on warm sunny days. The young hoppers spend the latter in groups basking in the sun, while at night or on cool days they crawl under stones or other shelter, or climb some plant. In either case they may be found in dense collections. Analyses to date do not show the relative importance of heat and of light in these reactions; but both, and particularly the former, are known to be of decided importance. Movement stops at night when the body temperature falls below the threshold for torpor, and begins in the morning when it rises above that level. A further rise in temperature will send the hoppers into greater activity, which results in such a scattering that the surface covered by a band in full daytime activity is three times that occupied by the same numbers at night. Comparison with the observations of Boyer and Buchsbaum, which have been summarized in chapter iv, in connection with slumber aggregations of insects, suggests that the temperature threshold of activity will be found to be lower on sunny than on dull days. On cool days the hoppers have been observed to keep their nighttime aggregation throughout the day. Uvarov (1928), from whose book much of the present account is summarized, cites observations which indicate that the temperature threshold for activity is higher in older nymphs.

With still further increase in temperature the bands start on their irregular and apparently aimless wandering. Some become restless and make small irregular jumps. These seem to initiate jumping on the part of others which at first is aimless, but which at length settles into a definite direction. Uvarov considers favorably the suggestion by Loeb (1918), based upon the work of Lyon (1904) and others, that there is a tendency of an animal to move so as to stop the movement of images of surrounding objects on the retina. One hopper jumping thus starts others seeing it to jump in such a way that there will be no movement across the retina. In a band this is taken up continually and passes along as a sort of automatic restimulation. Further, Grassé (1923) has shown that even non-gregarious grasshoppers give greater activity when several individuals are experimented upon together than when one is taken singly.

If some such explanation holds, we have the common direction of the movement of the band determined by chance or by environmental factors. La Baume (1918) thinks that the bands of the Moroccan locust move downhill because of a positive geotropism. Most of the obvious environmental factors, such as direction of the sun's rays, direction of wind, location of lush vegetation, do not seem to be definitely related to the direction of movement.

The appearance of one of these bands on the march is shown in the accompanying figure from Uvarov, from which it may be seen that the line of movement tends to be broad and shallow and with an irregular front (Fig. 31). When two such wandering bands meet, they usually fuse and go off in the direction formerly taken by the larger band. Such migrating bands recognize no obstacles other

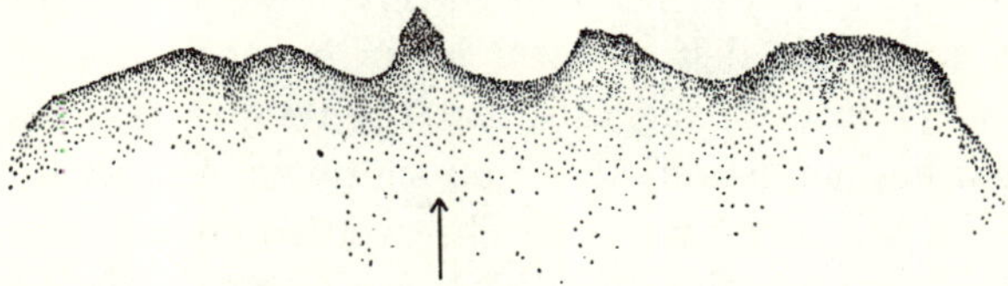

FIG. 31.—A band of locust hoppers on the march. (From Uvarov 1928, by permission of the Imperial Bureau of Entomology.)

than smooth vertical walls, and these merely cause a deflection. Inequalities of the surface are filled by the bodies of the first comers, and those following pass over the smoothed surface. Rivers are crossed by swimming with the same hopping motions that carry the insects along on land. The bands stop at noon if the heat becomes sufficient to produce heat torpor, and at night when cold torpor sets in.

We have already seen that the beginning of the wandering, as well as other movements of these hoppers, is largely determined by the temperature. We shall see later that, according to the phase theory of Uvarov, living together in dense bands tends to cause an alteration in the coloration, so that black pigment develops. Buxton (1924) reported that the body temperatures of a black form of *Calliptamus coelesyriensis*, a grasshopper of Palestine, was from 4° to 5° C. warmer than buff individuals of the same species under the same conditions. If Uvarov's theory is correct, it would appear that the collection in bands produces a coloration which increases the internal heat by increased absorption of the sun's rays, and that this in turn increases activity of the animals and is responsible, at least in part, for a greater tendency to wander. Apparently, the behavior and the coloration may be found to be inextricably intermixed with various other physiological processes and their morphological expressions, which have as yet escaped analysis.

The transition from a wingless band of hoppers to a winged swarm is gradual. The first winged forms to appear continue to move as hoppers with the wingless nymphs. Later, when more are molting to the adult form, the behavior of the band is modified; they rest much and are easily disturbed. After some days, isolated winged forms take off, fly in a circle, and again settle. When they pass over other winged individuals, these too may take to the air, perhaps as a response to air vibrations, as suggested by Vayssière (1921), since blinded locusts will respond, although those with eyes intact fail to do so if they are inclosed in glass (De Lepiney, 1928).

Faure (1923) gives almost the same account for the brown locust, *Locustana pardalina*, of South Africa. He reports that the bands of hoppers may be composed of three or four distinct nymphal stages, although they are more usually of the same stage. The members of a band do not all molt to form flying insects on the same day, but the winged males and females remain with the main band, probably until there are enough winged individuals to make a separate swarm. These precocious flyers camp with the main band at night even though they have ranged widely through the day.

Very large flying swarms travel for hundreds of miles. Small ones tend to remain near where they became winged. The distance covered is greater during the first few weeks of adult life. Later, as the females become heavy with eggs, the swarm tends to break up into sections. At night these flying swarms collect in clusters which are not so dense as those of the hopper bands. In South Africa brown locust swarms may fly on moonlit nights, particularly if harassed by birds. It is noteworthy that compact swarms leave large deposits of eggs behind.

Ordinarily, these locusts which Faure describes feed upon sweet grasses; but if food is scarce, they will eat almost anything. They readily become cannibalistic, eating injured members of the swarm. Mating does not begin until a day or two after the insects become winged; it continues at intervals thereafter until death. Nymphal aggregations are evidently not due to sex attraction; nor is the aggregation of the newly emerged winged adults, either with each other or with the nymphs which have not yet molted. Mating takes place immediately following egg-laying during the daytime; hence overnight aggregations, even of the mature adults, are not primarily due to mating reactions.

Uvarov cites cases which demonstrate that the migration of the adult swarms is not related to flood supply, since they will leave dense stands of vegetation upon which they normally feed and migrate out into arid regions. Neither is it due to a search for suitable nesting sites, for they will leave the regular nesting grounds and deposit their eggs wherever the physiological urge becomes sufficiently strong, regardless of the fact that the place may be entirely unsuitable for the development of the eggs. Further, he does not believe that the migration is a negative reaction to high parasitization, since the heavily parasitized individuals do not migrate, and since the others carry along with them their destructive red-mite and fly-larva parasites. Swarms have been known to stop and deposit their eggs on a barren hillside, where their eggs will develop poorly, if at all, and within sight of dense growths of one of their principal food plants growing in the type of habitat where their eggs would develop well. Uvarov believes that the emigration flight is both induced and regulated mainly by internal physiological factors.

The non-gregarious grasshoppers are solitary, not in the sense that there is but one or, at most, a few in a considerable area, but in the sense that for some unknown reason these Acrididae lack the tendencies which lead to mass movements. Their individual reactions to different environmental stimuli seem approximately like those of the individuals from the gregarious locusts, with the exception that their reactions are not so closely dependent upon those given by nearby grasshoppers. In behavior there is no hard and fast line that can be drawn between the two types, and in matters of form and color it appears that they also intergrade to a considerable degree.

This lengthy introduction to Uvarov's theory of phases of locusts is needed in order to have a proper background to understand the relations considered in that theory. To this, another known relationship should be added. After a period when non-gregarious grasshoppers have been no more than the usual agricultural pest, taking a relatively light toll of the available plants, a locust outbreak may occur either suddenly or after building up for a year or so; and this may be so serious as to present a great agricultural problem for an entire district. Such an outbreak may disappear as suddenly as it appeared, leaving behind only the normal population of grasshoppers. These outbreaks do not appear to have any definite periodicity. They are probably conditioned by a favorable combination of climatic and biotic factors. Weese (1924), working with spiders and their parasites, found that the climatic conditions which favored the development and survival of the hosts differed from those which were most favorable for the development and low mortality of their parasites. Uvarov is probably right in saying that "neither climatic factors in themselves nor the activity of natural enemies can be regarded as sufficient for a satisfactory explanation of the rapid increase in numbers of locusts in their breeding grounds at the beginning of an outbreak"; but he probably underestimates the possibilities of the two factors working together to supplement and reinforce each other.

In order to account for locust outbreaks and their sudden subsidence, Uvarov has put forth his theory of locust phases (1921,1928). The essence of this theory is: "Various species of gregarious locusts cannot be considered absolutely stable in all their characters, either morphological and (*sic*) biological; on the contrary, there are good reasons for regarding each species as exceedingly plastic and liable to fluctuations in all essential characters. These fluctuations have, of course, certain limitations, but in some cases the bounds are so wide that the extreme forms have been recognized as distinct species." These diverse forms of apparently related stock Uvarov calls "phases."

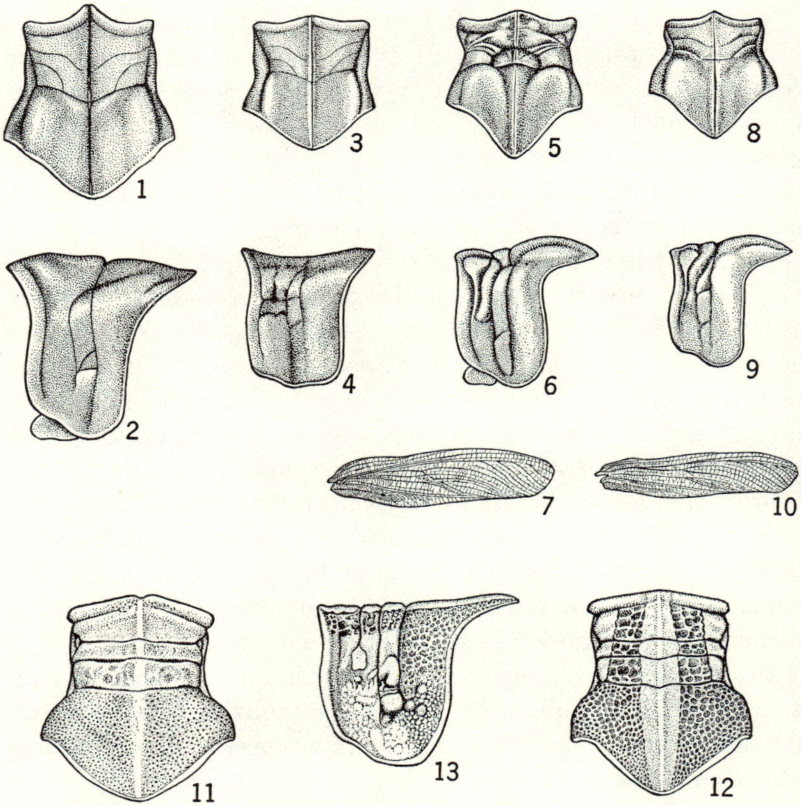

FIG. 32.—Showing morphological differences between phases of different species of acridid grasshoppers or locusts. 1, 2, Pronotum of *danica* phase of *Locusta migratoria* L. 3, 4, Ditto of *migratoria* phase of the same. 5, 6, 7, Pronotum and wing of *pardalina* phase of *L. pardalina* Walk. 8, 9, 10, Ditto of *solitaria* phase of the same. 11, Pronotum of *gregaria* phase of *Schistocerca gregaria* Forsk. 12, 13, Ditto of *flaviventris* phase of the same. (After Uvarov 1928, by permission of the Imperial Bureau of Entomology.)

Uvarov was led to this theory of phases by a consideration of the interrelations between the two supposedly good species, *Locusta migratoria* and *L. danica*. In his early work, on account of distinct differences in morphology, color, and ecological relations, he concluded that these are distinct species, as many others have regarded them. In hunting for delimiting structural characters, two were

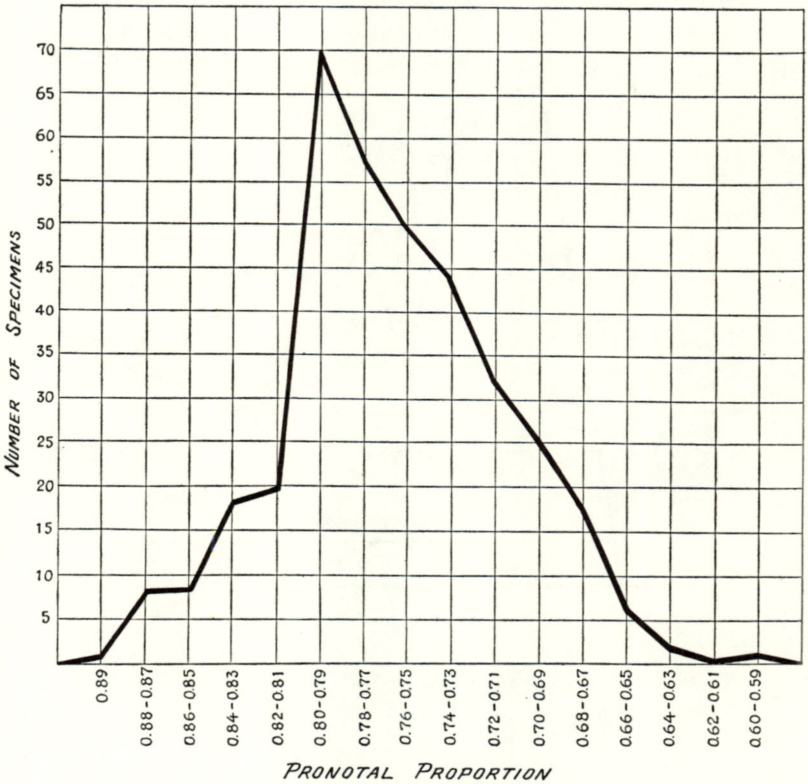

FIG. 33.—Graph of the range in variation in the pronotal proportion in 358 specimens of *Locusta migratoria* L. (After Uvarov 1928, by permission of the Imperial Bureau of Entomology.)

found—the proportions and shape of the pronotum and the relation between the length of the elytra and the hind femora. The pronotum differences of these two forms are shown in Figure 32. The differences may be summarized by saying that in *L. migratoria* (3, 4) the pronotum is relatively shorter and broader than in *L. danica* (1, 2) and has a low median keel. There is a more definite constriction in the middle of the pronotum, and the keel is straighter in profile. *Locusta danica* has the pronotum relatively longer and more compressed laterally; the midconstriction is less pronounced; the median keel is higher and is convex in outline. The elytra of the former are relatively longer and the hind femora relatively shorter than in

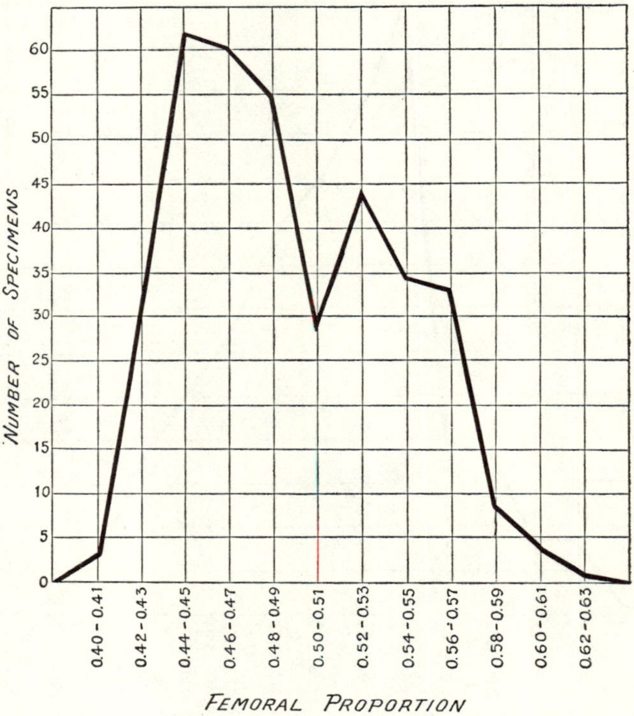

FIG. 34.—Graph showing the range of variation of the femoral proportion in 358 specimens of *Locusta migratoria* L. (After Uvarov 1928, by permission of the Imperial Bureau of Entomology.)

L. danica. These differences are summarized in numerical values in Figures 33 and 34. A study of Figure 33 shows that, so far as the pronotum characters are concerned, there is no evidence to indicate that here are two species, or even two forms of one species, but that with the femoral proportions there is a distinct difference in the grasshoppers shown by the two-humped curve. The hump to the right represents the *L. danica* type. Proportions taken for many individuals of both types show that there is no interval between figures for *L. migratoria* and for *L. danica.*

In *L. migratoria* both sexes are about the same size; the males are about 4 per cent smaller than the females. On the other hand, *L. danica* males are about 20 per cent smaller than females of that species. In both, the proportions between different parts of the body remain constant, independent of the absolute measurements.

The coloration of adults of the two forms is variable both in color and in pattern, so that they cannot be separated accurately on color characters. In general, *L. migratoria* is less variable in color than is *L. danica;* the color markings tend to be less sharp, and the general coloration tends to be more uniform. The hind tibiae are usually yellowish, though occasionally they are red. *Locusta danica* is more variable in adult coloration: bright-green forms are common; dark-brown and black forms are frequent; and the pattern is usually distinct, even if variable. The hind tibiae are frequently red, but this cannot be taken as absolutely diagnostic. During the breeding season males of *L. migratoria* become a bright yellowish, while males of *L. danica* show no color change when adult.

The situation regarding coloration of the hoppers is different. *Locusta danica* nymphs may be uniformly green, fawn, gray, brown, or even black. "Quite the opposite is the case with *migratoria,* in which each larval stage exhibits its constant color characters. Their coloration presents a combination of black and orange-red (or yellow), the earliest stages being almost entirely black, while orange or yellow appears later and extends with each molt without, however, entirely replacing the black." While there is variation even in this form, Uvarov says: "The main point is that this type of coloration never occurs in hoppers of *danica* in spite of the wide range of coloration in the latter." Somewhat similar conditions are exhibited in Figure 35 for the solitary and swarming phases of the South African locust, *Locustana pardalina,* and for the desert locusts, which have previously been known respectively as *Schistocerca flaviventris* and *S. gregaria.*

In *L. pardalina,* of South Africa, there are approximately the same differences between the swarming and solitary phases as have

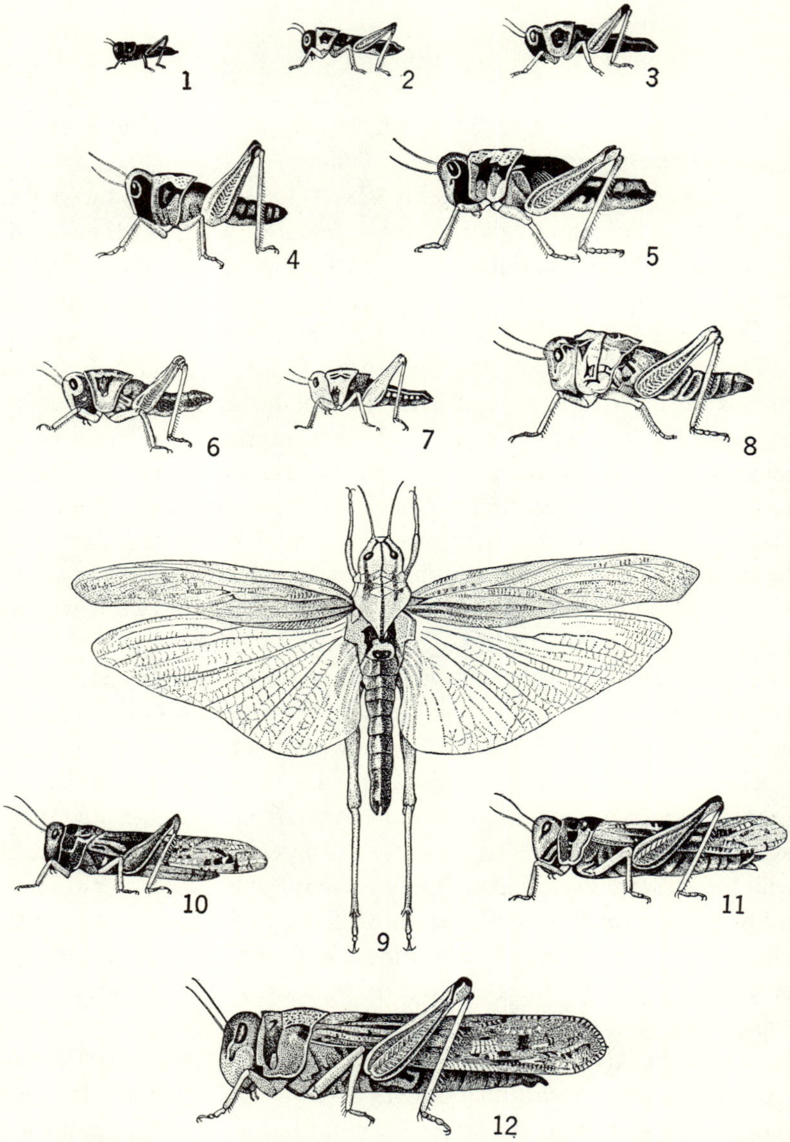

FIG. 35.—A black-and-white copy of a color plate by Faure (1923) showing solitary and swarm phases of the brown locust (*Locustana pardalina*). Black in this plate represents black or bluish black in the animals; heavy stippling represents dark brown; light stippling represents light or golden brown except on the head, prothrox, and metathoracic femora in Nos. 7 and 9 which are green.

The numbered drawings are: 1, 2, 3, 4, 5, swarm phase, first, second, third, fourth, and fifth stages of nymphs (or hoppers) respectively; 6, 7, 8, solitary phase, fourth, fourth, and fifth stages of nymphs (or hoppers) respectively; 9, adult female, solitary phase; 10, adult male, solitary phase; 11, adult female, solitary phase; 12 adult female, swarm phase, representing coloration of the young flier.

just been described for *Locusta migratoria* and *L. dancia.* In addition Uvarov, using specimens sent him by Faure, discovered differences in wing venation. With both *Locustana* and the *Locusta* we have been discussing, these characters are subject to great variation, and intermediate forms occur. Adults of the desert locust *Schistocerca flaviventris* have a distinctly higher crest of the pronotum than those of its supposed gregarious phase, usually known as *S. gregaria.* The coloration of the hoppers of the two phases for all three series is suggested in Figure 35 and may be compared with the description for the *Locusta* forms given above. The solitary nymphs are like *Locusta danica*, while the gregarious nymphs might be mistaken for *L. migratoria*, since they are of the same color and pattern.

The evidence that these phases of particular species may be transformed from one to the other does not as yet appear entirely conclusive. It does strongly suggest that such transformations may take place, and for that reason this relatively large amount of space has been devoted to the consideration of these locust phases. Uvarov observed in 1912 in the northern Caucasus that a swarm composed entirely of *L. migratoria* deposited eggs in nature in positions that were marked by those engaged in locust control work. The following year the hoppers were mainly of the parental type, but there was a considerable admixture of *L. danica* nymphs which showed the typical behavior of non-gregarious hoppers. There were many nymphs intermediate between the two. Similar field observations were made in the two following years in a different locality.

Plotnikov (1924; and, *vide* Uvarov, 1915, 1927) has carried on rearing experiments of the *L. migratoria* and *L. danica* nymphs. The account of the most convincing of his experiments that have come to my attention is summarized here. On May 26, he took 80 larvae of the third nymphal stage, and with typical migratoid coloration, from an ordinary cage of 0.02 cu. m. volume and transferred them to an open-air ground cage which covered 4 sq. m. of surface and was 60 cm. high. An equal number of similar larvae were left in the original cage for controls. After a time the larvae in the ground cage began to turn green, which Plotnikov regards as an intermediate color between migratoid and danicoid phases. They attained the fifth nymphal stage by June 11, when they had become quite green with the exception of 6 larvae, which were dark gray, almost black. In the original cage all the control larvae retained their typical migratoid coloration. The same experiment was repeated with danicoid nymphs, and only green forms appeared in the ground cage.

Plotnikov performed 15 experiments with larvae of the second brood of *L. danica* reared under conditions of crowding. When the larvae were kept in small cages or in glass jars, with 30—50 nymphs to 450–675 cc. space, a typical migratoid coloration was invariably obtained, but no dark specimens were to be found. When the nymphs were kept singly in the glass jars, they began turning green as early as the second stage. They were quite green by the fourth or fifth instar. When groups of 4 were put into 100 cc. of space in glass jars the fifth-instar nymphs were a mixture of migratoid and green, with some transitional forms.

In these experiments, in which the developing nymphs retain migratoid coloration only when they are crowded, Plotnikov reports that the nymphs having the migratoid color also lack the keeled pronotum typical of the *L. danica* nymphs, so that their structure as well as their color is affected by crowding.

In another series of experiments with non-migratoid nymphs, 86 second and third instar nymphs were placed in a 2,000 cc. cage for 12 days, at the end of which time they were in the fourth and fifth instars. Forty-five of these had definite migratoid coloration, while 9 more showed tendencies in that direction. Again, when 44 non-migratoid nymphs were put into a 1,000 cc. cage, a few typical migratoid animals were obtained. One regrets that there is no record of controls for these experiments reared under conditions such that crowding would be impossible.

Plotnikov considered the question of factors causing the migratoid color to appear in crowded cages. He eliminated cannibalism as a causal factor by feeding the isolated migratoid nymphs with killed hoppers. Under these conditions the isolated animals lost their typical migratoid coloration. He considers, too, that he has eliminated the humidity factor, since broods in highly humid jars gave the same results as those in an open outdoor cage with lower humidity. The same results were obtained in the hot midsummer as in the cooler autumn. This eliminated to some degree the factors of season, illumination, and temperature. Further, crowding experiments carried on in darkness produced migratoid coloration just as did rearing nymphs in open-air cages. Uvarov, in commenting on these experiments, thinks that the results are due to influences, as yet unknown, connected with the density of the larvae but not concerned with the density of the eggs or of the adults. One wonders whether the question of nematode infestation and its effects has been carefully checked in this work. Plotnikov (1927; *vide* Uvarov, 1928) has done work that confirms his earlier results that rearing under crowded or isolated conditions causes the changes above recorded; but he now regards these as aberrations, not as true transformations. The intermediate forms found in nature he interprets as hybrids between the two distinct races. These so-called "hybrids" do not differ in appearance from intermediate forms produced by cultural methods. In this connection we should record that Faure reports having seen a small, solitary-phase male of *L. pardalina* copulating with a large, swarm-phase female.

Faure (1923), working on the South African locust *Locustana pardalina*, observed transitions from the swarming phase to the solitary phase. Eggs collected from a field, where they had "in all probability" been deposited by a swarm, hatched out the usual black of the swarming nymphs of the first instar. If a number of these, some 60 or more, were kept together, swarm-phase colors held during the succeeding instars. If small numbers, not more than 10 or 12, were kept together in a cage, the color changed to that common for the solitary-phase nymphs after the first molt, and remained so during subsequent molts. The opposite change, from solitary to swarm phase, came on more gradually with crowding. Although without definite evidence, Faure believes, as a result of field and laboratory observations, that about 300 in a group will produce swarm-phase characteristics. Faure reconstructs the appearance of these two phases in nature as follows: After an epidemic of swarm-phase locusts, great numbers are killed off by man or other enemies, or by changed environmental conditions or overcrowding, and the species persists in the solitary phase. When conditions are again favorable, as at the end of a severe drought, these scattered grasshoppers multiply rapidly. As numbers permit, they gather into loose swarms, which deposit eggs in compact lots. From these compact deposits the swarming type of nymphs arises.

Another link in the chain of evidence connecting the two phases is furnished by Johnson (1926), working with the desert locust *Schistocerca gregaria* and *S. flaviventris*. Uvarov (1923), as a result of his studies on the phases of *Locusta migratoria*, suggested, from his inspection of museum specimens of these two supposedly good species of desert locust, that *Schistocera gregaria* was the swarm-phase and *S. flaviventris* the solitary phase of one and the same species. Field observations recorded by Johnson suggested that this was indeed the case, and these observations were followed by breeding experiments which gave the same results as have similar tests with other species, since the experimenter was able to control the appearance of either phase by regulating the density of the crowding. A part of this test was made in nature and on a large scale. Bands of hoppers thinned out by poison turned into the solitary phase so far as coloration and behavior were concerned. There is also preliminary evidence (Dampf, 1925, 1926) that the South American locust *S. paranensis* can be turned into the solitary phase usually known as *S. americana* by controlling the density of the population.

The phase theory of Uvarov remains to be tested even in a preliminary manner upon two other species, which, with the four discussed, make up the larger swarming locusts of the world. These are the African red locust *Nomadacris septemfasciata*, of which two types are known, differing in pronotal characters; and *Patangia succincta* of India, which apparently rarely swarms. Of the smaller swarming locusts, there is a suggestion that the Moroccan locust, *Dociostaurus maroccanus*, has a solitary phase which has been described as a pigmy race. Finally, there is the case of the Rocky Mountain locust, *Melanoplus spretus*, now apparently extinct in its typical form. In 1878, the United States Entomological Commission, composed of Riley, Packard, and Thomas, in their first annual report, record that they considered and rejected the possibility that the offspring from the breeding swarms of *M. spretus* change in the direction of a morphologically related species, *M. atlanis*, which differs from *M. spretus* by being less gregarious in its habits. The nymphs of the latter have coloration suggesting that of other typical swarming-hoppers, and the pronotum of the adult is proportionately shorter than that of *M. atlanis*. Somes (1914) questioned on morphological grounds the validity of the separation of the two species; and Parker (1925) and Hebard (1925) have suggested that the two are phases of the same species. This suggestion offers an interesting possibility of putting the whole phase theory to the test of critical experimentation without the necessity of making an expensive, time-consuming journey.

In final criticism of the phase theory, it does not seem to be clearly proved that the transformations from one phase to another do actually take place on a large scale and to a convincing degree. But as one reads through the descriptions of the different workers from different parts of the world and finds that independent students have thought that they have obtained these transformations, and when one learns that the same suggestion had been considered in 1878, long before Uvarov first stated his phase theory, it becomes impossible to dismiss the evidence entirely, although for the time being it must rest with a verdict of unproved but a promising opening for further work.

DROSOPHILA CULTURE EXPERIMENTS

The work of the *Drosophila* students probably presents the greatest mass of carefully controlled work upon the culture of a single animal species yet performed. Unfortunately, certain phases of the environment within the culture bottles cannot be controlled, but on the whole these workers have succeeded in creating standard conditions for their breeding experiments. For this reason one may accept their results without the mental reservations just indicated in the case of work on the transformation of locusts from one phase to another under laboratory conditions.

Bridges (1921) called attention to the necessity of having optimum environmental conditions in order to eliminate distortion in ratios in experimental work with *Drosophila*, and lists the effect of overcrowding as one of three main sources of disturbance. In the early *Drosophila* work, he says, this was the largest source of difficulty.

Eigenbrodt (1925) presents evidence that overcrowding the larvae of a homozygous race of bar-eyed *Drosophila* decreases the weight attained by the adult flies and decreases the size of the whole animal and of such individual parts as the thoracic length and length of wing. The number of eye facets, of hairs on or around the eye, and the number of teeth in the sex combs (found in the male only), are also decreased. Flies reared under overcrowded conditions do not exhibit the correlation of size with the temperature at which they are reared that is characteristic of normal flies. The rate of development is retarded by the overcrowding; variability tends to be increased, and sexual dimorphisms tend to disappear. In order to obtain standard results, Eigenbrodt concludes that *Drosophila* should never be reared under overcrowded conditions. He finds that the flies are normal in the foregoing relations if from 8 to 20 hatch out in an 8-dram vial which contains 9 grams of standard food. Under these conditions a higher population than 20 flies represents overcrowding.

Plunkett (1926), in his attack on the problem of how genes produce their effects, undertook to study experimentally the effects of various combinations of genetic and environmental factors upon the development of *Drosophila* bristles. These have the advantage for this sort of work of being discrete units that can be counted with a minimum of observational error. Their number is known to be affected by different genes, and also by such environmental factors as temperature and nutrition. The bristles have been plotted and named for the normal wild-type *Drosophila* and for various other stocks. Plunkett, in his experiments, used a mutant stock known as Dichaete, selected for low bristle number for many generations, and took care to insure uniformity in the different culture bottles, except for the factors under observation.

Under these conditions it was found that the number of flies developing in a bottle, other conditions being equal, has a pronounced effect on the mean bristle number. This effect is summarized in Table XXXVII for populations from eggs laid during 4 days for the first 7 groups, and during 8 days for the last 5 groups. The temperature was held at 25° C. throughout.

In commenting on these results, Plunkett says: "It is evident from the table that, under these conditions, there is no correlation between bristle number and density of population up to about forty

TABLE XXXVII

SHOWING DATA FOR THE EFFECTS OF DENSITY OF POPULATION UPON BRISTLE NUMBER IN DICHAETE *Drosophila*

(After Plunkett)

Flies per Bottle	Mean Number of Posterior Dorsocentral Bristles per Half-Fly
11.3	0.234±0.030
24.2	0.284±0.017
33.4	0.284±0.012
43.5	0.183±0.009
55.0	0.136±0.011
65.7	0.120±0.011
76.0	0.056±0.009
95.5	0.055±0.008
103.5	0.051±0.007
148.0	0.027±0.006
226.0	0.022±0.005
351.0	0.017±0.003

flies per bottle; but above this the bristle number falls off rapidly with increasing population, reaching almost zero (for these bristles) when the population is much in excess of 100 flies per bottle, as in ordinary 'stock' bottles. This 'crowding effect' makes it unsafe to draw quantitative conclusions as to the effects of other genetic factors or other environmental factors (e.g., temperature) on bristle numbers in flies raised under these conditions; i.e., more than forty offspring per bottle from eggs laid over a period of several days." In later experiments, when parents were kept in the bottles for not over 24 hours, there were no obvious effects of crowding up to about 80 or 100 offspring flies per bottle.

Plunkett continues: "Experiments designed to analyze the factors responsible for these 'crowding' and 'age of culture' effects, indicate that they are due largely, perhaps entirely, to competition for food. This factor seems to affect especially the younger larvae when in competition with older ones."

When virgin F_1 females from a cross of a wild type by vestigial winged *Drosophila* were backcrossed to vestigial males, an equal number of eggs destined to produce hybrid and homozygous vestigial flies was laid. Under conditions of overcrowding, Harnly (1929) found that the proportion of vestigial to the wild-type hybrid flies was reduced, thus demonstrating a pronounced selective effect of overcrowding. Similar results were obtained whether crowding was produced by using a reduced surface area of food or by increasing the length of the egg-laying period with the food area (and depth) remaining constant. Clausen (1924) reported a similarly low survival value for vestigial flies reared under crowded conditions.

Evidence from these students of genetics concerning the effectiveness of an environmental factor is the more trustworthy since they do not have the reputation of being easily convinced of the effectiveness of environmental factors upon morphological characters.

CONCLUSION

CHAPTER XIX

ANIMAL AGGREGATIONS AND SOCIAL LIFE

We have seen something of the kinds and the extent of the aggregation phenomena in nature and in the laboratory, among members of the animal kingdom and among specialized cells such as bacteria and spermatozoa, as well as cells under the artificial conditions that obtain in tissue cultures. We have reviewed the action of the direct and indirect environmental forces which control the formation of such aggregations. We have found that methods and different degrees of integration exist, which may serve to organize a group closely, although it lies well below the division-of-labor level of social life.

The well-known harmful effects of crowding have been summarized, as has the newer evidence that, despite the menace of overcrowding, many aggregations have survival values for their members and may even produce morphological changes as well as the more easily induced physiological effects. A further discussion of the extent and implication of the co-operation involved in these loosely integrated groupings is reserved for the final chapter. It is sufficient to state here that aggregations formed without sexual stimuli and at the lowest level of group integration may have survival value for their members and, under certain conditions, are essential for the survival of the race.

Although the matter has not been discussed, it has been the open as well as the implied suggestion throughout that these aggregations are important in social evolution. This question must now be faced directly. One avenue of approach to such a consideration might lie through an extensive review of the evidence concerning the origin of the social habit, but we shall limit this aspect of the discussion by citing summaries of generally accepted points of view and testing these against some of the known facts.

We have already suggested that sex may have evolved from the mutual stimulation which has been demonstrated to occur under certain conditions when two similar asexual cells are crowded together within a limited volume of medium. If this suggestion is sound, it would indicate that mass physiology of animals is much more primitive and fundamental than can be considered to be the case under the assumption that the gregarious or social habit in animals is at bottom an outgrowth only of the association of young individuals with one or both parents, and therefore usually a result of sexual reproduction. It is assumed that in special cases or at critical periods in social evolution the period of the association becomes lengthened and the family comes to react as a unit under many conditions. Students of social life in insects, especially as it exists among wasps, bees, and ants, usually adopt this explanation in some form for the origin of the social habit. Wheeler, in his summaries of studies on ants (1913*a*, 1918), and more recently in his review of social life in insects (1923, 1928), regards the insect colony as the result of an extension of the affiliation of mother and offspring. Wheeler's particular contribution is his theory that mutual feeding, which he calls "trophallaxis," is the bond that unites parent and offspring in the social insects; the mother feeds on secretions from the larvae and so is bound to them through self-interests, while they receive food from the mother to their own advantage. Wheeler shows that the social habit, meaning thereby a more or less prolonged association of young and adults, has arisen *de novo* at least thirty different times among insects alone, in nearly that many natural taxonomic families or subfamilies, belonging to five different orders. The gradual development of the mother-offspring family from the solitary insects is shown almost diagrammatically by the growth of the social habit among the solitary wasps (Wheeler, 1923).

Herbert Spencer's suggestion that colony life arose from the consociation of adult individuals for nonsexual, co-operative purposes was an early recognition of that type of social unit and, at the time it was advanced (1893–94), was not well grounded on proved fact. Spencer suggests that in some cases permanent swarms arise from such consociation and that natural selection establishes such of these groupings as are advantageous. In terms of human society, this view would stress the importance of the gang, rather than that of the family, as a preliminary step in the evolution of the social habit. It is important to note that the gang cuts across family lines in its formation, just as do the sleeping collections of male robins. which occur in our parks and orchards during the breeding season. Unfortunately for Spencer's principle, which may be correct in many instances, he limited his theory by a concrete example in ants, where we now know that social development is probably due to an extension of mother-offspring relations; but the easy dismissal of his illustration does not necessarily wreck the underlying principle.

Wheeler (1913*a*) expressed the usual attitude toward these consociations in relation to social origins when, after describing some cases of aggregations in ants, he dismissed them as entirely fortuitous instances, which would occur wherever ants might be abundant and places of refuge scanty; or as the manifestation of highly developed social proclivities, and not of such proclivities in the process of development. More recently (1930) he said: "Societies really represent very different emergent levels from the associations and have arisen in a different way, though, of course, ancient aggregative or associative proclivities may have been retained by many species and may serve to reinforce their specifically social behavior." Highly developed social life demands well-developed sense organs, central nervous system and muscular apparatus, and in addition to these, according to Wheeler, there must be a development of the family. "All the societies of insects," he says, "are merely single families in origin. . . . . The family origin of the flocks and herds of birds and mammals and hordes and tribes of primitive man is also apparent; though in these societies the family is open and not closed as in insects, and there is a retention in the flocks, herds and hordes of primitive aggregative or associative tendencies which seem to hark back to the ancestral fish and tadpole stages."

This attitude may be entirely correct so far as the highly integrated societies are concerned; but even in these closely knit organizations there is evidence, as has just been recognized, that the numerous aggregations, an account of whose formation, integration, and physiological effects have made up the body of the present discussion, have social significance. In more loosely organized social life we would expect the cohesiveness shown by animals at the aggregation level to make up a greater part of the total unifying forces which operate to produce a social unit. Further, we must recognize that at least a part of the forces which operate to bind members of families together are similar to those which operate in the case of other kinds of social groupings.

It is worth re-emphasizing, as Child (1924) and Alverdes (1927) have recognized, that both the congregation and the family bases of societies are in fact but two different types of aggregations, so that at all events aggregations of some sort are essential for the development of the social habit. In other words, this phase of the problem of social origins is not the question whether the social habits as seen among birds, ants, men, and others arose from aggregations such as we have been discussing in earlier sections of this book or from some sort of family; but rather it becomes a question as to the kind of aggregation which gave immediate rise to them, since the family type is only one of a number of kinds of aggregations, as we have seen from Deegener's outline.

The main value of the detailed classification of social organization given by Deegener lies in his recognition and elaboration of the essential unity of sex-conditioned and asexually conditioned social tendencies. Throughout his two main categories of loose accidental unions, or associations, and essential groupings, or societies, Deegener has recognized important subdivisions, differentiated according to whether the groups are formed on a sexual basis or prolongation of some sort of family relations, or whether by the gathering of individuals from different parents into more or less well-integrated groups. The former type, while common enough in homotypic groupings, does not compose all such groups; while the latter is most characteristic, though not necessarily exclusively so, of social bands or flocks which contain more than one species.

In order to appreciate the importance of social organizations which are not fundamentally united on a sexual basis, one should review Deegener's classification to find the extent to which he recognizes these asexually conditioned groupings. Among the homotypic categories alone are to be found such aggregations as the following: all collections in a favorable locality of limited extent; hibernating groups; animals collected about food; individuals joined in migrating bands; aggregations brought together by tropistic reactions which lead the responding organisms into a limited space; collections due to unfavorable conditions, whether passive, as in drift lines of beetles formed by wind action, or active, due to the moving together of stimulated individuals. From heterotypic categories one finds such social mutualism as exists between flocks of cowbirds and herds of cattle; one animal living upon the shell or covering of another without being parasitic; two different species occupying the same runways, even though these runways are made by only one of the species involved; the well-known relationship between ants and aphids, where the former feed on excretions of the latter and in turn afford them some protection; the same sort of relationship where narcotic material is supplied rather than food; the relationship between so-called "robber guests" and their hosts; or that of harmless guests which feed upon fragments dropped by a larger or more active species; or of the animals which make their small nests in the large nest of a larger individual; or of the small individuals which remain in the neighborhood of a larger one without being attacked, and thus avoid attacks from others; or the cases of animals cemented into the built-up covering of another, as caddis-fly larvae use small mussels; or those which live within the body of other animals without becoming parasitic; and, finally, the different types of parasitism.

All the foregoing list must be dismissed, wholly or in part, as falling outside the range of social phenomena, if the latter is to be regarded as limited solely to those relations which depend upon some sort of a sexual or family basis. While there are many who would be willing to dismiss a part of these as falling without the field of the social life of animals, I know of no student of social life who would dismiss them all. From such considerations as these we are drawn to the conclusion that, important as sex and the family are as integrating social factors, they do not form the sole outlet for the expression of the fundamental social appetites, nor are they the only foundation upon which social structures have arisen.

Observations on bird behavior (Allee, 1923*a;* Sherman, 1924) furnish interesting information concerning the problem of the extent to which social groups originate through individual, and to what extent through family, behavior. The question at hand is, Do these annually or semiannually recurring bird flocks form by the coming together of individuals or by the collections of some sort of family groups? The answer is that both methods occur. There is much evidence that ducks and geese migrate in flocks in which family units can be recognized (McAtee, 1924) and that in the tropical rain-forests parrakeet flocks are made up of pairs rather than of individuals. With the whistling swan and the Canada goose supposed family-groups have been identified in mid-winter (Miner, 1923). The large flocks of bronze grackles make a conspicuous feature of summer bird life in the Mississippi Valley. On July 3, M. Nice[1] reports seeing one of the birds of a flock beg from another as young ones do from their parents, and interprets this as evidence that a family joined the flock before the young were entirely independent.

On the other hand, heterotypic flocks are frequent in which a species may be represented by a single specimen. An extreme instance of such a flock is furnished by Beebe (1916), who records a flock of 28 birds composed of 23 different species. He comments upon the common occurrence of heterotypic flocks in British Guiana. Such extremely heterotypic groups could scarcely have been formed by family rather than by individual units.

Sherman (1924), known, like Nice, to be a careful observer of bird habits, gives much detailed information to show that not all flocking of birds is on a family basis. Proof of this is easily given by calling attention to the flocking habits of the cowbirds. The female cowbird deposits her egg in the nest of some other bird, usually smaller, and leaves it there to be cared for by the latter. This socially parasitized foster-mother frequently hatches and rears the young

[1] Personal communication.

cowbird. When able to fly, these young cowbirds, reared separately and by foster-parents of different species, join in the well-known cowbird flocks. Such flocks can form only by the collection of individuals.

Bobolinks and goldfinches begin the formation of their autumn flocks by the congregating of old males. Chimney swifts, a pre-eminently flocking species, leave their nests by one's and two's to join the immense late summer flock. Many birds rear more than one brood a season; and Nice writes that the only young of the first brood that she ever knew to stay with their parents throughout the raising of the second brood were one set of bluebirds.

Family ties apparently sit more lightly with birds than some would have us believe. Passerine birds may change mates for the second brood. Nice reviews the literature on banded birds and finds that in 7 pairs of 3 species there was a known shifting of mates in one season, as contrasted with 20 pairs of 11 species in which there was none.[1]

Nice writes further that she does not believe that song sparrows retain the family unit when flocking or for flocking, and she cites trapping experience with a robin to show that the female outstayed her mate and her three sets of offspring, and so evidently could not have joined in a postbreeding-season flock. Much similar evidence exists that with such relatively highly social individuals as birds, which are capable of forming well-integrated flocks that exhibit definite physiological division of labor and concerted group action, the flocking may occur by the congregation of individual birds, just as has been observed for the sleeping aggregations, where the male birds may come together to sleep even during the breeding season.

It has just been stated that these bird groups form well-integrated social units. Something of the intricacy of the social organization possible among them has been revealed by the work of Schjelderup-

[1] In this connection Miss Sherman contributes the following pertinent observation in a personal letter: "My chimney swifts led a perfectly upright life for ten seasons, but on the eleventh a shocking scandal occurred. An unprincipled female supplanted the mate that had helped build the nest and had begun to lay her eggs, which were thrown from the nest in which the interloper raised her brood."

Ebbe (1922) in his analysis of the intragroup relationships shown by a flock of domestic hens. Such a group forms not a closed but an open society; that is, new members are admitted as occasion demands. The flock is organized by what Schjelderup-Ebbe calls the "peck-right," or the "peck-order."

The rank of individuals within the group is indicated by their reaction when another member pecks or threatens to peck them. A given hen will submit to pecking by certain individuals without expressing resentment, and will in turn have the right to treat others similarly without their showing protest reactions. Hens with this power are said to have the "peck-right" over those submitting to the pecking. The "peck-order" decides which birds may peck others without being pecked in return. The ranking is determined by combat or by passive submission. A newcomer can win a position above the bottom of the peck-order only by fighting.

Sometimes the peck-order within the flock is in a simple continuous series, thus: A pecks B; B pecks C; C pecks D; and so on down to the humblest member of the flock. But it may happen that the peck-order is more confused. A may peck B; B may peck C; and still C may have the peck-right over A. Frequently, strong hens are pecked by weaker ones. This is due to the fact that young ones are attacked by older members of the flock, newcomers by old-timers, sick by healthy; and the order, once established, tends to remain permanent. A revolt or a fight may either change or confirm a previously existing peck-order. If the original order was accepted passively rather than as the result of a combat, a rebellion is more likely to occur. A hen revolting against a previously recognized superior fights less fiercely than at other times. This indicates a psychic obstacle to such an attack. Once a hen has accepted an inferior position, it is more difficult for her to return to superiority than if she had fought for the position at the start.

Position in the peck-order is associated with certain behavior traits. The hen which is entitled to peck all others is usually the least malicious, and a threatening note usually suffices for a peck. A hen low in the scale is usually cruel to the remaining hens. Katz and Toll (1923) tested the intelligence of hens and found that their position in the social scale corresponds roughly with their IQ.

During breeding time a hen is more easily irritated by other hens, and frequently against her superiors. A hen with chicks is very courageous; but if the chicks are removed, she may become timid and retiring in her behavior. Cocks are said by Schjelderup-Ebbe to behave in a manner similar to hens, but more ferociously. A cock may interfere with the fighting of two hens, or even of two other cocks, if he is superior to both. When among hens, he stands at the top of the peck-order.

This is not the picture of a simple society, though many of the complexities of the organization have only been suggested in the preceding summary. It may be objected that we have turned from the formation of flocks in nature to the examination of the integration of artificial flocks of domestic fowls. It does not necessarily follow that the details of flock organization are the same in nature, but fortunately there is evidence bearing exactly on this point. The same observer reports (1923) that in flocks of wild ducks essentially similar group organizations exist.

Another complex type of organization of a bird community in nature has been worked out in detail, particularly by Allen (1911, 1913) and by Howard (1920). This is the matter of territory in bird life. The evidence shows clearly that the males pre-empt fairly definite spaces before the breeding season begins, and maintain their position during the breeding season, driving off intruding males before and after a female has appeared to accept the territory and the male as her mate. Such spaced community organizations are apparently widespread among birds and again indicate clearly a distinct social development. These territorial relations are not limited to birds, but are also known for fishes (Reighard, 1920), as well as for mammals.

Among the fishes we have further evidence of the kind we have been reviewing for birds. This is well known in the case of the black catfish minnows observed by Bowen (1930). While it is true that the group of young originates from the eggs guarded and fertilized by a single male, yet, at least in the absence of the male, the young separate each night and come together one by one to form a group as the light increases the next day. Further, the group does not keep its original character of being made up of the young of one male; but the groups readily mix, either by the fusing of two separate groups or by the junction of individuals with groups with which they have not previously had acquaintance. In both cases, the group, once formed, is fully as well integrated as though all belonged to the original family, and shows a high degree of group protection, particularly through the multiplicity of eyes, which may detect danger, and the transmission of stimuli, which lead to quick separation and flight. The members of the group are safer than if they swam alone. This case is the more instructive since the group re-forms daily, and since, as Bowen has shown, the coming-together is largely conditioned by the possession of a weak social appetite combined with certain reciprocal reflexes.

These illustrations suggest the possibility of the formation of groups of decided social integration as the result of the congregation of individuals, without a family aggregation of any sort appearing in this particular type of organization.

Social organizations among birds include phenomena of leadership, of group integration, of division of labor, at least to the establishing of sentinels, of joint action in common defense by spreading an alarm and by joint attacks, so that we are warranted in ranking them as well-developed social groups. Even so, we repeat, they may arise from the congregation of individuals as well as by the coming-together of families. In so emphasizing the possible social development from the aggregations of individuals with which this book is mainly concerned, I have no intention of underestimating the other well-known method of social development by the extension of the family type of aggregation, which may be seen in process of development among the solitary wasps and which comes to a high state of perfection in the social wasps, bees, ants, and termites.

Students of human sociology are generally agreed that our complex modern social organization rests primarily on family groups (Thomas, 1909; Lowie, 1920; Malinowski, 1927). Miller in 1928 reviewed a part of the literature appearing just preceding that date dealing with the sex relations of non-human anthropoids and of man. Miller concludes that a precultural gregariousness, not sexually conditioned, but including sexual promiscuity, existed among early man and has influenced the course of development of human institutions. He emphasizes observations on the higher primates which indicate sexual promiscuity in horde life rather than family organization whether of the polygamous or monogamous type.

Such observations give point to the strongly intrenched system of human taboos and laws which are directed against promiscuous horde sex-relations, and bear out the theory that these taboos and laws are in themselves a universal, tacit recognition of the fact that such promiscuous tendencies are strong enough to be a menace to human society as now organized. The firm hold of the tendency toward promiscuity among modern men even under these taboos and laws is witnessed by the prosperous condition of prostitution, which from this point of view is a lusty survivor from primitive horde life rather than a recent development conditioned by modern economic or other social pressure. The high percentage of occurrence of gonorrhea and syphilis, diseases not excessively easy to acquire, also bears evidence that the sexual behavior of man, at least as shown under the prevailing Europeo-American civilization, consists in part of elements which cannot be distinguished from the activities which Miller thinks are characteristic of the ape horde.

While there are many indications that horde life stands in the remote human background, it appears that other anthropoids than man have made the transition to some sort of family life. Yerkes (1929), after commenting frequently on the incompleteness and uncertainty of our knowledge of the social traits of the great apes, concludes his detailed presentation of existing evidence as follows:

"Gregariousness and degree of dependence of the individual on the group tend from lemur to man to diminish, and at the same time to give place to a more definite and stable social unit, the family. There is a great diversity within the several types. Lemurs and their kind may live in bands or as mated pairs. Monkeys almost invariably constitute bands, as do also gibbons and siamang, but by contrast the anthropoid apes live either as families or in groups constituted by temporarily associated families. The transition from pronounced gregariousness to family life seems to occur between gibbon and chimpanzee.

"Leadership, the dominance of one individual, and variation in social value and influence in accordance with individual traits, apparently tend to become increasingly important from lemur to man. The transition is from the leader of the herd to the patriarchal head of the family.

"Likewise, sociability and social dependence, in certain at least, of their social aspects tend to increase from lemur to ape, but so does individualism; and whereas the chimpanzee is extremely sociable and dependent therefore upon its social environment, both orang-outan and gorilla are markedly less so than are certain monkeys and lemurs. Obviously knowledge does not permit of safe generalization. With respect to mutual aid and like expression of sympathetic interest there can be no doubt of marked increase: lemur, monkey, ape, man.

"Permanency of mating, although not definitely established for any infrahuman primate, is rendered increasingly probable from lemur, through monkey, to ape, by the nature and abundance of pertinent evidence. The same may be said of monogamy, for although it may exist in any of the four groups which we are comparing, so also, according to pertinent observations, may polygamy. Whether there is definite phylogenetic tendency toward the one or the other type of family it is impossible to say. But in any event the family as a social unit seems to become more prevalent and also more stable as we progress from lemur to man."

Men have other aspects of group life in common with these nonhuman anthropoids. Thus the monosexual human gangs and clubs seem to have their counterpart in the sleeping group of male gibbons which Spaeth (see Yerkes, 1929) found in the Siamese jungle. Such human and gibbon behavior appears to be an expression of the widespread phenomenon of the formation of monosexual groups which have been recorded notably among various other mammals (seals, for example), among birds, and among the so-called solitary bees and wasps. As in man, the young orang-outans congregate in gangs; and Yerkes records various observations of groups of chimpanzees engaging in play.

The high degree of group integration and the division of labor which may exist in social groups of monkeys are well illustrated by the following extract from a personal communication from J. F. V. Phillips, stationed in Tanganyika territory, East Africa. In writing about the social habits of baboons, he quotes from Sclater (1900), who says that these animals associate in groups of varying numbers up to about a hundred individuals; that, when moving from place to place, the old males are usually seen on the outskirts and always form a rear guard; and that, when resting, a sentinel or two is always placed on top of a rock to warn the troop of approaching danger.

Phillips comments: "This is entirely correct; the sentinel is exceedingly sharp and detects the least noise, scent, or appearance of man or leopard. In East Africa I have seen another species of baboon behaving in the same manner. The sentinels are often the largest, strongest males, that is, with the exception of the real leader of the group; they will remain faithfully at their post 'waughing' (the typical note of danger is 'waugh,' 'waugh,' very guttural and somewhat alarming) despite the proximity of danger. Upon these notes of warning reaching the ear of the leader, he will immediately assemble the leaders of the group, marshaling the males at the rear and along the sides, the females and the young at the forefront, or within the cordon of the males; he himself will alternately lead or bring up the rear, according to the plan of flight or the degree of danger. When things get too hot for the sentinels, they scamper off a short distance, mount some high position, and give further warning to the leader. In times of slaughter, the young are protected by the parents, often with great danger to the latter."

The comparison of the behavior of this baboon horde with certain aspects of human behavior of which we are justly proud is too marked to need emphasis. Like behavior is frequently exhibited in some degree by herds of various sorts of mammals. The group defense of eggs or young by birds or of the nest by social insects shows similar behavior elements. The baboon example is particularly valuable in that it dramatizes the possibilities of co-operation under group organization, even in non-human communities.

Restating the general argument of the present chapter, we may say that it seems quite possible that sex arose originally from the beneficial stimulation received as a result of the aggregation of two, or more, simple asexual organisms. Sex, once originated, became one of the integrating factors in further social development. In sex-conditioned society the offspring of one pair or of one parent may have remained in association with their parents for the immediate mutual benefit of all concerned, or there may have intervened a sexually promiscuous horde life from which the consociation of offspring with their individual parents arose as a further protective evolution. When these lengthened associations of parent and offspring continued long enough, a division-of-labor type of society could evolve. At first this would occur between parents and monomorphic offspring; later, dimorphic and even polymorphic offspring, as in ants and termites, might develop. However it arose, the family and the highly integrated division-of-labor society which may originate through it is only one type of expression of the fundamental tendency of animals to aggregate. There are other social phases of animal life which have developed on this same foundation of animal aggregations as a result of forces not centering about sex; these have produced social units of importance in animal life.

A part of the difficulties we have encountered in discussing the rôle of different types of animal aggregations in the evolution of social groupings may be avoided if we recognize that there are many levels of social organization and that these overlap. Among the groups which we may fairly call "social" there are: (1) those that show their social habit merely through the toleration of the close proximity of other similar individuals in the same restricted space—these may exist without any positive mutual attraction and may be called the toleration level; (2) those that form groups which react more or less definitely as units—the group integration level; (3) those which show physiological division of labor; finally (4), those that show morphologically distinct castes, each associated with some phase of the division of labor. The animals on the higher planes of social development continue to show the group attributes characteristic of the lower levels. Survival values have been demonstrated throughout this whole series and extend well below the toleration social level to the threshold of primitive life.

The first indication of structurally modified castes is to be found in the dimorphism accompanying sex, and the division of labor associated with sex runs not only through all the distinctly social levels but also through the majority of all types of animals as well. As we have suggested, sex itself, and therefore the divisions of labor associated with it, may have arisen as an outgrowth of certain benefits associated with primitive asexual aggregations. Sex phenomena aside, we have just seen that the different types of social organization, including even the physiological division of labor type, may have developed from aggregations formed by the coming-together of individuals without direct sexual causation as well as from those collections which have resulted from sexual appetite or from reproductive activities. The most permanent societies appear to have arisen when sexual and parental integration have operated in addition to the more elemental aggregation tendencies.

CHAPTER XX

THE PRINCIPLE OF CO-OPERATION

Espinas (1878) approached his task of organizing the materials available concerning animal societies with much the same point of view that we have developed from a behavioristic and ecological approach to the problem, the study of which has led to the present volume. Espinas says in his introduction to *Des sociétés animales*, "No living being is solitary. Animals, especially, sustain multiple relations with the organisms of their environment; and, without mentioning those that live in permanent intercourse with their kind, nearly all are driven by biological necessity to contract, even if only for a brief moment, an intimate union with some other member of their species. Even among organisms devoid of distinct and separate sexes, some traces of social life are manifested, both among animals that remain, like plants, attached to a common stock, and among the lowly beings which, before separating from the parental organism, remain for some time attached to it and incorporated in its substance. Communal life, therefore, is not an accidental fact in the animal kingdom; it does not arise here and there fortuitously and, as it were, capriciously; it is not, as is so often supposed, the privilege of certain isolated species in the zoölogical scale, such as the beavers, bees, and ants, but, on the contrary—and we believe we are in a position to prove this statement abundantly—a normal, constant, universal fact. From the lowest to the highest forms in the series, all animals are at some time in their lives immersed in some society; the social medium is the condition necessary to the conservation and renewal of life. This is, indeed, a biological law which it will be expedient to elucidate. Moreover, from the lowest to the highest stages in the series, we detect in the development of social habits a progression which, if not uniform, is at least constant, so that each social group carries the perfecting of these habits a little farther in one or another direction. Finally, social facts are subject to laws, and these are the same everywhere that such facts appear, so that they constitute a considerable and uniform domain in nature, a homogenous whole thoroughly integrated in all its parts."[1]

More recently Kropotkin, Deegener, Alverdes, and Wheeler have contributed to the development of this thesis. The latter, in *Social Life among Insects* (1923) expressed his point of view as follows: "All living things are genetically related as members of one great family, one vast living symplasm, which though fragmented into individuals in space, is nevertheless absolutely continuous in time. In the great majority of organic forms each generation arises from the co-operation of two individuals. Most animals and plants live in associations, herds, colonies or societies and even the so-called 'solitary' species are obligatory, more or less co-operative members of groups or associations of individuals of different species. Living beings not only struggle and compete with one another for food, mates and safety, but they also work together to insure to one another these same indispensable conditions for development and survival."

We are not concerned here, as some recent writers have been (Wheeler, 1911, 1930; Child, 1924), with renewing the Spencerian analogy between the living and the social "organism" however much we are impressed by the remarkable similarities between the interrelations of the organelles of a cell or the organs of the body, and the individuals composing a heteromorphic hydroid colony or a consociation of free individuals. Rather, it has been our task to present material gathered by numerous workers in the field of ecological physiology, particularly that collected within the last decade, and to focus it upon the present problem. The results obtained show that the generalizations of Espinas, of Deegener, and of Wheeler rest upon a broader base than that furnished by observational behavior studies concerned primarily with the struggle for existence as conceived in the nineteenth century, and with the more obvious survival values, or upon the lives of insects admittedly social in habit.

Certain of these problems with which we have been dealing be-

[1] Translated by Wheeler, 1928.

come more clear if presented in the simplest possible form. In order to free our minds of the perplexities brought in by the special requirements of our present-day species, with their long history of past adjustments to environments, or of mutational changes, let us try to consider conditions existing when living molecules first evolved from their non-living antecedents.

—However, whenever, and wherever life first appeared on this planet, considerations which we have given in detail in the preceding chapters make it highly probable that, unless the first living molecules appeared in considerable numbers approximately simultaneously within a limited microhabitat, there would have been little chance of survival; a single isolated living particle must have succumbed to the unconditioned unfavorable environment. If this occurred, a certain slight modification of the environment would result as the particle disintegrated. In doing so, it might free some X-substance, such as various workers, from Semper down to Robertson, Drzewina and Bohn, and Burnet, have assumed to be necessary for the well-being of living organisms; or the decomposing protoplasmic molecule might fix, by adsorption or otherwise, some of the elements of the environment harmful for a living system. In other words, the living protoplasm itself or the products of its metabolism during life, or freed by death and disintegration, would probably condition the immediate environment in such a manner that if another particle of living matter appeared soon in that niche it would have a better chance of survival.

If, on the other hand, several of these living molecules appeared approximately simultaneously in the same restricted microhabitat, then by the processes of metabolism they would tend to condition their environment similarly, and by fixation of toxic substances, or by some one of the other communal activities, such as the production of an X-substance, or the modification of the electrical conditions, this primitive aggregation of living particles would show the survival value which we have demonstrated is frequently exhibited by present-day animal aggregations of approximately the same integration level.

It may be that numerous transitions from the non-living to the living would occur one after the other in the same micro-niche, with a successive conditioning and a progressively greater longevity of some or all of the particles, until finally conditions would become sufficiently favorable for permanent survival. Whatever the details, it seems probable that this mechanism was operative from the very beginning of life and is a fundamental trait or property of living matter.

In order to discuss this trait more easily, it should be named. A few years ago it might have been called "unconscious co-operation"; but since many modern psychologists have discarded the concept of consciousness, the idea of lack of consciousness is less helpful than formerly. It may be regarded as an automatic mutual interdependence among organisms, or, for the sake of simplicity, as the principle of co-operation. The only trouble with calling this relationship one of co-operation, which it is, lies in the fact that the word carries with it an idea of conscious effort (cf. Durkheim, 1922) possible only after long ages of organic evolution, and then only in certain favored types of animals, while the evidence appears to be clear that the sort of co-operation of which we are speaking is a fundamental trait of living matter. As in all the other fundamental properties of living organisms, there is probably no hard and fast line to be drawn here between the living and the non-living. The mutual interdependence of the living must have grown out of similar but simpler interdependence in antecedent non-living matter, and may, in fact, be merely a highly specialized biological application of the mass law of chemistry.

If this analysis be sound, as it appears to be, the potentiality of social life is inherent in living matter, even though its first manifestations are merely those of a slight mutual interdependence, or of an automatic co-operation which finds its first biological expression as a subtle binding link of primitive ecological biocoenoses. Lest we be accused of having been carried too far by enthusiasm, it may be well to pause for a moment to examine the extent to which this automatic co-operation has been demonstrated to exist among animals. Are we, in fact, dealing with a phenomenon known to be sufficiently widespread to be thought of as having general rather than special biological significance? Chart I gives a diagrammatic list of the portions of the animal kingdom in which this principle of automatic co-operation has been demonstrated to date.

CHART I

Diphyletic Tree of Relationships within the Animal Kingdom*

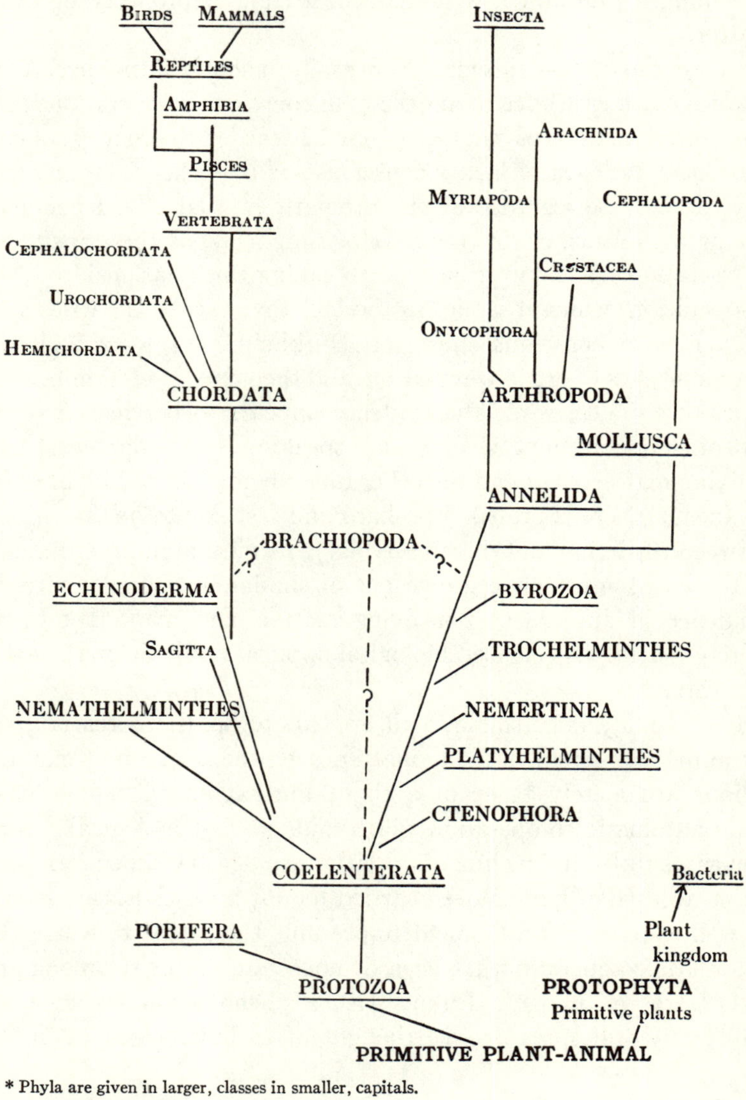

* Phyla are given in larger, classes in smaller, capitals.

This chart shows a diphyletic tree giving the relationships within the animal kingdom (Allee, 1926*a*). The distance from the base represents the relative degree of specialization. The phyla and classes underscored are those in which survival values from aggregations have been demonstrated, other than those known to occur in connection with normal sexual reproduction. Without reasonable doubt proper tests would reveal that aggregations of animals in all of the divisions still unchecked also possess survival value, at least when the animals are exposed to unusual or unfavorable conditions, such as those which would be furnished by hypotonic sea-water for the marine forms or by distilled water for the Trochelminthes. Exposure to various toxic agents would undoubtedly reveal group survival, providing the group were not too large nor the concentration too great. In addition to the group survival values known to be so widely distributed among animals, taxonomically considered, we have seen that similar survival values have been demonstrated for such diverse organisms as bacteria, for the spermatozoa of several kinds of aquatic animals and of mammals, and for tissue-culture cells. Evidently mutual interdependence, or automatic co-operation, is sufficiently widespread among the animal kingdom to warrant the conclusion given above that it ranks as one of the fundamental qualities of animal protoplasm, and probably of protoplasm in general.

Even if we are prepared to grant the foregoing conclusion, it does not necessarily follow that the principle of automatic co-operation is of great importance, though it may be exhibited by all known major groups of animals. Before we can satisfy ourselves that we are dealing with an important as well as a universal principle, it is necessary to find how commonly it is exhibited in nature. Running through the preceding pages and building up a summary of the various organisms whose aggregations have been discussed in these pages, we find, even when the different species of such animals as planarians and grasshoppers are lumped together, about 125 such. Aggregations of all these have been found in nature, with the exception of echinoderm larvae, which have not been reported in the density in which they may be found in laboratory containers. With about 14 exceptions, these aggregations are exhibited in addition to the congregation of the two sexes during the breeding season. Less than one-third of these forms possess sufficient social appetite to allow them to be classed as social animals in the usual sense of the term. Definite racial survival values have been demonstrated for about one-half, again excepting ordinary bisexual breeding relations; and of this half, about two-thirds are usually called "non-social." Even this brief survey shows that in addition to being a widespread phenomenon, taxonomically considered, survival values frequently accrue from animal aggregations in a state of nature, and often much below the level of group integration usually called "social."

The field naturalist, interested in observing a wide range of animal life, is familiar with the widespread occurrence of aggregations. Inland waters are notoriously poorer in population than is the sea; but in California, during the breeding season, I have seen ponds paved with the pebble-like clusters of salamander eggs. In mid-Great Salt Lake our boat ploughed through surface-covering masses of aggregated *Ephydra* flies that rose in choking numbers. Aldrich (1912) calculated 370,000,000 of these were to be found along every mile of Salt Lake beach. In the nearby mountain ponds of Utah aggregations of ostracods of the size of a walnut were to be found, at times occupying a portion of each cow track with which the bottom of the ponds were stippled; and similar collections of annelid worms occur in Indiana ponds. The collections of *Hydra* in favorable spots along Lake Michigan remind one of the abundance of marine organisms; and in some portions of spring-fed watercress swamps the supply of *Planaria dorotocephala* seems exhaustless.

Along the seashore, in such favorable locations as part of the California coast, the supply of animal life is appalling. One cannot step on the rocks exposed at low tide without crushing sea urchins, sea anemones, barnacles, or mollusks. Even in the less prolific regions around Cape Cod every available rock or solid timber washed by the tidal currents is the base for a densely packed aggregation, composed of many or of few species. Favorable bottom areas are similarly packed; and *Mytilus* and *Crepidula fornicata*, if proper substratum be wanting, form chains of animals, attaching to each other in the absence of solid objects. A suitable bit of mud flat may be packed with *Mya* so that the openings of their siphons fairly crowd the surface of the mud. At times and in favorable locations jelly fishes, or even minute copepods, may discolor the sea for miles; the entrance to the White Sea may be covered by red streaks produced by the presence of multitudes of starfish eggs (Mesiacev, 1927).

Brues (1926) estimated that the *Hymenorus* beetle population of a single panicle of Florida yucca would be about 15,000, and cites case after case of well-established insect aggregations. Some of the more striking include the hibernating aggregations of ladybird beetles (*Hippodamia convergens*) in northern California, of which Carnes (1912) records that two men in a single day can gather from 1,200,000 to double that number from the hibernating masses among the pine needles. A thousand chinch bugs have been found in the shelter of a single tuft of grass 3 inches in diameter (Headlee, 1910). Howard (1898, 1901) records flights of a chrysomelid beetle, one of which formed a belt 15 feet thick and a hundred yards wide over the course of the Gila River. The flight continued for 2 days. Cicadas, monarch butterflies, migratory locusts, and many Diptera, including *Polenia rudis*, *Muscina*, house flies, midges, and other insects, are known to collect in great numbers. In this survey I have not mentioned the collections of insects about electric lights, or the insects in the shore drift of lakes, or the vast collections of the more strictly social species, or the type of relationship usually called "symbiosis."

The recapitulation we have just made summarizes the evidence on two points. Aggregations of animals with little or no group organization, which possess survival values for the aggregants, have been demonstrated sufficiently widely throughout the animal kingdom to indicate that if studied in the other taxonomic units with the proper methods they can probably be demonstrated to occur in all the larger taxonomic divisions. Certainly they have already been demonstrated in groups sufficiently widespread to indicate that the absence of such group protection, other things being equal, is to be regarded as an exception rather than the rule. Such aggregations are ecologically as well as taxonomically widespread, and they are abundant in nature, as well as being widely distributed. It is upon such evidence that we may conclude that the mutual interdependence, or automatic co-operation, of which we are speaking is a fundamental and important principle in biology.

There is nothing in this recent work which displaces the earlier conclusion that overcrowding is harmful; but this newer evidence which we have been interested in presenting does show that under proper conditions, and entirely apart from breeding or hibernation, beneficial results may follow aggregations, in many organisms of the same or of different species, within a limited space. This means that in groupings caused by the tropistic reactions of individuals to environmental factors there may be a natural co-operation effective long before the physiological organization of the group has reached the level of development which occurs in the groupings usually designated as being truly social.

Symbiosis, commensalism, and intra-organismal relations aside, such unconscious co-operation was unknown to Espinas or to Kropotkin, who were much impressed by the evidences of mutual aid among insects and the larger animals. It was unknown to Wheeler when he wrote the 1923 conclusion quoted above, to which he was led by the studies of the ecologists and by his own knowledge of the behavior of ants and other social insects. The knowledge which we have summarized, showing that such general co-operation exists among loosely organized, or among apparently unorganized, groups of animals living even temporarily in the same region, gives us much clearer evidence than has been available to these students of social life, that their conclusion that co-operation is one of the major biological principles is correct, and that its roots extend far below the level of well integrated social activity.

From this point of view the first step toward the development of societies had probably already been taken when life came into existence on this planet. These first living particles were probably dependent on each other for the final adaptation of their physical environment so that they could continue to live. In the course of evolution they became more independent of close proximity to each other. A further advance was made when such more or less solitary animals developed, in addition to the general automatic co-operation inherent in living matter, a new toleration for close aggregation in a limited area, where they had collected not as a result of a social appetite but on account of their individual reactions to the surrounding environmental conditions. Such collections occur frequently as the result of forced movements in which the animal reacts, apparently mechanically, to the forces operating upon it, and may persist only because of the inertia of toleration. These tropistically conditioned groupings show survival values in addition to those resulting from the general co-operation of which we have spoken so frequently. Such additional survival values may be shown either by the effect of the group upon the individuals, rendering them more resistant to adverse environmental conditions, or conversely by so effecting the environment by the removal of toxic materials, or by some other ameliorating device, that it becomes more favorable for the continued existence of the animals. Group survival values can slip into the background as animals become well adjusted to the environment, to reappear apparently afresh when conditions of existence become again less favorable. These new survival values may be qualitatively as well as quantitatively different from those shown previously.

The last advance in this series comes when individuals cease to react as separate units and respond only as members of a group—when, as in the case of ants or termites and, rarely, with men, they are largely group-centered rather than self-centered. Many of the so-called "altruistic" drives in man apparently are the development of these innate tendencies toward co-operation, which find their early physiological expression in many simpler animals.

With the development of the nervous system, closer co-operation becomes possible and larger numbers are affected. There is much reason for thinking that many of the advances in evolution have come about through the selection of co-operating groups rather than through the selection of individuals. This implies that the two great natural principles of struggle for existence and of co-operation are not wholly in opposition, but that each may have reacted upon the other in determining the trend of animal evolution.

As a result of the working of these two principles, man has developed social groups, the scope of whose organization has been constantly extended until at the present time we are confronted with the problems centering about national versus international organization. Now, as in each stage of the social evolution of man, the proponents of the narrower organization maintain that the type of groupings they advocate satisfies the natural instinctive and traditional drives of man, while the more inclusive grouping is an abnormal desire for an idealistic utopia. So might the conservative primitive-living molecules, the protozoans, flatworms, isopods, or ants have argued, had they the wit, at each stage of their co-operative evolution. It may be helpful, and restful as well, to remember that the great majority of the evolution of social life has been brought about, not by conscious effort on the part of those undergoing evolution, but by the natural working-out of these two fundamental principles of struggle and co-operation.

We have been concerned in this book in tracing the earliest beginnings of these secondary (group) reactions (whether shown in overt acts or more subtly revealed), exhibited only under restricted conditions in nature which may be mimicked by properly controlled laboratory conditions. We have found that the physiology of the group considered independently from that of the individuals of which it is composed, begins simply and shows stages in development which can be arranged in various sorts of ascending series and which culminate in the group-centered, division-of-labor type of society that at first glance seems impossibly remote from the life of the so-called "solitary" animals.

Brilliant students of the highly social life of insects, like Wheeler, have found evidence that the behavior of these societies, taken together with observations on ecological associations and the various activities that center about reproduction, indicate the existence of a fundamental tendency toward co-operation. It has been much more easy for a student beginning with the humbler group levels to follow, from the social beginnings which he learns to recognize in almost unintegrated animal aggregations, the possibilities of the development of great social structures; and to trace their growth slowly and as yet imperfectly, but surely.

BIBLIOGRAPHY

BIBLIOGRAPHY

Ackerman, L. 1926. The physiological basis of wing production in the grain aphid. *Jour. Exp. Zoöl.* **44**:1–63.

Agar, W. E. 1914. Parthenogenetic and sexual reproduction in *Sinocephalus velutus* and other Cladocera. *Jour. Genetics* **3**:179–94.

Aldrich, J. M. 1912. The biology of some western species of the dipterous genus *Ephydra. N.Y. Jour. Ent. Soc.* **20**:77–99.

Alexander, C. P. 1920. Crane flies of New York. *Cornell University Memoirs* **28**:711.

Allard, H. A. 1916. The synchronal flashing of fireflies. *Science* **44**:710.

———. 1917. Synchronism and synchronic rhythm in the behavior of certain creatures. *Amer. Nat.* **51**:438–46.

———. 1918. Rhythmic synchronism in the chirping of certain crickets and locusts. *Ibid.* **52**:548–52.

Allee, W. C. 1912. An experimental analysis of the relation between physiological states and rheotaxis in Isopoda. *Jour. Exp. Zoöl.* **13**:270–344.

———. 1914. The ecological importance of the rheotactic reaction in stream isopods. *Biol. Bull.* **27**:52–66.

———. 1920. Animal aggregations. *Anat. Rec.* **17**:340.

———. 1923. Studies in animal aggregations: the temperature relation with isopods. *Ibid.* **24**:405–6.

———. 1923*a*. Animal aggregations: a request for information. *Condor* **25**: 129–31.

———. 1926. Studies in animal aggregations: causes and effects of bunching in land isopods. *Jour. Exp. Zoöl.* **45**:255–77.

———. 1926*a*. The evolution of the invertebrates. Chapter in *The Nature of the World and of Man*, pp. 260–303. Chicago: University of Chicago Press.

———. 1927. Studies in animal aggregations: some physiological effects of aggregation in the brittle starfish, *Ophioderma brevispina. Jour. Exp. Zoöl.* **48**:475–95.

———. 1927*a*. Animal aggregations. *Quar. Rev. Biol.* **2**:367–98.

———. 1928. Studies in animal aggregations: mass protection for *Planaria* from ultra-violet radiation. *Physiol. Zoöl.* **1**:509–30.

———. 1928*a*. Studies in animal aggregations: mass protection from fresh water for *Procerodes*, a marine turbellarian. *Jour. Exp. Zoöl.* **50**:61–84.

———. 1929. Studies in animal aggregations: natural aggregations of the isopod, *Asellus communis. Ecology* **10**:14–36.

———. 1929*a*. Mass protection from hypotonic sea water for *Procerodes*, a marine turbellarian. *Jour. Exp. Zoöl.* **54**:349–79.

ALLEE, W. C., and J. F. SCHUETT. 1927. Studies in animal aggregations: the relation between mass of animals and resistance to colloidal silver. *Biol. Bull.* **53**:301–17.

ALLEN, ARTHUR A. 1911–13. The red-winged blackbird: a study in the ecology of a cat-tail marsh. *Abstr. Proc. Linnean Soc. N.Y.*, Nos. 24–25.

———. 1921. Banding bats. *Jour. Mammal.* **2**:53–58.

ALLEN, EZRA. 1928. A note on the chromosomes of *Moina macrocopa*. *Science* **67**:18.

ALLEN, EZRA, and A. M. BANTA. 1929. Growth and maturation in the parthenogenetic and sexual eggs of *Moina*. *Jour. Morph. and Physiol.* **48**: 123–52.

ALLEN, G. M. 1925. *Birds and Their Attributes.* Boston: Marshall Jones Co. 338 pp.

ALVERDES, FRIEDRICH. 1927. *Social Life in the Animal World.* New York: Harcourt, Brace & Co. 216 pp.

ANDREWS, E. A. 1910. Conjugation in the crayfish, *Cambarus affinis*. *Jour. Exp. Zoöl.* **9**:235–65.

———. 1929. Population of ant mounds. *Quar. Rev. Biol.* **4**:228–57.

ARMBRUSTER, L. 1923. *Der Warmhaushalt im Bienenvolke.* Berlin: Pfennigstorff. 120 pp.

ASCHEMEIER, C. R. 1921. On the gorilla and chimpanzee. *Jour. Mammal.* **2**:90–94.

———. 1922. Beds of the gorilla and chimpanzee. *Ibid.* **3**:176–78.

ASHBY, E. 1929. The interaction of factors in the growth of *Lemna*. III. The interrelationship of intensity and duration of light. *Ann. Bot.* **43**:332–54.

———. 1929*a*. The interaction of factors in the growth of *Lemna*. IV. The influence of minute quantities of organic matter upon growth and reproduction. *Ibid.* **43**:805–16.

ASHBY, E., B. D. BOLAS, and F. Y. HENDERSON. 1928. The interaction of factors in the growth of *Lemna*. I. Methods and technique. *Ibid.* **42**:771–82.

ASHBY, E., and M. T. SU. 1929. The interaction of factors in the growth of *Lemna*. II. Method of dry weight estimation. *Ibid.* **63**:329–32.

ATKIN, E. E., and A. BACOT. 1917. The relation between the hatching of the eggs and the development of the larvae of *Stegomyia fasciata* (*Aedes calopus*), and the presence of bacteria and yeasts. *Parasitology* **9**:482–536.

AVEBURY, SIR JOHN LUBBOCK. 1892. *Ants, Bees and Wasps.* Bibliography. New York: Appleton. 448 pp.

———. 1899. *On the Senses, Instincts and Intelligence of Animals, with Special Reference to Insects.* London: K. Paul, Trench and Trübner. 292 pp.

BABÁK, E. 1906. Experimentelle Untersuchungen über die Variabilität der Verdauungsröhre. *Arch. für Entw. Mech.* **21**:612–702.

BACOT, A. 1917. The effect of the presence of bacteria or yeasts on the hatching of the eggs of *Stegomyia fasciata* (the yellow fever mosquito). *Jour. Roy. Micr. Soc.* 1917:173–74.

BAKER, L. E., and A. CARREL. 1926. Action of fibroblasts of the protein fraction of embryonic tissue extract. *Jour. Exp. Med.* **44**:387–95.

BALBIANI, G. 1860. Observations et expériences sur les phénomènes de reproduction fissipare chez les Infusoires ciliés. *Comp. Rend. d. l'Acad. d. Sci. Paris* **50**:1191–95.

BALDWIN, S. P. 1921. Recent returns from trapping and banding birds. *Auk* **38**:228–37.

BALTZER, F. 1914. Die Bestimmung und der Dimorphismus des Geschlechtes bei *Bonellia*. *Sitzungber. d. Physik-med. Gesell. Würzburg* 1914: 14–19.

———. 1914*a*. Die Bestimmung des Geschlechts nebst einer Analyse des Geschlechtsdimorphismus bei *Bonellia*. *Mitt. Zool. Sta. Neapol.* **22**:1–44.

———. 1923. Zur Entwicklungsgeschichte und Auffassung des Männchens der *Bonellia*. *Verhandl. d. Deutsch. Zool. Gesellsch.* **28**:65–68.

———. 1924. Über die Giftwirkung weiblicher *Bonellia* Gewebe auf das *Bonellia* Männchen und andere Organismen und ihre Beziehung zur Bestimmung des Geschlechts der *Bonellien* larvae. *Mitt. d. Naturf. Gesellsch. Bern* **8**:1–20.

———. 1925. Untersuchungen über die Entwicklung und Geschlechtsbestimmung der *Bonellia*. *Publ. d. Staz. Zool. d. Napoli* **6**:223–285.

———. 1926. Über die Vermännlichung indifferenter *Bonellia* Larven durch *Bonellia*-Extrakte. *Rev. Suisse d. Zool.* **33**:359–74.

———. 1928. Über metagame Geschlechtsbestimmung und ihre Beziehung zu einigen Problemen der Entwicklungsmechanik und Vererbung (Zusammenfass. Schrift.) *Verh. d. Deutsch. Zool. Gesellsch.* **32**:273–325.

———. 1928*a*. Zucht und Versuchsmethoden (Geschlechtsbestimmung) bei *Bonellia* Zuchtmethoden bei *Thalassene* und *Echiurus*. *Handb. biol. Arbeitsmeth. Lief. 278*, Abt. **9**:431–42.

———. 1928*b*. Neue Versuche über die Bestimmung des Geschlechts bei *Bonellia viridis*. *Rev. Suisse. d. Zool.* **35**:225–31.

BANKS, N. 1902. Sleeping habits of certain Hymenoptera. *Jour. N.Y. Ent. Soc.* **10**:209–14.

———. 1908. Sleeping habit of a bee. *Ent. News* **19**:340.

BANTA, A. M. 1914. Sex recognition and mating behavior of the wood frog. *Biol. Bull.* **26**:171–83.

———. 1925. The relation between previous sexual reproduction and the production of male offspring in *Moina*. *Amer. Nat.* **59**:50–61.

BANTA, A. M., and L. A. BROWN. 1923. Some data on control of sex in Cladocera. *Eugenics, Genetics and the Family: Scientific Papers, Second Int. Cong. of Eugenics* **1**:142–49.

BANTA, A. M., and L. A. BROWN. 1924. Rate of metabolism and sex determination in Cladocera. *Proc. Soc. Exp. Biol. and Med.* **22**:77–79.

———. 1929. Control of sex in Cladocera. I. The mothers as a means of controlling production. *Physiol. Zoöl.* **2**:80–92.

———. 1929*a*. Control of sex in Cladocera. II. The unstable nature of the excretory products involved in male reproduction. *Ibid.* **2**:93–98.

———. 1929*b*. Control of sex in Cladocera. III. Location of the critical point for control of sex. *Proc. Nat. Acad. Sci.* **15**:71.

———. 1929*c*. Control of sex in Cladocera. IV. Relation between the rate of the mother's development and the sex of the young. *Physiol. Zoöl.* **2**:302–8.

———. 1929*d*. Control of sex in Cladocera. V. Experimentally accelerated development of mothers and sex of young: mammalian endocrine substances without specific effect on cladocerans. *Ibid.* **2**:309–22.

BANTA, A. M., and T. R. WOOD. 1928. Genetic evidence that the Cladocera male is diploid. *Science* **67**:18–19.

BARBOUR, T. 1926. *Reptiles and Amphibians: Their Habits and Adaptations.* Boston: Houghton Mifflin. 125 pp.

BARKOW, H. C. L. 1846. *Der Winterschlaf nach seinen Erscheinungen im Tierreich.* Berlin. 525 pp.

BARON, M. 1930. Analyse der mitogenetischen Induktion und deren Bedeutung in der Biologie der Hefe. *Archiv. Wiss. Bot.* **10**:28–84.

BARTHÉLÉMY, H. 1923. Action de l'eau, NaCl, BrCl, et $CaCl_2$ sur les spermatozoïdes *de Rana fusca* et de *Bufo vulgaris. Comp. Rend. Acad. Sci.* **177**:654–57.

———. 1923*a*. Polyspermie dite physiologique et polysperme expérimentale des œufs utérins de *Rana fusca. Ibid.* **177**:1068–70.

———. 1926. Influence de la dilution du sperme sur la durée de survie des spermatozoïdes de la grenouille rousse. *Ibid.* **182**:1418–20.

BATES, A. F. C. 1895. A swallow roost in Maine. *Auk* **12**:48–51.

BEEBE, WILLIAM, G. I. HARTLEY, and P. G. HOWES. 1916. *Tropical Wild Life in British Guiana.* New York: New York Zoöl. Soc. 504 pp.

BELT, THOMAS. 1874. *The Naturalist in Nicaragua.* London: Bumpus. 403 pp.

BERNARD, C. 1865. *Introduction à l'étude de la médecine expérimentale.* Paris: Baillière. Translated in 1927 by H. C. Greene, New York. 400 pp.

BETHE, A. 1898. Dürfen wir den Ameisen und Bienen psychische Qualitäten zuschreiben? *Arch. f. d. ges. Phys.* **60**:15–100.

BILSKI, F. 1921. Uber den Einfluss des Lebensraumes auf das Wachstum der-Kaulquappen. *Pflüger's Archiv.* **188**:254–72.

———. 1926. Über den Einfluss verschiedener Bedingungen, besonders der Alkoholvergiftung auf Regeneration des Kaulquappenschwanzes. *Arch. f. Entw. d. Org.* **107**:219–40.

BLAIR, K. A. 1915. Luminous insects. *Nature* **96**:411–15.

BLAISDELL, F. S. 1912. Hibernation in *Cicindela seminis*. *Ent. News* **23**:156–59.

BLANCHARD, F. N. 1930. The stimulus to the breeding migration of the spotted salamander *Ambystoma maculatum*. *Amer. Nat.* **64**:154–67.

BODENHEIMER, F. S. 1929. Studien zur Epidemiologie, Ökologie und Physiologie der afrikanischen Wanderheuschrecke. *Zeitschr. f. angew. Entomol.* **15**:435–537.

BÖLSCHE, W. 1909. Daseinkampf und gegenseitige Hilfe in der Entwicklung. *Kosmos* **6**:14–16; 42–46.

BOGARDUS, E. S. 1922. The world as a group concept. *Jour. Applied Sociol.* **7**:31–38.

———. 1922*a*. The principle of group priority—the individual and the group. *Ibid.* **7**:784–87.

BOHN, G., and A. DRZEWINA. 1920. Variations de la sensibilité à l'eau douce des *Convoluta* suivant les états physiologiques et le nombre des animaux en expérience. *Comp. Rend. Acad. Sci.* **171**:1023–25.

———. 1928. Nouvelles recherches sur les effets de masse dans les agglomérations animals et essais d'interprétation. *Soc. Biol.* **99**:1567–70. (*See also* Drzewina, A., and G. Bohn.)

BOND, FRANK. 1921. The later flights of the passenger pigeon, 1872, 1875, 1876. *Auk* **38**:523–27.

BONNEVIE, K. 1907. Untersuchungen über Keimzellen der Bryozoen. *Jena Zeitschr. f. Naturw.* **42**:567–98.

BORODIN, D. N. 1930. Energy emanation during cell division processes (M-rays). *Plant Physiol.* **5**:119–30.

BOTTOMLEY, W. B. 1915. Some accessory factors in plant growth and nutrition. *Proc. Roy. Soc. London* B. **88**:237–47.

———. 1915*a*. A bacterial test for plant food accessories (auximones). *Ibid.* B. **89**:102–8.

———. 1917. Some effects of organic growth promoting substances (auximones) on the growth of *Lemna minor* in mineral culture solutions. *Ibid.* B. **89**:481–507.

———. 1920. The effect of nitrogen fixing organisms and nucleic acid derivatives on plant growth. *Ibid.* B. **91**:83–95.

BOULENGER, G. A. 1912. Some remarks on the habits of British frogs and toads. *Proc. Zoöl. Soc. London* 1912:19–22.

BOUVIER, E. L. 1919. *La vie psychique des insects*. Paris: E. Flammarin. 299 pp. (*See* Howard, L. O.)

BOWEN, E. S. 1930. The rôle of the sense organs in aggregations of *Ameiurus melas*. *Ecol. Monographs* **1**.

BRADLEY, J. C. 1908. Gregarious sleeping habits among aculeate Hymenoptera. *Ann. Ent. Soc. Amer.* **1**:127–30.

BRANDT, A. 1897. Vergesellschaftung und gegenseitiger Beistand bei Tieren. *Virchow's Sammlung Vorträge* **279**:1–48.

BRANNER, J. C. 1912. Geologic work of ants in tropical America. *Ann. Rept. Smithsonian Inst. 1911*, pp. 303–33.

BRASSLER, K. 1926. Massenflüge von *Amphimallus solstitialis L. Zeitschr. f. wiss. Insektenbiol.* **21**:267–68.

BRESSLAU, E. 1922. Die Ausscheidung entgiftender Schutzstoffe bei Ciliaten. *Zentralblatt f. Bakteriol., Parasitenk.* I. Abt. Orig. Bd. **89**, Beiheft: *87–*90.

———. 1924. Die Ausscheidung von Schutzstoffen bei einzelligen Lebewesen. *Ber. Senckenberg. Naturf. Ges.* **54**:49–67.

BREWSTER, WILLIAM. 1890. Summer robin roosts. *Auk* **7**:360–73.

———. 1895. A flight of pine grosbeaks. *Ibid.* **12**:245–56.

BRIDGES, C. B. 1921. Gametic and observed ratios in *Drosophila. Amer. Nat.* **55**:51–61.

BRINDLEY, H. H. 1910. Gregarious caterpillars. *Cambridge Phil. Soc.* **15**:576–87.

BRINTON, D. G. 1895. The progressive depopulation of northern regions. *Science* **2**:66–67.

———. 1902. *The Basis of Social Relations: A Study in Ethnic Psychology.* New York: Putnam's Sons. 204 pp.

BROOKS, W. K. 1885. The life history of the *Hydromedusae. Boston Soc. Nat. Hist. Mem.* **3**:359–430.

BROWN, C. R., and M. HATCH. 1929. Orientation and "fright" reactions of whirligig beetles (*Gyrinidae*). *Jour. Comp. Psych.* **9**:159–84.

BROWNE, F. BALFOUR. 1929. On the habits of certain social caterpillars. *Br. Assoc. Adv. Sci. 96th Mtg. (Glasgow, 1928)*, p. 564.

BROWNLEE, J. 1915. Studies in the meaning and relationships of birth and death rates. *Jour. Hygiene* **15**:11–35.

BRUES, C. T. 1903. On sleeping habits of some aculeate Hymenoptera. *Jour. N.Y. Ent. Soc.* **2**:228–30.

———. 1926. Remarkable abundance of a cistelid beetle with observations on other aggregations of insects. *Amer. Nat.* **60**:526–51.

BUCHANAN, R. E. 1918. Life phases in a bacterial culture. *Jour. Infect. Dis.* **23**:109–25.

BUCHANAN, R. E., and E. I. Fulmer. 1928–30. Vol. I, *Physiology and Biochemistry of Bacteria.* 516 pp. Vol. II, *Effects of Environment upon Microorganisms.* 709 pp. Vol. III, *Effects of Micro-organisms upon Environment.* 575 pp. Baltimore: Williams & Wilkins.

BUCHNER, PAUL. 1928. Die Grenzen des symbioseprinzipes. *Naturwissenschaften.* **16**:517–21.

BÜCHNER, LUDWIG, 1879. *Liebe und Liebes-Leben in der Thierwelt.* Berlin. 368 pp.

BURNET, F. M. 1925. Hydrogen peroxide and bacterial growth. *Australian Jour. Exp. Biol. and Med. Sci.* **2**:65–76.

———. 1925*a*. A note on the effect of dyes on bacterial growth. *Ibid.* **2**: 77–82.

BURRILL, A. C. 1912. Anon the reasons for bird migration: a favorite food theory. *Bull. Wis. Nat. Hist. Soc.* **10**:101–6.

———. 1913. Economic and biologic notes on the giant midge. *Ibid.* **10**: 124–63.

———. 1913*a*. Notes on Lake Michigan swarms of chiromids. *Milwaukee Bull. Wis. Nat. Hist. Soc.* **11**:52–69.

BURROWS, M. T. 1913. The tissue culture as a physiological method. *Trans. Cong. Amer. Physic. and Surg.* **9**:77–90.

———. 1913*a*. The tissue culture *in vitro. XVII Inter. Cong. Med. Gen. Path. and Path. Anat. London*, pp. 217–37.

———. 1921. Reserve energy of actively growing embryonic tissues. *Proc. Soc. Exp. Biol. and Med.* **18**:133–36.

———. 1921*a*. Problems in cancer research. *Jour. Cancer Research* **6**:131–38.

———. 1924. Problems in cancer research. *Southern Med. Jour.* **17**:233–41.

———. 1926. Studies on the nature of the growth stimulus in cancer. *Jour. Cancer Research.* **10**:239–51.

BURROWS, M. T., J. E. BURNS, and Y. SUZUKI. 1917. Studies on the growth of cells. The cultivation of bladder and prostatic tumors outside the body. *Jour. Urol.* **1**:3–15.

BURROWS, M. T., and C. G. JOHNSON. 1925. The action of oils in the production of tumors. *Arch. Int. Med.* **36**:293–332.

BURROWS, M. T., and L. H. JORSTAD. 1926. On the source of vitamin B in nature. *Amer. Jour. Physiol.* **77**:24–37.

———. 1926*a*. On the source of vitamin A in nature. *Ibid.* **77**:38–50.

BURROWS, M. T., and Y. SUZUKI. 1919. The study of problems of immunity by the tissue culture method. III. A method for determining the resistance to diphtheria infection. *Jour. Immunol.* **4**:1–18.

BUTTEL-REEPEN, H. VON. 1903. *Die stammesgeschichtliche Entstehung des Bienenstaates sowie Beiträge zur Lebensweise der solitären und sozialen Bienen (Hummels, Meliponinen, etc.).* Leipzig: G. Thieme. 138 pp.

———. 1905. Soziologisches und Biologisches, von Ameisen und Bienenstaat. Wie ensteht eine Ameisenkolonie? *Arch. Rass. Gesellsch. Biol.* **2**:20–35.

———. 1906. Das Schwärmen der Bienen. *Zeitsch. f. wiss. Insektenbiol.* **2**: 359–61.

———. 1907. Zur Psychobiologie der Hummeln. *Biol. Centbl.* **27**:579–604.

———. 1907*a*. Psychobiologische und biologische Beobachtung an Ameisen, Bienen und Wespen. *Naturwiss. Wochenschr.* **22**:465–78.

———. 1915. *Leben und Wesen der Bienen.* Braunschweig: Vieweg u. Sohn. 300 pp.

BUXTON, P. A. 1923. Further cases of co-ordinated rhythm in insects. *Entomologist* **56**:271–73.

———. 1924. Heat, moisture and animal life in insects. *Proc. Roy. Soc. London* **96**:123–31.

CALKINS, G. N. 1926. *Biology of the Protozoa*. Philadelphia: Lea & Febinger. 623 pp.

CANDOLLE, A. P. DE. 1832. *Physiologie végêtale*. 3 vols. Paris: Béchet jeune.

CARNES, E. K. 1912. Collecting ladybirds (*Coccinellidae*) by the ton. *Monthly Bull. Comm. Horticulture, Calif.* **1**:71–81.

CARPENTER, KATHLEEN E. 1927. The lethal action of soluble metallic salts on fishes. *Brit. Jour. Exp. Biol.* **4**:378–90.

———. 1930. Further researches on the action of metallic salts on fishes. *Jour. Exp. Zoöl.* **56**:407–22.

CARREL, A. 1924. Tissue culture and cell physiology. *Physiol. Rev.* **4**:1–20.

———. 1928. Modern techniques of tissue culture and results. *Arch. Exp. Zellforsch.* **6**:70–81.

CARREL, A., and BAKER, L. E. 1926. The chemical nature of substances required for cell multiplication. *Jour. Exp. Med.* **44**:503–21.

CARREL, A., and A. H. EBELING. 1923. Action of serum on lymphocytes *in vitro*. *Ibid.* **38**:513–19.

CASTELLANI, ALDO. 1926. Observations on the importance of symbiosis or close association of different species of organisms in the production of certain biochemical phenomena and in the causation of certain diseases and certain symptoms of disease. *Jour. Trop. Med. and Hyg.* **29**:217–26.

CASTLE, W. A. 1928. An experimental and histological study of one life cycle of *Planaria velata*. *Jour. Exp. Zoöl.* **51**:417–85.

CAULLERY, M. 1921. *Le parasitisme et la symbiose*. Paris: G. Doin. 400 pp.

———. 1922. Parasitism and symbiosis in their relation to the problem of evolution. *Ann. Rept. Smithsonian Inst. 1920*, pp. 399–409.

CAULLERY, MAURICE, and MARGUERITE COMAS. 1928. Le determinisme du sex chez un nématode parasite des larves de chironomes. *Comp. Rend. Acad. Sci.* **186**:646–47.

CHADBOURNE, ARTHUR P. 1889. An unusual flight of killdeer plover along the New England Coast. *Auk* **6**:255–63.

CHAMBERS, R. 1908. Einfluss der Eigrösse und der Temperatur auf das Wachstum und die Grösse des Frosches und dessen Zellen. *Arch. Mikros. Anat. u. Entw. Gesch.* **72**:601–61.

CHAMPY, C. 1914. La présence d'un tissu antagoniste maintient la différenciation d'un tissu cultivé en debois de l'organisme. *Soc. Biol.* **76**:31–32.

CHAPMAN, R. N. 1928. The quantitative analysis of environmental factors. *Ecology* **9**:111–23.

CHESNEY, A. 1916. The latent period in the growth of bacteria. *Jour. Exp. Med.* **24**:387–418.

Chidester, F. E. 1911. The mating habits of four species of the *Brachyura*. *Biol. Bull.* **21**:235–48.

Child, C. M. 1911. Studies on the dynamics of morphogenesis and inheritance in experimental reproduction. II. Physiological dominance of anterior over posterior regions in the regulation of *Planaria dorotocephala*. *Jour. Exp. Zoöl.* **11**:187–220.

———. 1911*a*. Studies on the dynamics of morphogenesis and inheritance in experimental reproduction. IV. Certain dynamic factors in the regulatory morphogenesis of *Planaria dorotocephala* in relation to the axial gradient. *Ibid.* **13**:103–52.

———. 1915. *Senescence and Rejuvenescence*. Chicago: University of Chicago Press. 481 pp.

———. 1924. *Physiological Foundations of Behavior*. New York: Holt & Co. 330 pp.

———. 1928. Axial development in aggregates of dissociated cells from *Corymorpha*. *Physiol. Zoöl.* **1**:419–61.

Chittenden, F. H. 1898. Insects that affect asparagus. *U.S. Bur. Entom. Bull.* **10**:54–62.

Christie, J. R. 1929. Some observations on sex in the *Mermitihdae*. *Jour. Exp. Zoöl.* **53**:59–76.

Church, Frances. 1927. The effect of crowding on the rate of growth in fishes. Master's thesis, University of Chicago Library. 32 pp.

Churchman, J. W. 1920. Communal activity of bacteria. *Proc. Soc. Exp. Biol. and Med.* **18**:22–23.

———. 1921. Further studies on the behavior of bacteria toward gentian violet. *Jour. Exp. Med.* **33**:569–81.

Churchman, J. W., and Morton C. Kahn. 1921. Communal activity of bacteria. *Ibid.* **33**:583–91.

Clark, N. A. 1926. Plant growth-producing substances, hydrogen-ion concentration and the reproduction of *Lemna*. *Plant Physiol.* **1**:273–79.

Clark, N. A., and E. M. Roller. 1924. "Auximones" and the growth of the green plant. *Soil Sci.* **17**:193–98.

Clausen, R. E. 1924. The inheritance of cinnabar eye color in *Drosophila melanogaster*, including data on the locus of jaunty. *Jour. Exp. Zoöl.* **38**:423–36.

Cobb, N. A. 1928. Initial survey of nemas in upper twenty millimeters of marine beach sand near low tide mark. *Proc. Philos. Soc. Wash.* **19**:199–200.

Cobb, N. A., G. Steiner, and J. R. Christie. 1923. *Agamermes decaudata:* a nema parasite of grasshoppers and other insects. *Jour. Agric. Res.* **23**:921–26.

———. 1927. When and how does sex arise? *Off. Rec. U.S. Dept. Agric.* **6**: No. 43, p. 6.

Cohen, D. A. 1900. Casual observations of a colony of black-crowned night herons' colonial nesting habits. *Condor* **2**:10–12.

Cohn, E. J. 1917. Relation of H-ion concentration of sperm suspensions and their fertilizing power. *Anat. Rec.* **11**:530.

———. 1918. Studies in the physiology of spermatozoa. *Biol. Bull.* **34**:168–218.

Colton, H. S. 1908. Some effects of environment on growth of *Lymnaea. Proc. Acad. Nat. Sci. Phila.* **60**:410–48.

Conklin, E. G. 1898. Environment and sexual dimorphism in *Crepidula. Ibid.* 1898: 435–44.

Conn, H. W. 1914. *Social Heredity and Social Evolution.* New York: Abingdon Press. 348 pp.

Cornetz, V. 1912. L'illusion de l'entraide chez la fourmi. *Rev. des Idées* **9**:292. **9**:292.

Courtis, S. A. 1907. Response of toads to sound stimuli. *Amer. Nat.* **41**: 677–82.

Cox, Philip. 1889. A bird wave. *Auk* **6**:241–43.

Crabb, E. D. 1929. Growth of a pond snail, *Lymnaea stagnalis*, as indicated by increase in shell size. *Biol. Bull.* **56**:41–64.

Craig, Wallace. 1908. The voices of pigeons regarded as a means of social control. *Amer. Jour. Sociol.* **14**:86–100.

———. 1911. Emotions in the pigeons. II. The mourning dove. *Auk* **28**: 398–408.

———. 1911*a*. Emotion in the pigeons. III. The passenger pigeon. *Ibid.* **28**:408–26.

———. 1914. Male doves reared in isolation. *Jour. An. Behav.* **4**:121–38.

———. 1916. Synchronism in the rhythmic activities of animals. *Science* **44**: 784–86.

Crampton, G. C. 1912. Experiments performed upon Protozoa confined in capillary tubes. *Arch. f. Protistenkunde* **27**:9–15.

———. 1912*a*. Inhibition of cell division in *Paramoecium. Science* **35**:634–35.

Crozier, W. J. 1922. An observation on the "cluster-formation" of the sperms of *Chiton. Amer. Nat.* **56**:478–80.

Cummins, Harold. 1920. The rôle of voice and coloration in spring migration and sex recognition in frogs. *Jour. Exp. Zoöl.* **30**:325–43.

Curran, H. R. 1925. The growth of bacteria. Thesis, Cornell University, Ithaca.

Cuthbertson, A. 1926. Studies on the Clyde crane flies. *Ent. Monthly Mag.* **62**:36–38, 84–88.

Cutler, D. W., and L. M. Crump. 1923. The rate of reproduction in artificial cultures of *Colpidium colpoda.* I. *Biochem. Jour.* **17**:174–86.

———. 1923*a*. The rate of reproduction in artificial cultures of *Colpidium colpoda.* II. *Ibid.* **17**:875–86.

———. 1924. The rate of reproduction in artificial cultures of *Colpidium colpoda*. III. *Ibid.* **18**:903–12.

———. 1925. The influence of washing on the reproductive rate of *Colpidium colpoda*. II. *Ibid.* **19**:450–53.

CUTLER, D. W., L. M. CRUMP, and H. SANDON. 1922. A quantitative investigation of the bacterial and protozoan population of the soil, with an account of the protozoan fauna. *Phil. Trans. Roy. Soc. London* **211**:317–50.

DALL, W. H. 1882. Notes on the cluster fly, *Pollenia rudis*. *Proc. U.S. Nat. Mus.* **5**:635–37.

DAMPF, A. 1925. (*See* Hoffmann, Dampf, and Varela.)

———. 1926. Der Färbungswechsel bei den Wanderhauschreckenlarven, ein biologisches Rätsel. *Verh. III. Int. Ent. Kongr. Zurich* **2**:276–90.

DAUGEARD, P. A. 1928. Le determinisme des mouvements chez les organismes inférieurs. *Ann. Protistol.* **1**:3–10.

DAVENPORT, C. B. 1897, 1899. *Experimental Morphology*. New York: Macmillan. 509 pp.

DAVIDSON, J. 1914. The host plants and habits of *Aphis rumicus* L., with some observations on the migration of, and infestation of, plants by aphides. *Ann. Applied Biol.* **1**:118–41.

DAVIS, W. T. 1894. Staten Island crows: their roosts. *Auk* **2**:228–31.

DAY, F. 1880. *Fishes of Great Britain and Ireland*. Part I. London: William & Norgate. 326 pp., 93 pl.

DEEGENER, P. 1917. Versuch zu einem System der Assoziations- und Sozietätsformen im Tierreiche. *Zool. Anzeig.* **49**:1–16.

———. 1918. *Die Formen der Vergesellschaftung im Tierreiche. Ein systematisch-soziologischer Versuch*. Leipzig: Veit. 420 pp.

———. 1919. Beiträge zur Kenntnis sozialer Raupen. *Deutsch. Ent. Zeitschr.* 1919: 65–115.

———. 1919*a*. Der psychische Konnex sozialer Raupen. *Ibid.* 1919: 416–17.

———. 1919*b*. Tiersoziologische Beobachtungen und Versuche. *Sitzenber. Gesell. Naturf. Berlin* 1919: 1–34.

———. 1920. Beobachtungen an sozialen Blattwespenlarven. *Deutsch. Ent. Zeitschr.* 1920: 310–14.

———. 1920*a*. Die Tiergesellschaftkunde. *Aus der Heimat* **33**:113–26.

———. 1920*b*. Soziologische Studien an Raupen und Bemerkungen über Licht und statischen Sinn. *Arch. f. Naturgesch.* **86**:91–154.

———. 1921. Der sogenannte Phototropismus der Raupen und sein biologischer Wert. *Zeitschr. f. Allg. Physiol.* **19**:119–32.

———. 1922. *Ein Lehrjahr in der Natur*. 2 vols. Jena: Fischer.

———. 1922*a*. Soziologische Beobachtungen an *Hyponeuta cognatellus*. *Hb. Biol. Centralbl.* **42**:241–53.

———. 1922*b*. Kokongesellschaften und Gesellschaftskokons. *Zeitschr. f. wiss. Insektenbiol.* **17**:176–98.

DEEGENER, P. 1923. Weitere Mitteilungen über Gesellschaftskokons bei Lepidopteren. *Ibid.* **18**:51–56.

———. 1925. *Lebensgewohnheiten der Insekten.* Leipzig: Quelle und Meyer. 135 pp.

DELAGE, Y., and M. GOLDSMITH. 1911. *Les théories de l'évolution.* Paris: Flammarion. 371 pp.

DEMUTH, G. S. 1921. Swarm control. *U.S. Dept. Agric. Farmers Bull.* 1198. 47 pp.

DEVARIGNY, H. 1894. Recherches sur le nanisme expérimentale. Contribution à l'étude de l'influence du milieu sur les organismes. *Jour. Anat. et Phys.* **30**:147–88.

DICKEY, DONALD R. 1923. Evidence of interrelation between fox and caribou. *Jour. Mammal.* **4**:121–23.

DITMARS, R. L. 1907. *The Reptile Book.* New York: Doubleday Page & Co. 465 pp.

DOLBEAR, A. E. 1897. The cricket as a thermometer. *Amer. Nat.* **31**:970–71.

D'OMBRAIN, E. A. 1906. Visit to an Ibis rookery. *Emu* **5**:185–89.

DREW, A. H. 1923. Growth and differentiation in tissue cultures. *Brit. Jour. Exp. Path.* **4**:46–52.

DRZEWINA, A., and G. BOHN. 1920. Variations de la sensibilité à l'eau douce des *Convoluta*, suivant les états physiologiques et le nombre des animaux en expérience. *Comp. Rend. Acad. Sci.* **171**:1023–25.

———. 1921. Variations de la susceptibilité aux agents nocifs avec le nombre des animaux traités. *Ibid.* **172**:485–87.

———. 1921*a*. La defense des animaux groupés vis-à-vis des agents nocifs. *Ibid.* **172**:779–81.

———. 1921*b*. Action nocive de l'eau sur les Stentors, en fonction de la masse du liquide. *Compt. rend. Soc. de biol.* **84**:917–20.

———. 1921*c*. Variations dans le temps de la résistance aux agents physiques et chimiques chez *Rana fusca*. *Ibid.* **84**:963–65.

———. 1921*d*. Sur des phénomènes d'auto-protection et l'auto-destruction chez des animaux aquatiques. *Comp. Rend. Acad. Sci.* **173**:107–9.

———. 1922. Sur des phénomènes d'auto-destruction et d'auto-agglutination chez les *Convoluta*. *Ibid.* **174**:330–32.

———. 1922*a*. Immunisation des *Convoluta* contre l'action du chlorure de potassium par des doses plus fortes que la dose rapidement mortelle. *Compt. rend. Soc. de biol.* **86**:252–54.

———. 1923. Effets tardifs de la dilution du sperme sur le développement de l'œuf d'Oursin. *Comp. Rend. Acad. Sci.* **176**:79–81.

———. 1923*a*. Variations dans le temps du pouvoir fécondant du sperme d'Oursin traité par des substances nocives, en rapport avec son degré de dilution. *Compt. rend. Soc. de biol.* **89**:383–86.

———. 1923*b*. Action toxique du rouge neutre en présence de la lumière. *Ibid.* **89**:386–88.

———. 1924. Expulsion des algues vertes symbiotes chez les *Convoluta roscoffensis*, sous l'influence de l'acide carbonique. *Comp. Rend. Acad. Sci.* **178**: 876–78.

———. 1924*a*. Action mortelle ou nulle de quelques acides sur *Convoluta roscoffensis*, de part et d'autre d'un point critique. *Compt. rend. Soc. de biol.* **90**:625–26.

———. 1925. Au sujet de la régulation du pH de l'eau de mer. *Ibid.* **93**:917–19.

———. 1925*a*. Sur l'acidification du milieu par cytolyse cellulaire. *Comp. Rend. Acad. Sci.* **181**:692–94.

———. 1926. Influence de la température sur le sperme d'Oursin, en fonction de la masse. *Compt. rend. Soc. de biol.* **95**:376–78.

———. 1926*a*. Action de l'argent métallique sur le sperme et les larves d'Oursin. *Comp. Rend. Acad. Sci.* **182**:1651–52.

———. 1926*b*. Action antagoniste de l'argent et de l'étain metallique sur les êtres vivants. *Ibid.* **183**:571–72.

———. 1926*c*. Activation par la lumière des effets de l'argent sur *Convoluta. Ibid.* **183**:677–79.

———. 1926*d*. Influence de l'acide carbonique sur le sperme d'Oursin, en fonction de la masse. *Ibid.* **185**:317–18.

———. 1927. Influence des parois des vases sur les réactions des animaux. *Ibid.* **185**:875–77.

———. 1927*a*. Effets tardifs de l'argent metallique sur les œufs d'Oursin. *Compt. rend. Soc. de biol.* **97**:1522.

———. 1928. Les *Convoluta. Ann. Sci. Nat. Zool.* X Ser. **11**:299–398.

DURKHEIM, E. 1922. *De la division du travail social.* Paris: Alcan.

EBELING, A. H., and A. FISCHER. 1922. Mixed cultures of pure strains of fibroblasts and epithelial cells. *Jour. Exp. Med.* **36**:285–89.

EDDY, M. W. 1925. Metabolic rate and bunching in catfish. *Anat. Rec.* **31**: 332–33.

EDDY, S. 1928. Succession of Protozoa in cultures under controlled conditions. *Trans. Amer. Micros. Soc.* **47**:283–304.

EDDY, W. H., H. L. HEFT, H. C. STEVENSON, and R. JOHNSON. 1921. The yeast test as a quantitative measure of vitamin. *Proc. Soc. Exp. Biol. and Med.* **18**:138–40.

EDDY, W. H., H. L. HEFT, and H. C. STEVENSON. 1922. A reply to Fulmer, Nelson and Sherwood concerning medium F. *Jour. Biol. Chem.* **51**:83–85.

EIDMAN, H. 1918. Zeile und Aufgaben der Tiersoziologie. *Biol. Zentralbl.* **48**:101–5.

EIGENBRODT, H. J. 1925. The somatic effects of certain environmental conditions on a homozygous race of *Drosophila*. Ph.D. thesis in University of Illinois Library. 126 pp.

EIJKMAN, C. 1904. Über thermolabile Stoffwechselprodukte als Ursache der natürlichen Wachstumhemmung der Mikroorganismen. *Centralbl. f. Bakt.* I Orig. **37**:436–49.

———. 1906. Über natürliche Wachstumshemmung der Bakterien. *Ibid.* **41**: 367–69.

———. 1907. Über die Ursache der Wachstumhemmung in Bakterienkulturen. *Deutsch. Med. Woch.* **33**:265–66.

ELLICOTT, E. L. 1880. Bundles of snakes. *Amer. Nat.* **14**:206–7.

ELTON, CHARLES. 1927. *Animal Ecology.* London: Sidgwick & Jackson. 207 pp.

EMERSON, A. E. 1926. Development of soldier termites. *Zoölogica* (New York Zoölogical Society) **7**:69–100.

———. 1928. Communication among termites. *Intern. Cong. Ent. 1928.* **2**: 722–26.

EMERSON, A. E., and ROBERT SIMPSON. 1929. Apparatus for the detection of substratum communication among termites. *Science* **59**:648–49.

EMERY, C. 1910. Il polimorfismo e la fondazione della società hegli insetti social. *Scientia (rivista di Scienza)* **7**:336–49.

ENGLISH, P. F. 1923. Gregarious wood rats. *Jour. Mammal.* **4**:1–9.

ENTEMANN, M. 1901. Colonies in *Polistes.* Reported from the Zoölogical Club of the University of Chicago. *Science* **13**:112–13.

ESPINAS, A. V. 1878. *Des sociétés animales.* Paris: Librairie Germer Baillière. 588 pp.

ESSENBERG, CHRISTINE. 1915. The habits of the water strider, *Gerris remiges. Jour. An. Behav.* **5**:397–402.

EVERMANN, B. W. 1901. Birds in dry seasons. Proc. Biol. Soc. Wash. reported by F. A. Lucas. *Science* **14**:896.

———. 1921. The Año Nuevo Steller sea lion rookery. *Jour. Mammal.* **2**:16–19.

EWING, H. E. 1916. Eighty-seven generations in a parthenogenetic pure line of *Aphis avenea* Fabre. *Biol. Bull.* **31**:53–102.

———. 1925. The factors of inheritance and parentage as affecting the ratio of alate to apterous individuals in Aphids. *Amer. Nat.* **59**:311–26.

———. 1926. Wing production in plant lice. *Ibid.* **60**:576–79.

FABRE, J. H. 1879–1907. *Souvenirs entomologiques.* 10 vols. Paris: Delagrave.

———. 1912. *Social Life in the Insect World.* Translated by B. Miall. London: Unwin. 327 pp.

———. 1915. *The Hunting Wasps.* New York: Dodd Mead. 427 pp.

FAGE, LOUIS, and RENÉ LEGENDRE. 1927. Sur le rhythme lunaire des assaimayes du *Nereis irrorata* Malm. *Bull. Soc. Zool. France.* **52**:84–86.

FARR, W. 1843. Causes of the high mortality in town districts. *5th Ann. Rept. Reg. Gen. of Births, Deaths and Marriages in England* (2d ed.), pp. 406–35.

———. 1875. Effects of density of population on health. *Suppl. to 35th Ann. Rept., ibid.* xxiii–xxv.

FAURE, J. C. 1923. The life history of the brown locust. *Jour. Dept. Agric. So. Africa* **7**:205–24.

FERRIÈRE, A. 1915. *La loi du progrès en biologie et en sociologie et la question de l'organisme social.* Paris: Giard et Brière. 680 pp.

FIELDE, A. M. 1905. Progressive order of ants. *Biol. Bull.* **10**:1–16.

———. 1905*a*. Sense of smell in ants. *Independent* **59**:375–78.

———. 1905*b*. Odor in ants. *Ann. N.Y. Acad. Sci.* **16**:304.

FIELDE, A. M., and G. H. PARKER. 1905. The reactions of ants to material vibrations. *Proc. Acad. Nat. Sci. Phila.* **56**:642–50.

FINLEY, W. L. 1902. Among the sea birds of the Oregon coast. *Condor* **4**:53–57.

———. 1904. A winter with the birds of Costa Rica. *Ibid.* **6**:152–60.

———. 1905. Among the sea birds off the Oregon coast. *Ibid.* **7**:120–29, 161–69.

———. 1906. Herons at home. *Ibid.* **8**:36–40.

———. 1907. Among the gulls on Klamath Lake. *Ibid.* **9**:12–16.

———. 1907*a*. Among the pelicans. *Ibid.* **9**:35–40.

FISCHEL, WERNER. 1927. Beiträge zur Soziologie des Haushuhns. *Biol. Zentralbl.* **47**:678–96.

FISCHER, A. 1923. Contributions to the biology of tissue cell. I. The relation of cell crowding to tissue growth *in vitro. Jour. Exp. Med.* **38**:667–72.

———. 1924. The interaction of two fragments of pulsating heart *in vitro. Ibid.* **39**:577–83.

———. 1925. *Tissue Culture.* Copenhagen: Levin and Munksgaard. 305 pp.

———. 1925*a*. Cytoplasmic growth principles of tissue cells. *Arch. f. Exp. Zellforsch.* **1**:375–77.

FISCHER-SIGWART H. 1897. Biologische Beobachtungen an unseren Amphibien. *Vierteljahrschr. d. Natur. Ges. Zurich* **42**:238–316; **43**:279–316.

FISHER, R. A. 1925. Application of "Student's" distribution. *Metron* **5**:3–17.

FISHER, WALTER K. 1904. Home life of a *Buccanus,* man o' war bird. *Condor* **6**:57–61.

———. 1904*a*. Albatross pictures. *Ibid.* **6**:151–59.

FLATTELY, F. W. 1921. Rhythm in nature. *Ann. Rept. Smithsonian Inst. 1920*, pp. 389–97.

FLETCHER, ROBERT K. 1926. Notes on a migration of the snout butterfly. *Ent. News.* **37**:106–7.

FLOERSCHEIMS, CECIL. 1906. On some enemies of the diurnal Lepidoptera. *Tutt's Ent. Rec.* **18**:36–39.

FORBES, S. A. 1887. The lake as a microcosm. *Bull. Peoria Acad. Sci. 1887.* 17 pp. Reprinted in *Ill. State Nat. Hist. Survey Bull.* **15**:537–50.

FORD, H. A. 1852. On the characteristics of *Troglodytes* gorilla. *Proc. Acad. Nat. Sci. Phila.* **6**:30–33.

FORD, H. J. 1915. *The Natural History of the State.* Princeton University Press. 188 pp.

FOREL, A. 1921–23. *Le monde social des fourmis.* 6 vols. Genève: Librairie Kundig.

———. 1929. *The Social World of the Ants Compared with That of Man.* Translated by C. K. Ogden. 2 vols. New York: Albert and Charles Boni.

FOWLER, J. R. 1927. Concerning group protection in aquatic animals. Master's thesis, University of Chicago Library.

———. 1931. The relation of numbers of animals to survival in toxic concentrations of electrolytes. *Physiol. Zoöl.* **4**.

FOX, K. MUNRO. 1923. Lunar periodicity in reproduction. *Proc. Roy. Soc. London* B. **95**:523–51.

FRAENKEL, G. 1929. Untersuchungen über Lebensgewohnheiten, Sinnesphysiologie und Sozialpsychologie der wandernden Larven der afrikanischen Wanderheuschrecke. *Biol. Zentralbl.* **49**:657–80.

FRANK, G., and M. KUREPINA. 1930. Die gegenseitige Beeinflussung der Seeigeleier als mitogenetischer Effekt betrachtet. *Arch. f. Entw. d. Org.* **121**: 634–38.

FRAZER, J. G. 1910. *Totemism and Exogamy.* Vol. IV. New York: Macmillan. 379 pp.

FRIEDMANN, HERBERT. 1929. *The Cow Birds: A Study in Social Parasitism.* Springfield, Ill.: C. C. Thomas. 421 pp.

FRISCH, K. VON. 1918. Beitrag zur Kenntnis sozialer Instinkte bei solitären Bienen. *Biol. Zentralbl.* **38**:183–88.

FRISON, T. H. 1927. The development of castes of bumblebees (*Bremidae*). *Ann. Ent. Soc. Amer.* **20**:156–81.

FUCHS, H. M. 1914. Action of egg secretions on fertilizing power of sperm. *Arch. für Entw. Mech.* **40**:205–53.

———. 1915. Studies in the physiology of fertilization. *Jour. Genetics* **4**:215–301.

FULMER, E. I. 1921. The acclimatization of yeast to ammonium fluoride and its inversion in wort. *Jour. Phys. Chem.* **25**:455–72.

FULMER, E. I., and V. E. NELSON. 1922. Water soluble B and "bios" in yeast growth. *Jour. Biol. Chem.* **51**:77–82.

FULTON, B. B. 1925. Physiological variation in the snowy tree cricket. *Ann. Ent. Soc. Amer.* **18**:363–83.

———. 1928. A demonstration of the location of auditory organs in certain Orthoptera. *Ibid.* **21**:445–48.

———. 1928*a*. Sound perception by insects. *Science Monthly* **27**:552–56.

GAIN, L. 1913. The penguins of the antarctic regions. *Ann. Rept. Smithsonian Inst. 1912*, pp. 475–82.

GALTON, F. 1883. *Inquiries into Human Faculty and Its Development.* New York: Macmillan. 387 pp.

GALTSOFF, P. S. 1925. Regeneration after dissociation. (An experimental study on sponges.) *Jour. Exp. Zoöl.* **42**:183–251.

GATES, B. N. 1914. Temperature of the bee colony. *Bull. U.S. Dept. Agric. 96.* 25 pp.

GEE, W. P. 1916. Effects of acute alcoholization on the germ cells of *Fundulus heteroclitus. Biol. Bull.* **31**:379–406.

GEMMILL, J. F. 1900. On the vitality of the ova and sperm of certain animals. *Jour. Anat. and Physiol.* **34**:163–81.

GESENIUS, H. 1929. Über stoffwechselwirkungen mitogenetischen Strahlen. *Biochem. Zeitschr.* **212**:240.

GIBSON, A. 1902. Some interesting habits of lepidopterous larvae. *Rept. Ent. Soc. Ontario* **33**:74–78.

———. 1903. Basswood, or linden, insects. *Ibid.* **34**:50–61.

GLASER, O. 1915. Can a single spermatozoön initiate development in *Arbacia? Biol. Bull.* **28**:149–51.

GOETSCH, W. 1923. Die Abhängigkeit sozialer Insekten vom Nest. *Sitz. Ber. Ges. Morph. u. Physiol. München* **34**:19–38.

———. 1924. Versuch über den Einfluss des Lebensraums auf Körpergrösse von Hydren, Planarien und Kaulquappen. *Verhandl. d. Deutsch. Zool. Ges.* **29**:60–62.

———. 1924*a*. Lebensraum und Körpergrösse. *Biol. Zentralbl.* **44**:529–60.

GOLDMAN, E. A. 1920. Mammals of Panama. *Smithsonian Misc. Coll.* **69**:1–309.

GOLDSCHMIDT, R. 1926. Bemerkungen zum Problem der Geschlechtsbestimmung bei *Bonellia. Biol. Zentralbl.* **46**:441–52.

GOODWIN, S. H. 1904. Pelicans nesting at Utah Lake. *Condor* **6**:126–29.

GOULD, G. M. 1895. The katydid's orchestra. *Science* **2**:380–81.

GOULD, H. N. 1917. Studies on sex in the hermaphrodite mollusc, *Crepidula plana.* I. *Jour. Exp. Zoöl.* **23**:1–52.

———. 1917*a*. Studies on sex in the hermaphrodite mollusc, *Crepidula plana.* II. *Ibid.* **23**:225–50.

GOUNELLE, E. 1900. Observation fait en 1889 dans le Sertâo méridional de l'État de Bahia. *Bull. Ent. Soc. France* 1900: 168–69.

GRAHAM-SMITH, G. S. 1920. The behavior of bacteria in fluid cultures as indicated by daily estimates of the numbers of living organisms. *Jour. Hygiene* **19**:133–204.

GRASSÉ, P. 1923. Sur la phototropisme des quelques criquets. *Compt. rend. Soc. de biol.* **89**:898–99.

GRASSI, B., and ANNA FOÀ. 1907. Ricerche sulle filossere, etc. esequite nel R. osservatorio antifilosserico di Fauglia, etc. *Rend. dell' Accad. d. Lincei*, l Sett.

GRASSI, B., and A. SANDIAS. 1897–98. The constitution and development of the society of termites. Observation on their habits with appendices on the parasite Protozoa of the *Termitidae* and on the *Embiidae*. *Quar. Jour. Micr. Sci.* **39**:245–322.

GRAVE, B. H. 1922. An analysis of the spawning habits and spawning stimuli of *Chaetopleura apiculata* Say. *Biol. Bull.* **42**:234–56.

———. 1927. An analysis of the spawning habits and spawning stimuli of *Cumingia tellinoides*. *Ibid.* **52**:418–36.

———. 1928. Vitality of the gametes of *Cumingia tellinoides*. *Ibid.* **54**:351–62.

GRAVE, B. H., and R. C. DOWNING. 1928. The longevity and swimming ability of spermatozoa. *Jour. Exp. Zoöl.* **51**:383–88.

GRAY, JAMES. 1913. The electrical conductivity of fertilized and unfertilized eggs. *Jour. Mar. Biol. Assoc.* N.S. **10**:50–59.

———. 1913*a*. The effects of hypertonic solutions upon the fertilized eggs of *Echinus* (*E. esculentus* and *E. acutus*). *Quar. Jour. Micr. Sci.* **58**:447–81.

———. 1915. Note on the relation of spermatozoa to electrolytes and its bearing on the problem of fertilization. *Ibid.* **61**:119–26.

———. 1916. The electrical conductivity of echinoderm eggs and its bearing on the problems of fertilization and artificial parthenogenesis. *Phil. Trans. Roy. Soc. London* B. **207**:481–529.

———. 1920. The relation of spermatozoa to certain electrolytes II. *Proc. Roy. Soc. London* B. **91**:147–57.

———. 1922. A critical study of the facts of artificial fertilization and normal fertilization. *Quar. Jour. Micr. Sci.* **66**:419–37.

———. 1928. Effect of dilution on the activity of spermatozoa. *Brit. Jour. Exp. Biol.* **5**:337–45.

———. 1928*a*. Senescence of spermatozoa. *Ibid.* **5**:345–61.

GREENLEAF, W. E. 1924. The influence of volume of culture media and cell proximity on the rate of reproduction of Protozoa. *Proc. Soc. Exp. Biol. and Med.* **21**:405–6.

———. 1926. The influence of volume of culture medium and cell proximity on the rate of reproduction of Infusoria. *Jour. Exp. Zoöl.* **46**:143–69.

GREGORY, L. H. 1917. Effect of starvation on wing development of *Microsiphon destructa*. *Biol. Bull.* **33**:296–303.

GROSVENOR, C. H., and G. SMITH. 1913. The life cycle of *Moina rectirostris*. *Quart. Jour. Micr. Sci.* **58**:511–22.

GROTE, W., C. VOGT, and B. HOFER. 1909. *Die Süsswasserfische von Mitteleuropa.* Leipzig: Engelmann. 558 pp.

GUDGER, E. W. 1919. A historical note on the synchronous flashing of fireflies. *Science* **50**:188–90.

———. 1929. Some instances of supposed sympathy among fishes. *Sci. Mon.* **29**:266–71.

GURWITSCH, A. 1923. Ausstrahlung einer spezifischen mitotischen Energie. *Arch. f. Entw. d. Or.* **180**:11–41.

———. 1926. *Das Problem der Zellteilung, physiologisch betrachtet.* Berlin: J. Springer. 221 pp.

GURWITSCH, A., and NINA GURWITSCH. 1924. Fortgesetzte Untersuchungen über mitogenetische Strahlung und Induktion. *Arch. Mikrosk. Anat. w. Entwicklungsmech.* **103**:68–80.

GURWITSCH, LYDIA. 1924. Untersuchungen über mitogenetische Strahlung. *Ibid.* **103**:483–90.

HABERLANDT, G. 1919. Zur Physiologie der Zellteilung. *Sitzungsber. preuss. Akad. Wiss. Berlin.* **20**:322–48.

———. 1921. Wundhormone als Erreger von Zellteilungen. *Beitr. allg. Bot.* **2**:1–53.

———. 1922. Über Zellteilungshormone und ihre Beziehungen zur Wundheilung, Befruchtung, Parthenogenesis, und Adventorembryologie. *Biol. Zentralbl.* **42**:145–72.

HAJÓS, K. 1922. Beiträge zur Frage der wachstumhemmenden Wirkung von Bouillonkulturen. *Centralbl. f. Bakt.* I. Orig. **88**:583–85.

HAMILTON, G. V. 1914. A study of sexual tendencies in monkeys and baboons. *Jour. An. Behav.* **4**:295–318.

HARNLY, M. H. 1929. An experimental study of environmental factors in selection and population. *Jour. Exp. Zoöl.* **53**:141–70.

HARRISON, R. G. 1928. On the status and significance of tissue culture. *Arch. Exp. Zellforsch.* **6**:4–27.

HARTLAND, E. S. 1921. *Primitive Society: The Beginnings of the Family and the Reckoning of Descent.* New York: E. P. Dutton. 180 pp.

HARTMANN, O. 1919. Über das Verhalten der Zell-, Kern-, und Nucleolengrösse und ihrer gegenseitigen Beziehungen bei Cladoceren während des Wachstums, des Generationzyklus und unter dem Einfluss äusserer Faktoren. *Arch. f. Zellforsch.* **15**:1–94.

HARTMANN, R. 1880. *Anthropoid Apes.* New York: Appleton. 326 pp.

HARTOG, M. 1895. On certain habits and instincts of social insects. *Science* **1**:98–101.

HAVILAND, MAUD. 1926. *Forest, Steppe and Tundra.* Cambridge University Press. 217 pp.

HAWKES, O. A. M. 1926. On the massing of the lady bird, *Hypodermia convergens* in the Yosemite Valley. *Proc. Zoöl. Soc. London* 1926: 693–705.

HEADLEE, T. J. 1910. Burning chinch bugs. *Circ. Kan. State Exp. Sta. 16.* 7 pp.

HEATH, HAROLD. 1902. The habits of the California termites. *Biol. Bull.* **4**:44–63.

———. 1924. Certain features of the termite caste problem. *Anat. Rec.* **29**:115.

HEATON, T. B. 1926. The nutritive requirements of growing cells. *Jour. Path. and Bact.* **29**:293–306.

HEBARD, M. 1925. The Orthoptera of South Dakota. *Proc. Acad. Nat. Sci. Phila.* **77**:33–155.

HEGNER, R. W., F. M. ROOT, and D. L. AUGUSTINE. 1929. *Animal Parasitology.* New York: Century Co. 640 pp.

HEMPELMANN, F. 1911. Zur Naturgeschichte von *Nereis dumerilii* Aud. et Edw. *Zoologica.* **25**:1–135.

HENKING, H. 1891. Untersuchungen über die ersten Entwicklungsvorgänge in den Eiern der Insekten. II. Über Spermatogenese und deren Beziehung zur Eientwicklung bei *Pyrrhocoris apterus* L. *Zeitschr. f. wiss. Zoöl.* **51**:685.

HENRICI, A. T. 1928. *Morphologic Variation and the Rate of Growth of Bacteria.* Springfield, Ill.: C. C. Thomas. 194 pp.

HENTIC, H. VON. 1923. Reaction of animals to changes in physical environment. *Jour. Comp. Psych.* **3**:61–71.

HERBST, C. 1928. Untersuchungen zur Bestimmung des Geschlechtes. I. Ein neuer Weg zur Lösung des Geschlechtsbestimmungsproblems bei *Bonellia viridis. Sitzungsber. d. Heidelberger Akad. d. wiss. math.-naturw. Klasse.* 1928: 2 Abh., 19 pp.

HERING, M. 1926. *Biologie der Schmetterlinge.* Berlin: Springer. 480 pp.

HESS, C. VON. 1913. Gesichtsinn. *Winterstein's Handb. d. vergl. Physiol.* **4**:555–840.

HESS, W. N. 1920. Notes on the biology of some common *Lampyridae. Biol. Bull.* **38**:39–77.

HESS, W. R. 1926. Die Temperaturregulierung im Bienenvolk. *Zeitschr. f. vergl. Physiol.* **4**:465–87.

HESSE, R. 1924. *Tiergeographie.* Jena: Fischer. 613 pp.

HEYMONS, R. 1915. Vierfüssler, Insekten und Spinnenkerfe. *Brehms Tierleben* **2.** Leipzig und Wien: Bibliographisches Institut.

HILL, R. B. 1926. The estimation of the number of hookworms harbored by the use of the dilution egg count method. *Amer. Jour. Hygiene* **6**:19–41 (Suppl.).

HIMMER, A. 1925. Körpertemperaturmessungen an Bienen und anderen Insekten. *Erlanger Jb. Bienenkde.* **3**:(H 1) 144.

———. 1926. Der soziale Wärmhaushalt der Honigbiene. I. Die Wärme im nichtbrütenden Wintervolk. *Ibid.* **4**:1–50.

———. 1927. Der soziale Wärmhaushalt der Honigbiene. II. Die Wärme der Bienenbrut. *Ibid.* **5**:(H 1).

———. 1927*a*. Ein Beitrag zur Kenntnis des Wärmehaushalts im Nestbau sozialer Hautflüger. *Zeitschr. f. vergl. Physiol.* **5**:375–89 (citations).

HINE, J. S. 1906. Habits and life histories of some flies of the family Tabanidae. *Techn. Ser. U.S. Dept. Agric. Ent.* No. 12: 19–38.

HINRICHS, MARIE. 1926. Modification of development on the basis of differential susceptibility to radiation. III. *Arbacia* germ cells and (*a*) ultraviolet radiation, (*b*) visible radiation following sensitization. *Biol. Bull.* **50**:455–72.

———. 1926*a*. The effect of ultraviolet radiation on the fertilizing power of *Arbacia* sperm. *Ibid.* **50**:473–89.

———. 1927. Ultraviolet and *Arbacia* fertilization. *Ibid.* **53**:416–37.

HOFFBAUER, C. 1902. Uber den Einfluss des Wasser volumens auf das Wachsthum der Fische. *Allg. Fisch. Zeit.* **27**:103–4, 119–22.

HOFFMANN, C. C.; A. DAMPF; and VARELA. 1925. Informe de la comision cientifica exploradora de la plaga de la langosta en el Estado Veracruz. *Monog. Inst. Hyg. Mexico* 3. 140 pp.

HOGG, J. 1854. Observations on the development and growth of *Lymnaeus stagnalis*. *Quar. Jour. Micr. Soc.* **2**:91–103.

HOLMES, S. J. 1903. Sex recognition among amphipods. *Biol. Bull.* **5**:288–92.

———. 1905. Reactions of *Ranatra* to light. *Jour. Comp. Neur. and Psychol.* **15**:305–49.

———. 1911. Reactions of mosquitoes to light in different periods of their life history. *Jour. An. Behav.* **1**:29–32.

———. 1926. *Life and Evolution.* New York: Harcourt Brace. 449 pp.

HOLMQUIST, A. M. 1926. Studies in arthropod hibernation. *Ann. Ent. Soc. Amer.* **19**:395–429.

———. 1928. Life history and habits of the mound building ant. *Formica ulkei*. *Ecology* **9**:70–87.

———. 1928*a*. Studies in arthropod hibernation. II. The hibernation of the mound building ant, *Formica ulkei*. *Physiol. Zoöl.* **1**:325–58.

HOLT, E. W. L. 1898. On the breeding of the dragonet (*Callionymus lyra*). *Proc. Zoöl. Soc. London.* 1898: 281–315.

HOPKINS, A. D. 1898. Mutual aid among beetles. *Canadian Entom.* **30**:21–29.

HORNADAY, W. T. 1879. On the species of Bornean apes, with notes on their habits. *Proc. Amer. Assoc. Adv. Sci.* 1879: 438–56.

———. 1885. *Two Years in the Jungle.* New York: Scribners. 512 pp.

HORSFALL, R. B. 1920. Remarkable habits of the sage grouse. *Zoölogica* (Sci. Con. N.Y. Zoöl. Soc.) **2**:243–50.

HOWARD, H. E. 1920. *Territory in Bird Life.* London: Murray. 208 pp.

HOWARD, L. O. 1898. Swarming of western flea beetle. *Bull. Div. Entom. U.S. Div. Agric.* N.S. **18**:100.

HOWARD, L. O. 1901. Migration of the western willow flea beetle. *Ibid.* N.S. **30**:97.

———. 1922. *English Translation of Bouvier's "La vie psychique des insectes."* New York: Century Co. 377 pp.

HOWARD, L. O., H. G. DYAR, and F. KNAB. 1912. The mosquitoes of North and Central America and the West Indies. *Carnegie Inst. of Wash. Publication No. 159.* **1**:120–39.

HOWELL, A. B. 1920. Contribution to the life history of the California mastiff bat. *Jour. Mammal.* **1**:111–17.

———. 1920*a*. Some Californian experiences with bat roosts. *Ibid.* **1**:169–77.

HOWITT, A. W. 1904. *The Native Tribes of Southeast Australia.* New York: Macmillan. 819 pp.

HOWLEY, JAMES P. 1884. The Canada goose. *Auk* **1**:310–13.

HUBER, JACOB. 1907. The founding of colonies by *Atta sexdens. Ann. Rept. Smithsonian Inst. 1906*, pp. 355–72. Translated from *Biol. Centralbl.* **25**:606–19, 624–35.

HUNT, CRESWELL J. 1906. Change of location of a crow roost. *Auk* **23**:429–31.

HUXLEY, JULIAN. 1922. Dedifferentiation in *Echinus* larvae and its relation to metamorphosis. *Biol. Bull.* **43**:210–38.

———. 1927. Editor's introduction to Elton's *Animal Ecology*, pp. ix-xvii. London: Sidgwick & Jackson.

HYMAN, LIBBIE H. 1923. Physiological studies on Planaria. V. Oxygen consumption of pieces with respect to length, level, and time after section. *Jour. Exp. Zoöl.* **37**:47–68.

IHERING, H. VON. 1894. Die Ameisen von Rio Grande do Sol. *Berlin Ent. Zeitsch.* **39**:321–446.

———. 1912. Biologie und Verbreitung der Brasilianischen Arten von Eciton. *Ent. Mittheil.* **1**:226–35.

IMMS, A. D. 1913. On *Embia major*, n.s. from the Himalayas. *Trans. Linn. Soc. London.* **11**:167–95.

ISELEY, F. B. 1914. Experimental study of the growth and migration of fresh-water mussels. *Rept. U.S. Comm. Fish. 1913.* Appendix III. 24 pp.

IVANITZKY, NADINE. 1922. Les institutions des primitif Australiens. *Rev. d. Inst. d. Sociol.* **2**:175–220.

IZUKA, A. 1903. Observations on the Japanese palolo. *Jour. Coll. Sci. Imp. Univ. Tokyo 17.* Art. 11. 37 pp.

JAHN, THEODORE L. 1929. Studies on the physiology of the euglenoid flagellates. I. The relation of the density of population to the growth rate of *Euglena. Biol. Bull.* **57**:81–106.

JENNINGS, H. S. 1901. Aggregations in *Paramecium* as tropisms. *Science* **13**:74–75.

———. 1906. *Behavior of Lower Organisms.* New York: Columbia University Press. 366 pp.

———. 1913. The effect of conjugation in *Paramecium*. *Jour. Exp. Zoöl.* **14**:280–391.

JEWELL, M. E., and H. BROWN. 1929. Studies on Northern Michigan bog lakes. *Ecology* **10**:427–76.

JOHNSON, H. B. 1926. A further contribution to our knowledge of the bionomics and control of the migratory locust, *Schistocerca gregaria*, in the Sudan. *Bull. Ent. Sect. Wellcome Trop. Res. Lab.* **22**:1–14.

JOHNSON, W. H. 1929. The reaction of *Paramecium* to solutions of known hydrogen-ion concentration. *Biol. Bull.* **57**:199–224.

JONES, J. W. L. 1903. *Sociality and Sympathy. An Introduction to the Ethics of Sympathy.* New York: Macmillan. 91 pp.

JONES, LYNDS. 1897. The Oberlin summer grackle roost. *Wilson Ornith. Chapter of Agassiz Assoc. Bull.* **15**:37–56.

JORDAN, H. 1881. Einfluss des bewegten Wassers auf die Gestaltung der Muscheln aus der Familie Najades, Lam. *Biol. Centralbl.* **1**:392–99.

JUST, E. E. 1914. Breeding habits of the heteronereis form of *Platynereis megalops* at Woods Hole, Mass. *Biol. Bull.* **27**:201–12.

———. 1915. Analysis of fertilization of *Platynereis megalops*. *Ibid.* **28**:93–114.

KAEDING, B. 1905. Bird life on the Farollone Islands. Nesting colonies. *Condor* **5**:121–27.

KAFKA, G. 1914. *Einführung in der Tierpsychologie auf experimenteller und ethnologischer Grundlage.* I. Bd. *Die Sinne der Wirbellosen.* Leipzig: Barth. 593 pp.

KALMUS, H. 1929. Versuche über die Teilung von *Paramecium caudatum* in der Kapillare nebst Bemerkungen über den sogenannten Raumfaktor. *Arch. f. Protistenk.* **66**:402–8.

KAMMERER, P. 1909. Allgemeine Symbiose und Kampf ums Dasein als gleichberechtigte Triebkräfte der Evolution. *Arch. f. Rass. und Ges. Biol.* **6**:585–608.

———. 1913. Pansymbiose. *Naturwissensch. Jahrg.* **1**:1222–25.

KARSCH, F. 1900. Päderastie und Tribadie bei den Tieren auf Grund der Literatur. *Jahrb. f. Sex. Zwischenstufen* **2**:126–60.

KATZ, D. 1926. Sozialpsychologie der Vögel. *Ergebn. d. Biol.* **1**:447–78.

KATZ, D., and A. TOLL. 1923. Die Messung von Charakter- und Begabungsunterschieden bei Tieren (Versuch mit Hühnern). *Zeitschr. f. Psychol. u. Physiol. d. Sinnesorg.*, Abt. I, *Zeitschr. f. Psych.* **93**:287–311.

KAWAJIRI, MINORU. 1928. On studies of the population density of cultured fishes. I. On the influence of population density of fishes upon the survival and the rate of growth. *Jour. Imp. Fish. Inst. Tokyo.* **24**:8–11.

KEASBEY, L. M. 1903. Co-operation, coercion and competition. *Science* **17**:922.

KELLOGG, VERNON. 1903. Some insect reflexes. *Ibid.* **18**:693–96.

———. 1907. Review of *The Flock* by M. Austin. *Ibid.* **25**:179–81.

KENDALL, W. C. 1902. Notes on the sticklebacks. *Ibid.* **15**:588.

KENNEDY, C. H. 1927. Some non-nervous factors that condition the sensitivity of insects to moisture, temperature, light and odors. *Ann. Ent. Soc. Amer.* **20**:87–106.

KENNEALY, A. E. 1906. An approximate law of fatigue in the speeds of racing animals. *Proc. Amer. Acad. Arts and Sci.* **42**:273–331.

KENYON, F. C. 1898. The occurrence in great abundance of insects ordinarily merely common. *Science* **8**:551–52.

KESSLER, KARL F. 1880. On the law of mutual aid. Lecture before Russian Congress of Naturalists (*see* Kropotkin).

KIRKALDY, G. 1903. Upon maternal solicitude in *Rhynchota* and other non-social insects. *Ann. Rept. Smithsonian Inst. 1903*, pp. 577–85. Revised from *Entomologist* **5**:113–20.

KNAB, F. 1906. The swarming of *Culex pipiens*. Bibliography of swarming aggregations in Diptera. *Psyche* **13**:123–33.

———. 1907. The swarming of *Anopheles punctipennis* Say. *Ibid.* **14**:1–4.

KÖHLER, W. 1925. *The Mentality of Apes.* Translated by E. Winter. New York: Harcourt Brace. 342 pp.

KOFOID, C. A., and OLIVE SWEZY. 1919. Studies on the parasites of the termites. *Univ. Calif. Publ.* **20**:Nos. 1–4, pp. 1–116.

KRAJNIK, B. 1922. Über eine Modifikation des Mikrorespirationsapparatus. *Biochem. Zeitschr.* **130**:286–93.

KRIZENECKY, J. 1923. Über ein homotypisches Synaporium bei den Enchytraeiden. *Zool. Anz.* **55**:80–88.

KROGH, A. 1916. *The Respiratory Exchange of Animals and Man.* New York: Longmans, Green & Co. 173 pp.

KROPOTKIN, P. 1914. *Mutual Aid, a Factor of Evolution.* 2d ed. New York: Knopf. 223 pp.

KUCZYNSKI, ROBERT E. 1928. *The Balance of Births and Deaths.* New York: Macmillan. 140 pp.

KUENKEL, K. 1916. *Zur Biologie der Lungenschnecken.* Heidelberg. 440 pp.

KUESTER, E. 1907. *Anleitung zur Kultur der Mikroorganismen.* Leipzig und Berlin: Teubner. 201 pp.

———. 1908. Keimung und Entwicklung von Schimmelpilzen in gebrauchten Nahrlösungen. *Ber. d. d. Bot. Gesellsch.* **26a**:246–48.

KULAGIN, N. M. 1899. Zur Biologie der Infusorien. *Physiol. russe.* **1**:269–79.

LA BAUME, W., H. BUCHER, and OTHERS. 1918. Die Heuschreckenplage und ihre Bekämpfung. *Monog. angew. Ent. Berlin.* **11**:1–20.

LAMEERE, A. 1922. The origin of insect societies. *Ann. Rept. Smithsonian Inst. 1920*, pp. 511–22.

LANESSAN, J. M. A. 1881. La lutte pour l'existence et l'association pour la lutte. From *Bibliothèque biologique*, edited by Doin (*see* Kropotkin and Lanessan).

———. 1903. *La lutte pour l'existence et l'évolution des sociétés.* Paris: Alcan. 280 pp.

LANG, ANDREW. 1903. *Social Origins.* New York: Longmans, Green. 311 pp.

LANGE, D. 1905. A remarkable flight of *Corixa* water boatmen. *Canadian Entomol.* **37**:364.

LANGHANS, V. H. 1909. Über experimentelle Untersuchungen zu Fragen der Fortpflanzung, Variation und Vererbung bei Daphniden. *Verh. deutsch. Zool. Ges.* **19**:281–91.

LAPICQUE, L. 1921. Note. *Compt. rend. Soc. de biol.* **84**:920.

LAURENT, PHILIP. 1917. The supposed synchronal flashing of fireflies. *Science* **45**:44.

LEGENDRE, R. 1907. Sur un facteur important du nanisme expérimental: les excreta *Comp. Rend. Assoc. Française p. l'Avance d. Sci.* 36. Session (*Rheims*) **2**:607–10.

———. 1908. Recherches sur le nanisme expérimental: influence des excreta. *Arch d. Zool. Exp. et Gen.* **4**:77–84.

———. 1925. La lune et les êtres marins. *Rev. Scientif. Année* **63**:225–36.

LEIDY, J. 1874. Parasites of termites. *Jour. Acad. Nat. Sci. Phila.* N.S. **8**: 425–47.

———. 1877. On intestinal parasites of *Termes flavipis. Proc. Acad. Nat. Sci. Phila.* 1877:146–49.

LEPINEY, J. DE. 1928. Sur le comportement de *Schistocerca gregaria. Forsk. Rev. Path. Vég. Ent. Agric. Mars* 1928: 1–15.

LEWIS, L. L. 1911. Vitality of reproductive cells. *Okla. Agric. Sta. Bull.* **96**:30–47.

LIBBY, O. G. 1899. Nocturnal flight of migrating birds. *Auk* **16**:140–46.

LIEBIG, J. 1851. *Chemische Briefe.* Heidelberg: Winter. 668 pp.

LILLIE, F. R. 1915. Sperm agglutination and fertilization. *Biol. Bull.* **28**: 18–33.

———. 1917. The free-martin: a study of the action of sex hormones in the foetal life of cattle. *Jour. Exp. Zoöl.* **23**:371–452.

———. 1919. *Problems of Fertilization.* University of Chicago Press. 278 pp.

———. 1921. Studies in fertilization. X. The effects of copper salts on the fertilization reaction in *Arbacia*, and a comparison of mercury effects. *Biol. Bull.* **41**:125–43.

LILLIE, F. R., and E. E. JUST. 1913. Breeding habits of the heteronereis form of *Nereis limbata*, at Woods Hole, Mass. *Biol. Bull.* **24**:147–60.

———. 1924. Fertilization, chap. viii in Cowdry's *General Cytology*, pp. 451–536. Chicago: University of Chicago Press.

LILLIE, R. S. 1901. Aggregation of *Arenicola* larvae due to common reactions. Reported from the Zoölogical Club of the University of Chicago. *Science* **13**:73–74.

LILLIE, R. S., and M. Baskervill. 1922. The action of ultra-violet rays on starfish eggs. *Amer. Jour. Physiol.* **61**:57–71.

LLOYD-JONES, ORREN, and F. A. Hayes. 1918. The influence of excessive sexual activity of male rabbits. *Jour. Exp. Zoöl.* **25**:463–98.

LOEB, J. 1918. *Tropisms, Forced Movements and Animal Conduct.* New York: Lippincott. 209 pp.

LONGLEY, W. H. 1918. Haunts and habits of tropical fishes. *Amer. Mus. Jour.* **18**:79–88.

LOOMIS, L. M. 1900. California water birds. *Proc. Calif. Acad. Sci. 1899* Part IV. Reviewed by F. A. Lucas in *Science* **11**:397.

LOWIE, R. H. 1920. *Primitive Society.* New York: Boni & Liveright. 572 pp.

LUCANUS, F. VON. 1922. Die Rätsel des Vogelzuges. Reviewed by J. Ritchie. *Nature* **109**:573.

LUCAS, F. A. 1899. Mental traits of the fur seal. *Science* **9**:785–86.

LUTZ, F. E. 1924. Insect sounds. *Bull. Amer. Mus. Nat. Hist.* **50**:333–72.

LYON, E. P. 1904. On rheotropism. I. Rheotropism in fishes. *Amer. Jour. Physiol.* **12**:149–61.

———. 1905. Rheotropism in fishes. *Biol. Bull.* **8**:238–39.

MCATEE, W. L. 1912. Certain phases of the theory of recognition marks. *Auk* **29**:226–33.

———. 1915. A remarkable flight of caddis flies and chironomids. *Science* **42**:694–95.

———. 1924. Do bird families have any permanency? *Condor* **26**:193–94.

MCCLENDON, J. F. 1910. On the effect of external conditions on the reproduction of *Daphnia. Amer. Nat.* **44**:404–11.

MCDERMOTT, F. A. 1916. Flashing of fireflies. *Science* **44**:610.

———. 1917. Observations on the light emission of American *Lampyridae. Canadian Entomol.* **49**:53–61.

MCDOUGALL, W. 1927. *The Group Mind.* Putnam: New York. 301 pp.

MCINDOO, U. E. 1920. The senses of insects. *Ann. Rept. Smithsonian Inst. 1919*, pp. 461–83.

MACKEY, G. H. 1897. Terns of Penikese Island. *Auk* **14**:278–84.

———. 1899. The terns of Muskaget and Penikese Islands, Mass. *Ibid.* **16**:259–66.

MALAQUIN, A. 1901. Le parasitism évolutif des *Monstrillidae. Arch. Zool. Exper. Gen.* **9**:81–232.

MALINOWSKI, B. 1927. *Sex and Repression in Savage Society.* New York: Harcourt, Brace. 285 pp.

MANN, B. P. 1882. Cluster flies, *Pollenia rudis. Psyche* **3**:378–79.

MANN, W. M. 1913. Homing behavior of ants. *Jour. An. Behav.* **3**:429–45.

MARSHALL, W. 1889. Leben und Treiben der Ameisen. *Zool. Vort.* **3** and **4**:1–144.

MARTIN, E. D. 1921. *Behavior of Crowds.* New York: Harpers Bros. 311 pp.

MASSART, J., and E. VANDERVELDE. 1893. Parasitisme organique et parasitisme social. *Bull. Sci. France Belg.* **25**:227–94.

MAST, S. O. 1911. *Light and the Behavior of Organisms.* New York: Wiley & Sons. 410 pp.

———. 1912. Behavior of fireflies (*Photinus puralis*). *Jour. An. Behav.* **2**:256–72.

MAUCK, A. V. 1901. On the swarming and variation in a myriapod (*Fontaria virginiensis*). *Amer. Nat.* **35**:477–78.

MAYER, A. G. 1900. On the mating instinct in moths. *Ann. and Mag. Nat. Hist.* **5**:183–90.

———. 1900*a*. An Atlantic "palolo," *Staurocephalus gregarius. Bull. Mus. Comp. Zoöl.* **36**:1–14.

———. 1902. The Atlantic palolo. *Sci. Bull. Brooklyn Mus. Arts and Sci.* **1**:93–103.

———. 1903. The Atlantic palolo. *Science* **17**:482.

———. 1903*a*. Reactions of lepidopterous larvae. Preliminary report before N.Y. Acad. Sci. *Ibid.* **17**:743.

MAYER, C. 1920. Trapping an elephant herd in Trenggonu. *Asia* **20**:1055–58.

MEDRKIEWICZOWNA, H. 1921. Le rôle de la surface libre du liquide dans le développement des cultures du *Colpidium colpoda. Soc. d. Sci. d. Varsovie.* **1**:1.

MEEK, ALEXANDER. 1920. The physiology of migrations in the sea. *Nature* **105**:197.

MESǏACEV, I. 1927. Einige zoogeographische und faunistische Ergebnisse der Expeditionen des wissenschaftlichen Meeresinstitutes in die nördlichen Meere. *Comp. Rend. Acad. Sci. U.R.S.S.* **14**:202–12.

METCALF, M. M. 1900. Hearing in ants. *Science* **11**:194.

MICHAEL, E. L. 1920. Marine ecology and the coefficient of association. A plea in behalf of quantitative biology. *Jour. Ecol.* **8**:54–59.

MICHAEL, E. L., and G. F. MCEWEN. 1916. Dependence of marine biology upon hydrography and the necessity of quantitative biological research. *Univ. of Calif. Publ. in Zoöl.* **15**:1–280.

MILLER, G. S. 1928. Some elements of sexual behavior in primates and their possible influence on the beginning of human social development. *Jour. Mammal.* **9**:273–93.

MILLER, H. A. 1921. The group as an instinct. *Jour. Amer. Sociol.* **27**:334–43.

MILLER, N. 1909. The American toad. *Amer. Nat.* **43**:641–88, 730–45.

MILLER, R. C. 1921. The flock behavior of the coast bush tit. *Condor* **23**:121–27.

———. 1921*a*. The mind of the flock. *Ibid.* **23**:183–86.

———. 1922. The significance of the gregarious habit. *Ecology* **3**:122–26.

MILNER, R. D., and G. S. DEMUTH. 1921. Heat production of honey bees in winter. *U.S. Dept. Agric. Bull. 988*, 14 pp.

MINER, JOHN THOMAS. 1923. *Jack Miner and the Birds, and Some Things I Know about Nature.* Chicago: Reilly & Lee. 178 pp.

MINNICH, D. E. 1920. Relation of phototropism to swarming in the honey bee, *Apis mellifera*. *Jour. Psychobiol.* **2**:177–80.

———. 1925. The reactions of the larvae of *Vanessa antiopa* to sounds. *Jour. Exp. Zoöl.* **42**:443–69.

MOCKERIDGE, F. A. 1920. The occurrence and nature of the growth-promoting substances in various organic manurial composts. *Biochem. Jour.* **14**:432–50.

———. 1924. The formation of plant growth-promoting substances by micro-organisms. *Ann. Bot.* **38**:723–34.

———. 1927. Auximone theory. *Bot. Gaz.* **83**:314–17.

MONTGOMERY, T. H., JR. 1910. The significance of courtship and secondary sexual characters of areneads. *Amer. Nat.* **44**:151–77.

MORGAN, L. H. 1868. Systems of consanguinity and affinity of the human family. *Smithsonian Contr. to Knowledge* **17**:487–90.

———. 1878. *Ancient Society*. New York: Holt. 560 pp.

MORRISON, T. F. 1929. Observations on the synchronous flashing of fireflies in Siam. *Science* **69**:400–401.

MORSE, E. S. 1916. Fireflies flashing in unison. *Ibid.* **43**:169–70.

———. 1916*a*. Fireflies flashing in unison. *Ibid.* **44**:387–88.

MORSE, MAX. 1903. Unusual abundance of a myriapod, *Parajulus pennsylvanius*. *Ibid.* **18**:59–60.

MOSIER, C. A., and T. E. SYNDER. 1918. Notes on gadflies in the Florida Everglades. *Proc. Ent. Soc. Wash.* **20**:115–26.

———. 1919. Further notes on Tabanidae in Florida Everglades. *Ibid.* **20**:182–84.

———. 1919*a*. Notes on the seasonal activity of Tabanidae in the lower Everglades of Florida. *Ibid.* **21**:186–96.

MOTTRAM, J. C. 1925. The *in-vitro* cultivation of tissues with reference to the production of cancer by means of radium and X-rays. *Brit. Jour. Exp. Path.* **6**:53–62.

MOZLEY, ALAN. 1928. The variation of *Lymnaea traskii* in pond and lake habitats. *Amer. Nat.* **62**:286–88.

MULLER, S., and H. SCHLEGEL. 1840. Beiträge zur Naturgeschichte des Orangs-autangs. *Forieps Notizen* **20**:336–43.

MUNDT, A. H. 1882. Migration of dragon flies. *Canadian Entomol.* **14**:56–57.

MYERS, E. C. 1927. Relation of density of population and certain other factors to survival and reproduction in different biotypes of *Paramecium caudatum*. *Jour. Exp. Zoöl.* **49**:1–43.

MYERS, J. G. 1927. Ethological observations on some *Pyrrhocoridae* of Cuba. *Ann. Ent. Soc. Amer.* **20**:279–301.

NATZMER, G. VON. 1913. Mixed colonies of ants. *Biol. Centralbl.* **33**:666–67.

NELSON, E. W. 1903. Notes on the Mexican cormorant nesting colonies. *Condor* **5**:139–45.

NELSON, G. 1911. Notes on Pelican Island. *Auk* **28**:393–97.

NEWMAN, H. H. 1917. Synchronic behavior. *Science* **45**:44.

NOBLE, G. K. 1929. The relation of courtship to the secondary sexual characters of the twolined salamander, *Eurycea bislineata* (Green). *Amer. Mus. Novitates* **362**:1–5.

NOBLE, G. K., and E. J. FARRIS. 1929. The method of sex recognition in the wood-frog *Rana sylvatica*. *Ibid.* **363**:1–17.

NOBLE, G. K., and RUTH C. NOBLE. 1923. The Anderson tree frog (*Hyla andersoni* Baird). Observations on its habits and life history. *Zoölogica* **11**: 417–55.

NORDQUIST, O. 1924. Times of entering of the Atlantic salmon, *Salmo salar* L., in the rivers. *Rapp. Proc. Verb. Con. Perm. Int. Exp. Mer.* **33**:1–58.

O'BRIEN, R. A. 1910. Remarks on the habits of the green tree ant of Australia. *Proc. Zoöl. Soc. London* **1910**: 138–50.

OHAUS, F. 1899–1900. Bericht über eine entomologische Reise nach Centralbrasilien. *Stettin Ent. Zeitg.* **60**:204–45; **61**:164–91, 193–274.

ORTON, J. H. 1926. On lunar periodicity in spawning of normally grown Falmouth oysters (*O. edulis*) in 1925, with a comparison of the spawning capacity of normally grown and dumpy oysters. *Jour. Marine Biol. Assoc.* N.S. **14**:199–225.

PALMER, R. H. 1916. A visit to Hat and Egg Island, Great Salt Lake. *Condor* **17**:113–23.

PAPANICOLAU, G. 1910. Über die Bedingungen der sexuellen Differenzierung bei Daphniden. *Biol. Centralbl.* **30**:430–40.

———. 1910*a*. Experimentelle Untersuchungen über die Fortpflanzungsverhältnisse der Daphniden. *Ibid.* **30**:689–92, 737–50, 752–74, 785–802.

PARHON, MARIE. 1909. Les exchanges nutritifs chez les abeilles pendant les quatres saisons. *Ann. d. Sci. Nat. Zool.* Ser. **9**:1–59.

PARK, R. E., and E. W. BURGESS. 1921. *Introduction to the Science of Sociology*. Chicago: University of Chicago Press. 1040 pp.

PARKER, G. H. 1902. Diurnal depth migration in the copepod *Labidocera aestiva*. From Proc. Boston Soc. Nat. Hist., reported by G. M. Allen, secretary, in *Science* **15**:471.

———. 1919. *The Elementary Nervous System*. New York: Lippincott. 229 pp.

———. 1919*a*. Organization of *Renilla*. *Jour. Exp. Zoöl.* **27**:499–507.

———. 1920. Activities of colonial animals. I. Circulation of water in *Renilla*. *Ibid.* **31**:343–65.

———. 1920*a*. Activities of colonial animals. II. Neuro-muscular movements and phosphorescence in *Renilla*. *Ibid.* **31**:475–513.

PARKER, J. R. 1925. Grasshopper migrations and modifications. *Ecology* **6**:458–59.

PARR, A. E. 1927. A contribution to the theoretical analysis of the schooling behavior of fishes. *Occ. Papers Bingham Oceanographic Coll.* **1**:1–32.

PARSHLEY, H. M. 1917. Insects in ocean drift. *Canadian Entomol.* **49**:45–48.

PASTEUR, L. 1869. Mémoire sur la fermentation alcoolique. *Ann. d. Chimie et d. Physique.* Ser. 3. **58**:323–426.

PATTEN, W. 1916. Co-operation as a factor in evolution. *Proc. Amer. Phil. Soc.* **55**:505–32.

———. 1920. *The Grand Strategy of Evolution.* Boston: Badger. 429 pp.

PAUL. 1899. Besitzen Raufen Verwandtschaftsinn? *Ent. Zeitsch. Intern. Ent. Ver.* **12**:141–42.

PAWLOW, P. N. 1925. Über die Adsorption. XI. Die Lebensdauer von Organismen in ihrer Abhängigkeit von der Konzentration und dem Volumen giftig wirkender Lösungen und von der Bevölkerungsdichtigkeit. *Kolloid. Zeitschr.* **37**:105–11.

PEARL, R. 1900. Studies on electrotaxis: I. On the reactions of certain Infusoria to the electric current. *Amer. Jour. Physiol.* **4**:96–123.

———. 1925. *The Biology of the Growth of Population.* New York: Knopf. 260 pp.

———. 1928. *The Rate of Living.* New York: Knopf. 185 pp.

PEARL, R., A. ALLEN, and W. H. D. PENNIMAN. 1916. A new synthetic culture medium and its influence on fertility at different densities of population. *Amer. Nat.* **60**:357–66.

PEARL, R., and FRANCES J. DUNBAR, 1905. Some results of a study of variation in *Paramecium.* *7th Ann. Rept. Mich. Acad. Sci.*, pp. 77–86.

PEARL, R., J. R. MINER, and S. L. PARKER. 1927. Experimental studies on the duration of life. XI. Density of population and life duration in *Drosophila.* *Amer. Nat.* **61**:289–318.

PEARL, R., and S. L. PARKER. 1922. On the influence of density of population upon rate of reproduction in *Drosophila.* *Proc. Nat. Acad. Sci.* **8**:212–19.

———. 1922*a*. Experimental studies on the duration of life. IV. Data on the influence of density of population on duration of life in *Drosophila.* *Amer. Nat.* **56**:312–21.

PEARL, R., and F. M. SURFACE. 1909. A biometrical study of egg production in the domestic fowl. I. Variation in annual egg production. *U.S. Dept. Agric. Bur. Animal Industry Bull. 110*, Part I, pp. 1–80.

———. 1911. Seasonal distribution of egg production. *Ibid.* Part II, pp. 81–170.

PEARSE, A. S. 1909. Observation on copulation among crawfishes, with special reference to sex recognition. *Amer. Nat.* **43**:746–53.

———. 1914. Habits of fiddler crabs. *Ann. Rept. Smithsonian Inst. 1913*, pp. 415–28.

———. 1926. *Animal Ecology.* New York: McGraw Hill. 417 pp.

PEARSE, A. S., and H. ACHTENBERG. 1920. Habits of yellow perch in Wisconsin lakes. *Bull. U.S. Fish.* **36**:293–366.

PEEBLES, F. 1929. Growth regulating substances in echinoderm larvae. *Biol. Bull.* **57**:176–87.

PEMBERTON, J. P. 1922. A large tern colony in Texas. *Condor* **24**:31–48.

PENFOLD, W. J. 1914. On the nature of bacterial lag. *Jour. Hygiene* **14**:215–41.

PESKETT, G. L. 1924. Allelocatalysis and the growth of yeast. *Jour. Physiol.* **59**:33.

———. 1924*a*. Allelocatalysis and the growth of yeast. *Biochem. Jour.* **18**:866–71.

———. 1925. Studies on the growth of yeast. I. The influence of the volume of culture medium employed. *Ibid.* **19**:464–73.

———. 1925*a*. Studies on the growth of yeast. II. A further note on allelocatalysis. *Ibid.* **19**:474–76.

PETERS, A. R. 1923. Substances needed for the growth of a pure culture of *Colpidium colpoda*. *Jour. Physiol.* **55**:1–32.

PETERSEN, C. G. J., and P. B. JENSEN. 1911. Valuation of the sea: I. Animal life of the sea bottom, its food and quantity. *Rept. Danish Biol. Sta.* **20**:1–76.

PETERSEN, W. 1929. The relation of density of population to rate of reproduction in *Paramecium caudatum*. *Physiol. Zoöl.* **2**:221–54.

PETRUCCI, R. 1906. Origine polyphilétique, homotypie et non-comparabilité directe des sociétés animales. *Notes et Mem. Inst. Solvay.* Paris: Giard et Brière. 126 pp.

PETRUNKEVITCH, A. 1910. Courtship of *Dysdera crocata*. *Biol. Bull.* **19**:127–29.

PHILIPS, J. C. 1916. Behavior and make-up of the migrating flocks of Canada geese. *Auk* **33**:24–27.

PHILLIPS, E. F. 1911. Bees. *U.S. Dept. Agric. Farmers Bull. 447.* 48 pp.

———. 1917. *Beekeeping.* New York: Macmillan. 457 pp.

PHILLIPS, E. F., and G. DEMUTH. 1914. Temperature of the honey bee cluster in winter. *Bull. U.S. Dept. Agric. 93.* 16 pp.

PHILLIPS, J. F. V. 1925. The Kuysna elephant: a brief note on their history and habits. *So. Africa Jour. Sci.* **22**:287–93.

———. 1926. "Wild pig" (*Potamochoerus choeropotamus*) at the Kuysna: Notes by a naturalist. *Ibid.* **23**:655–60.

PICKERING, S. 1917. The effect of one plant on another. *Ann. Bot.* **31**:183–87.

PIERON, H. 1921. Notes on the work of Drzewina and Bohn. *Compt. rend. Soc. de biol.* **84**:919–20.

PILSBRY, H. A., and J. BERQUEART. 1927. The aquatic mollusks of the Belgian Congo with a geographical and ecological account of Congo malacology. *Bull. Amer. Mus. Nat. Hist.* **53**:69–602.

PING, CHI. 1917. Observations on *Chironomus decorus* Johansen. *Canadian Entomol.* **49**:418–26.

PIPER, C. V. 1897. A remarkable sembling habit of *Coccinella transversoguttata*. *Ent. News* **8**:49–51.

PIRSCH, G. B. 1923. Studies on the temperature of individual insects with special reference to the honey bee. *Jour. Agric. Res.* **24**:275–87.

PITT, FRANCES. 1927. *Animal Mind.* London: Allen & Unwin. 340 pp.

PLATH, O. E. 1927. Notes on the hibernation of several North American bumblebees. *Ann. Ent. Soc. Amer.* **20**:181–93.

PLOTNIKOV, V. I. 1915. Reports on the work of the Turkestan Entomological Station for 1912, 1913, 1914, and part of 1915. (In Russian.) Tashkent. 60 pp.

———. 1924. Some observations on the variability of *Locusta migratoria* in breeding experiments. *Bull. Entomol. Res.* London **14**:241–43.

———. 1927. *Locusta (Pachytylus) migratoria*, L. and *danica*, L. as independent forms and their derivatives. (In Russian.) Tashkent. 33 pp.

PLUNKETT, C. R. 1926. The interaction of genetic and environmental factors in development. *Jour. Exp. Zoöl.* **46**:181–244.

POPOVICI-BAZNOSANU, A. 1921. L'influence de quelques facteurs sur l'accroissement des Gasteropodes d'eau douce. *Arch. d. Zool. Exp. et Gen.* **60**:501–21.

PREVOST, J. L., and J. B. DUMAS. 1824. Deuxième mémoire sur la génération. Rapports de l'œuf avec la liqueur fécondante Phénomènes appréciable résultant de leur action mutuelle. Développement de l'œuf des Batraciens. *Ann. Sci. Nat. II* 1824: 100–120, 129–49.

PYECRAFT, W. P. 1913. *The Courtship of Animals.* London: Hutchinson. 318 pp.

RABAUD, E., 1921. Sur le malentendu entre sociologues et biologistes. *Anthropologie* **31**:337–40.

———. 1927. Les rassemblements de mâles d'Halictes et le phénomène social chez les insects. *Bull. Biol. France Belg.* **61**:163–186.

———. 1929. Phénomène social et sociétés animals. *Ibid.* **63**:377–98.

RAHN, O. 1906. Über den Einfluss der Stoffwechselprodukte auf das Wachstum der Bakterien. *Centralbl. f. Bakt.* II. Abt. **16**:417–29, 609–17.

RAU, PHIL. 1915. Ability of mud-dauber to recognize its own prey. *Jour. An. Behav.* **5**:240–49.

———. 1917. Courtship of *Pieris protodice. Ibid.* **7**:143–44

RAU, PHIL, and NELLIE RAU. 1916. The sleep of insects: an ecological study. *Ann. Ent. Soc. Amer.* **9**:227–74.

———. 1929. The sex attraction and rhythmic periodicity in giant Saturniid moths. *Trans. Acad. Sci. St. Louis* **26**:81–221.

RAWIN, W. 1924. Weitere Beiträge zur Kenntnis der mitostischen Ausstrahlung und Induktion. *Arch. f. mikros. Anat. u. Entw.* **101**:53–61.

READE, W. W. 1867. The habits of the gorilla. *Amer. Nat.* **1**:177–80.

REAUMUR, R. 1734–42. *Mêmoires pour servir à l'histoire des insectes.* Paris: Roret. **2**:141–47, 170; **5**:667.

REEVES, CORA D. 1907. The breeding habits of the rainbow darter: a study in sexual selection. *Biol. Bull.* **14**:35–59.

REICHART, P. 1884. Schimpanse an Tanganika. *Arch. f. Naturgesch.* **1**:120–24.

REICHENOW, EDWARD VON. 1920. Biologische Beobachtungen an Gorilla und Schimpanse. *Sitzungber. d. Gesell. Naturf. Freunde* 1920: 1–41

REIGHARD, J. 1893. The ripe eggs and spermatozoa of the wall-eyed pike. *Bien. Rept. Mich. State Board Fish Comm.* **10**:93–171.

———. 1903. An experimental study of the spawning behavior of *Lampetra wilderi*. *Science* **17**:529.

———. 1903*a*. The natural history of *Amia calva* Linnaeus. Mark Anniversary Volume, pp. 57–109.

———. 1915. Breeding habits, development and propagation of the black bass. *Bien. Rept. Mich. State Board Fish Comm.* **7**:1–63.

———. 1920. The breeding behavior of the suckers and minnows. *Biol. Bull.* **38**:1–32.

REINHARD, H. J. 1927. The influence of parentage, nutrition, temperature and crowding on wing production in *Aphis gossypii*. *Texas Agric. Exp. Sta. Bull.* *353*, pp. 5–19.

———. 1921*a*. Responses of the large water strider, *Gerris remigis*, to contact

REINHEIMER, H. 1920. *Symbiosis, a Socio-physiological Study of Ecology.* London: Headly Bros. 425 pp.

REINKING, O. H. 1921. The synchronal flashing of fireflies. *Science* **53**:485–86.

REUTER, O. M. 1913. *Lebensgewohnheiten und Instinkte der Insekten bis zum Erwachen der sozialen Instinkte.* Berlin: R. Friedlander u. Sohn. 60 pp.

RIDDLE, C. D. 1928. Effect of crowding on the resistance of *Asellus* to KCl. Master's thesis, University of Chicago Library.

RILEY, C. F. C. 1912. Ecology of dragon fly nymphs: reactions to light and contact. *Ann. Ent. Soc. Amer.* **5**:273–92.

———. 1921. Distribution of the large water strider, *Gerris remigis* Say, throughout a river system. *Ecology* **2**:32–36.

———. 1921*a*. Responses of the large water strider, *Gerris remigis*, to contact and light. *Ann. Ent. Soc. Amer.* **14**:231–89.

ROBERTS, T. S. 1900. Nesting habits of Franklin's gull. *Auk* **17**:272–83.

ROBERTSON, T. B. 1921. Experimental studies on cellular multiplication. I. The multiplication of isolated Infusoria. *Biochem. Jour.* **15**:595–611.

———. 1921*a*. Experimental studies on cellular reproduction. II. The influence of mutual contiguity upon reproductive rate in Infusoria. *Ibid.* **15**:612–19.

———. 1922. Reproduction in cell communities. *Jour. Physiol.* **56**:404–12.

———. 1923. *The Chemical Basis of Growth and Senescence.* New York: Lippincott. 389 pp.

———. 1924. The nature of the factors which determine the duration of the period of lag in cultures of Infusoria. *Australian Jour. Exp. Biol. and Med. Sci.* **1**:105–20.

ROBERTSON, T. B. 1924*a*. The influence of washing upon the multiplication of isolated Infusoria and upon the allelocatalytic effect in cultures initially containing two Infusoria. *Ibid.* **1**:151–75.

———. 1924*b*. Allelocatalytic effect in cultures of *Colpidium* in hay infusion and in synthetic media. *Biochem. Jour.* **18**:1240–47.

———. 1925. The influence of cholesterol and oxycholesterol upon the multiplication of Infusoria (*Enchelys*). *Australian Jour. Exp. Biol. and Med. Sci.* **2**:83–90.

———. 1927. On some conditions affecting the viability of Infusoria and the occurrence of allelocatalysis therein. *Ibid.* **4**:1–24.

ROSSMAN, BRUNO. 1928. Untersuchungen über die Theorie der mitogenetischen Strahlung. *Arch. f. Entw. d. Organ.* **113**:346–406.

ROUBAUD, E. 1923. Les désharmonies de la fonction rénale et leurs conséquences biologiques chez les moustiques. *Ann. d. l'institut Pasteur.* **37**:627–79.

ROUBAUD, E., and J. COLAS-BELCOUR. 1927. Action des diastases dans le déterminisme d'éclosion de l'œuf chez le moustique de la fièvre jaune (*Stegomyia fasciata*). *Comp. Rend. Acad. Sci. Paris.* **184**:248–49.

ROUS, P., and F. S. JONES. 1916. A method for obtaining suspensions of living cells from the fixed tissues, and for plating out individual cells. *Jour. Exp. Med.* **23**:549–55.

ROUX, W. 1894. Über den Cytotropismus der Furchungszellen des Grasfrosches. *Arch. f. Entw. d. Organ.* **1**:161–202.

ROWAN, WILLIAM. 1922. Ecological note on the birds observed at the biological station of the University of Manitoba. *Ecology* **3**:43–69, 255–60.

RUSSELL, E. JOHN. 1927. *Soil Conditions and Plant Growth.* London: Longmans, Green. 516 pp.

RUTHVEN, A. G. 1908. Variations and genetic restrictions in garter snakes. *U.S. Nat. Mus. Bull. 61.* 201 pp.

RUTTER, CLOUDSLEY. 1903. Natural history of the Quinnat salmon. *Bull. U.S. Fish Comm.* **22**:65–141.

SALKIND, S., A. POTOZKY, and I. ZOGLINA. 1930. Die mitogenetische Beeinflussung der Eier von *Protodrilus* und *Saccocirrus*. *Arch. f. Entw. d. Org.* **121**:630–34.

SAMPSON, M. 1922. Iso-agglutination and hetero-agglutination of spermatozoa. *Biol. Bull.* **43**:267–83.

SARGENT, P. FLORENCE. 1926. *Overpopulation, Theory and Statistics.* London: Kegan Paul. 66 pp.

SARLES, M. P. 1929. The effect of age and size of infestation on the egg production of the dog hookworm, *Aneylostoma caninum*. *Amer. Jour. Hygiene* **10**:658–82.

SAUNDERS, A. A. 1914. An ecological bird study. *Auk* **30**:200–210.

SAVAGE, T. S. 1847. Notice of external characteristics and habits of *Troglodytes* gorilla. *Boston Jour. Nat. Hist.* **5**:417–43.

———. 1847*a*. On the habits of the driver or visiting ants of West Africa. *Trans. Ent. Soc. London* **5**:1–15, and *Proc. Acad. Nat. Sci. Phila.* **4**:195–203.

SAYLES, IRA. 1887. The sense of smell in *Cathartes aura*. *Auk* **4**:51–56.

SCHEERPELTZ, OTTO. 1926. Die Verdichtung des Vorkommens subterran lebender Insekten im Quellenniveau. *Koleopterolog. Rund.* **12**:55–63.

SCHIEFFERDECHER, P. 1904. Symbiose. *Sitz-Ber. niederrhein. Nat. Heilk. Bonn.* 1904B: 44–54.

SCHJELDERUP-EBBE, T. 1922. Beitrage zur Sozialpsychologie des Haushuhns. *Zeitschr. f. Psychol.* **88**:225–52.

———. 1923. Das Leben der Wildente (*Anas bosehas*) in der Zeit der Paarung. *Psychol. Forsch.* **3**:12–18.

SCHMEIL, OTTO. 1896. Deutschlands reilebende Süsswassercopepoden, III. Teil. *Centropagidae Bibliotheca Zoologica.* **8**: Heft 21, 1–143.

SCHRITTKY, C. 1922. Soziale Gewohnheiten bei solitären Insekten. *Zeitschr. f. wiss. Insektenbiol.* **17**:49–57.

SCHÜCKING, A. 1903. Zur Physiologie der Befruchtung Parthenogenese und Entwicklung. *Arch. f. d. ges. Physiol.* **97**:58–97.

SCHULZ, RICHARD. 1926. Untersuchung einer Tiergesellschaft. *Arch. Naturgesch.*, Abt. A, **92**:1–66.

SCHWARZ, E. A. 1896. Sleeping trees of Hymenoptera. *Proc. Ent. Soc. Wash.* **4**:24–26.

SCHWEMMLE, J. 1929. Mitogenetische Strahlen. *Biol. Zentralbl.* **49**:421–37.

SCLATER, W. L. 1900. *Mammals of South Africa.* 2 vols. London: R. H. Porter.

———. 1907. Bird islands of South Africa. *Condor* **9**:71–76.

SCOTT, H. 1916. Notes on the swarming of chloropid flies, *Psocidae*, etc., in houses. *Ent. Monthly Mag.* **52**:18–21, 43.

SCOTT, JOHN W. 1909. Some egg-laying habits of *Amphitrite ornata*. *Biol. Bull.* **17**:327–40.

SCOTT, W. E. D. 1887. The bird rookeries of southern Florida. *Auk* **4**: 135–44, 213–22, 272–84.

———. 1901. Data on song in birds: obervations on the song of the Baltimore oriole in captivity. *Science* **14**:522–26.

SEILER, J. 1927. Das Problem der Geschlechtsbestimmung bei *Bonellia*. *Naturwiss.* **15**:33–43.

SELOUS, F. C. 1907. *African Nature Notes and Reminiscences.* New York: Macmillan. 356 pp.

SEMPER, KARL. 1874. Über die Wachstumbedingungen der *Lymnaeus stagnalis*. *Zool. Inst. Würzburg.* **1**:138–67.

———. 1881. *Animal Life as Affected by the Natural Conditions of Existence.* New York: Appleton & Co. 472 pp.

SETON, E. T. 1919. Migrations of the gray squirrel (*Sciurus carolinensis*). *Jour. Mammal.* **1**:53–58.

SETON, E. T. 1922. A roving band of bats. *Ibid.* **3**:52.

———. 1922*a*. A magpie as sentinel for rabbits. *Ibid.* **3**:119.

———. 1929. *Lives of Game Animals.* 8 vols. New York: Doubleday Doran & Co.

SEVERIN, H. H. P., and H. C. SEVERIN. 1911. Habits of *Belostoma* (*Zaitha*) *flumineum* and *Nepa apiculata*. *Behavior Mono.* **1**:No. 3, 1–47

SHACKLEFORD, M. W. 1929. Animal communities of an Illinois prairie. *Ecology* **10**:126–54.

SHARP, M. A. D. 1927. Notes on the gorilla. *Proc. Zoöl. Soc. London* 1927: 1006–9.

SHAW, GRETCHEN. 1929. Further studies on the effect of crowding on the rate of growth in fishes. Master's thesis, University of Chicago Library.

SHELDON, C. 1921. A fox associating with mountain sheep on the Kenai peninsula, Alaska. *Jour. Mammal.* **2**:234.

SHELFORD, R. W. C. 1916. *A Naturalist in Borneo.* London: Unwin. 331 pp.

SHELFORD, V. E. 1911. Physiological animal geography. *Jour. Morph.* **22**: 551–618.

———. 1913. The reactions of certain animals to gradients of evaporating power of air. A study in experimental ecology. *Biol. Bull.* **25**:79–120.

———. 1913*a*. *Animal Communities of Temperate America.* University of Chicago Press. 362 pp.

———. 1929. *Laboratory and Field Ecology.* Baltimore: Williams & Wilkins. 608 pp.

SHELFORD, V. E., and E. D. TOWLER. 1925. Animal communities of San Juan channel and adjacent areas. *Publ. Puget Sound Biol. Sta.* **5**:33–73.

SHELLEY, F. C. 1929. The effect of crowding on sex in Cladocera. Master's thesis, University of Chicago Library.

SHERMAN, ALTHEA. 1924. Animal aggregations, a reply. *Condor* **26**:85–88.

SHIRAS, G. 1921. Notes on habits of deer, moose, beaver, wolves and muskrats. *Natl. Geog. Mag.* **40**:114–204.

SHUFELDT, R. W. 1896. The cormorant rookeries of the Lofoten Islands. *Auk* **13**:312–14.

SHULL, A. F. 1907. The stridulation of the snowy tree cricket (*Oecanthus nivens*). *Canadian Entomol.* **39**:213–25.

———. 1916. Letter to W. Craig. *Science* **44**:786.

———. 1918. Genetic relations of winged and wingless forms to each other and to sexes in the apid *Macrosiphon salarifoli*. *Amer. Nat.* **52**:507–20.

———. 1929. Determination of types of individuals in aphids, rotifers and Cladocera. *Biol. Rev.* **4**:218–48.

SIEBER-AUE, S. 1921. Masse und Persönlichkeit. *Grenzboten* **80** (Pt. 3):1–3.

SKENE, JAMES. 1825. On the emigration of a colony of caterpillars observed in Provence. From the MS reprinted in *Amer. Jour. Sci. & Arts* **9**:284–87. 1825. From Brewster's *Edinburgh Jour. of Sci.* **3**:93.

SMITH, ADAM. 1759. *Theory of Moral Sentiments*. London: Bell. 538 pp.

SMITH, G. 1915. The life cycle of Cladocera with remarks on the physiology of growth and reproduction in Crustacea. *Proc. Roy. Soc. London* B. **88**:418–35.

SMITH, H., and G. CLOWES. 1924. H-ion concentration on fertilization and development of *Arbacia, Asterias* and *Chaetopterous* eggs. *Biol. Bull.* **47**: 304–44.

SMITH, H. A. P. 1920. Evidence of interrelation of fox and caribou. *Jour. Mammal.* **4**:122–23.

SMITH, H. M. 1903. Breeding habits of the yellow catfish (*Ameiurus nebulosus*). *Science* **17**:243–44.

———. 1905. The drumming of the drum-fishes (*Sciaenidae*). *Ibid.* **22**:376–78.

———. 1909. A remarkable flight of bats in Luzon. *Ibid.* **29**:439.

SMITH, JOHN B. 1895. The Katydid's orchestra. *Ibid.* **2**:591.

SMITH, V. G. 1928. Animal communities of deciduous forest succession. *Ecology* **9**:479–500.

SNYDER, J. O. 1922. The return of marked king salmon grilse. *Calif. Fish and Game* **8**:102–7.

SNYDER, T. E. 1918. A peculiar habit of a horsefly (*Tabanus americanus*) in the Florida Everglades. *Proc. Ent. Soc. Wash.* **19**:141–45.

———. 1920. The colonizing reproductive adults of termites. *Ibid.* **22**:109–50.

———. 1924. Note on a mating flight of *Hexagenia bilineata* Say (Plectoptera). *Ibid.* **26**:24.

———. 1925. The origin of castes in termites. *Proc. Biol. Soc. Wash.* **38**:57–68.

———. 1926. The biology of termite castes. *Quar. Rev. Biol.* **1**:522–52

SOKOLOWSKY, A. 1910. *Genossenschaftsleben der Säugetiere: Eine psycho-biologische Studie über die Beziehungen der Säugetiere zu ihren Artgenossen.* Leipzig. 148 pp.

SOMES, M. P. 1914. The *Acrididae* of Minnesota. *Agric. Exp. Sta. Bull. 141.* 98 pp.

SOROKIN, P. 1928. *Contemporary Sociological Theories.* New York: Harper Bros. 785 pp.

SPALLANZANI, L. 1785. *Expériences pour servir à l'histoire de la génération des animaux et des plantes.* Genève: Chez B. Chirol. 413 pp.

SPELT, G. 1925. Beobachtung über Geselligkeit (Kolienbildung) bei überwinternden Käfern der Gattung Saprinas. *Zeitschr. f. wiss. Insektenbiol.* **20**:5–8.

SPENCER, BALDWIN, and F. J. GILLEN. 1899. *The Native Tribes of Central Australia.* London: Macmillan. 671 pp.

SPENCER, HERBERT. 1860. The social organism. *Westminster Rev.* **73**:90–121.

———. 1873. *Descriptive Sociology.* 8 vols. New York: Appleton.

SPENCER, HERBERT. 1877–97. *Principles of Sociology.* 3 vols. New York: Appleton.

———. 1893. The inadequacy of natural selection. *Contemp. Rev.* **63**:153–66, 439–56.

———. 1893*a*. Professor Weismann's theories. *Ibid.* **63**:743–60.

———. 1894. Weismannism once more. *Ibid.* **66**:592–608.

———. 1894*a*. Origin of classes among the parasol ants. *Nature* **51**:125–26.

SPRINGER, MARY G. 1922. The effect upon developing eggs of extracts of embryos of the same species. *Biol. Bull.* **43**:75–96.

STEINER, A. 1924. Über den sozialen Wärmhaushalt der Waldameise (*Formica rufa* var. *rufo-pratensis* For.). *Zeitschr. f. vergl. Physiol.* **2**:23–57.

———. 1925. Temperaturmessungen in den Nestern der Waldameise (*Formica rufa* var. *rufo-pratensis* For.) und der Wegameise (*Lasius niger* L.) während des Winters. *Mitt. naturf. Ges. Bern,* 1925:1–19.

———. 1929. Temperaturuntersuchungen in Ameisennestern mit Erdkuppeln, im Nest von *Formica exsecta* Nyl. und in Nestern unter Steinen. *Zeitschr. f. vergl. Physiol.* **9**:1–66.

———. 1930. Die Temperaturregulierung im Nest der Feldwespe (*Polistes gallica* var. *biglumis* L.) *Ibid.* **11**:461–502.

STEINER, G. 1925. Problem of host selection and host specialization of certain plant infesting Nemas and its application in the study of nemic pests. *Phytopathology* **15**:499–534.

STEP, EDWARD. 1913. *Mess Mates: A Book of Strange Companionships in Nature.* London: Hutchinson. 220 pp.

STEPHEN, A. C. 1929. Notes on the quantitative distribution of molluscs and polychaetes in certain intertidal areas on the Scottish Coast. *Proc. Roy. Soc. Edinburgh* **21**:205–16.

STETSON, G. R. 1899. The fauna of central Borneo. *Science* **5**:640–43.

STONE, CALVIN P. 1922. The congenital sexual behavior of the young male albino rat. *Jour. Comp. Psych.* **2**:95–153.

STONER, DAYTON. 1919. Swarming of the monarch butterfly in Iowa. *Ent. News* **30**:38.

STORROW, B. 1923. Herring investigations. I. Herring shoals. *Rept. Dove. Mar. Lab.* **12**:8–40.

STRONG, R. M. 1914. Habits and behavior of the herring gull. *Auk* **31**:22–50, 178–99.

STUART, C. A., and A. M. BANTA. 1929. Quantity of food as a factor in sex control in Cladocera. *Anat. Rec.* **44**:210.

———. 1931. Available bacteria and the sex ratio in *Moina. Physiol. Zoöl.* **4**:72–86.

STUART, C. A.; M. MCPHERSON; and H. J. COOPER. 1931. Studies on bacteriologically sterile *Moina macrocopa* and their food requirements. *Ibid.* **4**:87–100.

STUBBS, F. J. 1912. A contribution towards a solution of the problem of migration. *Zoölogist* **16**:441–49.

"STUDENT." 1925. New tables for testing the significance of observations. *Metron* **5**:18–21; Tables 26–32.

STUDER, T. 1913. Über *Putorius ermineus minimus* Cavazza. Eine Zwergform der Hermelius, *Putorius ermineus. Mitt. Natf. Ges. Bern* 1913:79–91.

STURTEVANT, A. H. 1915. Experiments on sex recognition in flies. *Jour. An. Behav.* **5**:351–66.

SUMNER, W. C., and A. G. KELLER. 1927. *The Science of Society.* 4 vols. New Haven: Yale University Press.

SUTTIE, J. D. 1922. Critique of the theory of herd instinct. *Jour. Mental Sci.* **68**:245–54.

SZYMANSKI, J. S. 1913. Analysis of the social instincts. *Biol. Centralbl.* **33**:644–58.

TANNER, F. W. 1925. The "bios" question. *Chem. Rev.* **1**:397–472.

TARDE, GABRIEL. 1903. Inter-psychology. *Internatl. Quar.* **7**:57–84.

TAVERNER, P. A. 1904. Discussion of the origin of migration. *Auk* **21**:322–33.

TAYLOR, H. S. 1924. *A Treatise on Physical Chemistry.* New York: D. Van Nostrand. 2 vols. 1323 pp.

TAYLOR, T. H. 1900. Additional note on the swarming and buzzing of Harlequin flies. In Miall, L. C., and A. R. Hammond: *The Structure and Life History of the Harlequin Fly*, Appendix, pp. 183–84. Oxford.

TERAO, ARATA. 1929. Relative rate of reproduction as a phase of population growth of organisms. *Jour. Imp. Fish. Inst.* [Tokyo] **24**:136–37. (Preliminary note.) *Proc. Imp. Acad. Tokyo* **5**:173–75.

TERAO, ARATA, and TOMOZO YANAKA. 1928. Influence of density of population upon the rate of reproduction in the water flea, *Moina macrocopa. Proc. Imp. Acad. Tokyo* **4**:556–58.

———. 1928*a*. Influence of density of population upon the egg laying in the fish *Oryzias latipes. Ibid.* **4**:559–60.

THOMAS, O. 1921. Bats on migration. *Jour. Mammal.* **2**:167.

THOMAS, W. I. 1909. *Source Book for Social Origins.* Boston: Badger. 932 pp.

THOMPSON, C. B. 1916. The brain and frontal gland of the castes of the "white ant" *Leucotermes flavipes* Kollar. *Jour. Comp. Neur.* **26**:553–603.

———. 1917. Origin of the castes of the common termite *Leucotermes flavipes* Kol. *Jour. Morph.* **30**:83–133.

———. 1918. The development of the castes of nine genera and thirteen species of termites. *Biol. Bull.* **39**:379–98.

———. 1922. The castes of *Termopsis. Jour. Morph.* **36**:495–531.

THOMPSON, C. B., and T. E. SNYDER. 1919. The question of the phylogenetic origin of the termite castes. *Biol. Bull.* **36**:115–29.

TOMSEN, C. 1921. Migrating muskrats. *Fins, Feathers and Fur* **28**:9.

TORRE-BUENO, J. R. 1911. The gerrids of the Atlantic states. *Trans. Amer. Ent. Soc.* **37**:243–52.

TOWNSEND, C. W. 1917. Labrador chickadee in southward migration. *Auk* **34**:160–62.

———. 1918. A winter crow roost. *Ibid.* **35**:405–16.

———. 1920. Courtship in birds. *Ibid.* **37**:380–93.

TROTTER, S. 1912. Relation of genera to the faunal areas. *Ibid.* **29**:159–65.

TROTTER, W. 1916. *Instincts of the Herd in Peace and War.* London: Unwin. 264 pp.

TRUE, F. W. 1895. Toads on the sea shore. *Science* **1**:166.

TURNER, C. H. 1907. The homing of ants: An experimental study of ant behavior. *Jour. Comp. Neur. and Psych.* **17**:367–429.

———. 1915. The mating of *Lasius niger* L. *Jour. An. Behav.* **5**:337–40.

TYLER, W. M. 1916. The call-notes of some nocturnal migrating birds. *Auk* **33**:132–41.

UICHANO, L. B. 1919. General facts in the biology of Philippine mound-building termites. *Philippine Jour. Sci.* **15**:59–65.

UVAROV, B. P. 1921. A revision of the genus *Locusta* with a new theory as to the periodicity and migration of locusts. *Bull. Entomol. Res. London* **12**: 135–63.

———. 1923. Notes on locusts of economic importance, with some new data on the periodicity of locust invasion. *Ibid.* **14**:31–39.

———. 1928. *Grasshoppers and Locusts.* London: Imp. Bur. Entom. 352 pp.

UVAROV, B. P., and B. N. ZOLOTAREVSKY, 1929. Phases of locusts and their interrelations. *Bull. Ent. Res.* **20**:261–65.

VAIHINGER, H. 1925. *The Philosophy "As If," a System of the Theoretical, Practical and Religious Fictions of Mankind.* Translated by C. K. Ogden. New York: Harcourt Brace. 370 pp.

VALLEY, G., and L. F. RETTGER. 1926. Carbon dioxide requirements of bacteria. *Jour. Bact.* (Abs.) **11**:78–79.

VAN NAME, W. G. 1910. Compound ascidians of the coasts of New England and neighboring British provinces. *Proc. Boston Soc. Nat. Hist.* **34**:339–424.

———. 1912. Simple ascidians of the coasts of New England and neighboring British provinces. *Ibid.* **34**:439–619.

VAUGHAN, T. W. 1919. Corals and formation of coral reefs. *Ann. Rept. Smithsonian Inst. 1917*, pp. 189–238.

VAYSSIERE, P. 1921. La lutte contre le criquet marocain (*Dociostaurus maroccanus Thunb.*). *Ann. Epiph. Paris* **7**:117–77.

VERHOEFF, K. W. 1902. Die zusammengesetzte Zirpvorrichtung von *Geotrupes. Sitzber. Gesell. Naturf. Freunde. Berlin* 1902: 149.

VERNON, H. M. 1895. Effects of environment on Echinoderm larvae. *Phil. Trans. Roy. Soc. London* B. **186**:577–632.

———. 1899. Effects of staleness of sex cells on development of echinoids. *Proc. Roy. Soc. London* B. **65**:350–60.

———. 1899*a*. The relation between marine animal and vegetable life. *Mitth. a. d. Zool. Sta. Neapel.* **13**:341–425.

———. 1903. *Variation in Animals and Plants.* New York: Holt. 415 pp.

WADLEY, F. M. 1923. Factors affecting the proportion of alate and apterous forms of aphids. *Ann. Ent. Soc. Amer.* **16**:279–303.

WAGNER, N. 1927. Über den von A. Gurwitsch entdeckten spezifischen Erreger der Zellteilung (mitogenetische Strahlen). *Biol. Zentralbl.* **47**:670–78.

WAGNER, V. A. 1907. *Psycho-biologische Untersuchungen an Hummeln mit Bezugnahme auf die Frage der Geselligkeit im Tierreiche.* Stuttgart: E. Schweizerbart'sche Verlagsbuchhandlung (E. Nägele). 239 pp.

WALLACE, A. R. 1872. *The Malay Archipelago.* London: Macmillan. 515 pp.

WALLIN, I. E. 1927. *Symbionticism and the Origin of Species.* Baltimore: Williams & Wilkins. 171 pp.

WALTER, E., 1925. Die Versuche 1924 in der bayerischen teichwirtschaftlichen Versuchsstation in Wielenbach. *Fischerei-Zeitung* **28**:458–64.

———. 1925*a*. Anmerkung zu Arens zur Ernährung der Forellen. *Ibid.* **29**:507.

WALTON, A. 1927. The relation of density of sperm-suspension and fertility as determined by artificial insemination of rabbits. *Proc. Roy. Soc. London* B. **101**:303–15.

WARD, H. B. 1921. Some of the factors controlling the migration and spawning of the Alaska red salmon. *Ecology* **2**:235–54.

———. 1926. Spawning habits of the bullhead catfish. *Bull. N.Y. Zoöl. Soc.* **29**:220.

WARD, H. L. 1905. Young of the red bat. *Science* **22**:20–21.

WARNER, L. H. 1927. A study of sex behavior in the white rat by means of the obstruction method. *Comp. Psych. Mono.* **4**:1–68.

———. 1928. A study of thirst behavior in the white rat by means of the obstruction method. *Jour. Genet. Psych.* **35**:178–90.

———. 1928*a*. A study of hunger behavior in the white rat by means of the obstruction method. A comparison of sex and hunger learning. *Jour. Comp. Psych.* **8**:273–99.

WARREN, E. 1900. On the reaction of *Daphnia magna* to certain changes in its environment. *Quar. Jour. Micr. Sci.* **43**:199–224.

WARREN, E. R. 1912. Some north central Colorado bird notes. *Condor* **14**:81–104.

———. 1922. Life of Yellowstone beaver. *Roosevelt Wild Life Bull.* **1**:187–221.

WASMANN, E. 1901–2. Neues über die zusammengesetzten Nester und gemischten Kolonien der Ameisen. *Allgem. Zeitschr. Ent.* **6**:353–55, 369–71; **7**:1–5, 33–37, 72–77, 100–108, 136–39, 169–73, 206–8, 235–40, 260–65, 293–98, 340–45, 385–90, 422–27, 441–49.

WASMANN, E. 1913. Ants and their guests. *Ann. Rept. Smithsonian Inst. 1912*, pp. 455–74.

WASMUND, E. 1928. Insekten-Massenschwärme am Bodensee und in Nürnberg. *Zeitschr. f. wiss. Insektenbiol. Berlin* **23**:234–43.

WAXWEILER, E. 1906. Esquisse d'une sociologie. *Notes et mem. Inst. Solvay*, Fasc. 2. Paris: Giard et Brière. 306 pp.

WEBB, G. B., W. W. WILLIAMS, and M. A. BARBER. 1909. Immunity production by inoculation of increasing numbers of bacteria, beginning with one living organism. *Jour. Med. Res.* **20**:1–25.

WEBSTER, F. M. 1909. Investigations of *Toxoptera graminum* and its parasites. *Science* **29**:278.

WEBSTER, H. M. 1913. The disastrous occurrence of *Vanessa* in California and Oregon during the years 1911–1912. *Canadian Entomol.* **45**:117–20.

WEESE, A. O. 1924. Animal ecology of an Illinois elm-maple forest. *Ill. Biol. Mono.* **9**:351–437.

WEISMANN, A. 1893. The all-sufficiency of natural selection. *Contemp. Rev.* **64**:309–38, 596–610.

WEISS, G. 1921. Wachstum und Zellgrösse bei *Cyprinus carpio*. *Zool. Jahrb., Abt. f. Allg. Zool. u. Phys.* **38**:137–68.

WELLMAN, G. B. 1917. Notes on the breeding of the American toad. *Copeia* No. 51: 107–8.

WENCKSTEIN, A. VON. 1891. Orang-utans von der Ostküste von Sumatra. *Cor. Bl. deut. Ges. f. Anthrop. München* **32**:30–32.

WESSEL, J. P. 1928. The effect of associated animals upon the rate of growth of aquarium fishes. Master's thesis, University of Chicago Library.

WESTWOOD, J. O. 1836. Nest of a gregarious Mexican butterfly, *Eucheira n.g. socialis* n.s. *Ent. Soc. Trans.* I, 1836: 38–44.

WHEELER, W. M. 1907. The polymorphism of ants, with an account of some singular abnormalities due to parasitism. *Bull. Amer. Mus. Nat. Hist.* **23**:1–93.

———. 1911. The ant colony as an organism. *Jour. Morph.* **22**:307–25.

———. 1913. Notes on the habits of certain Central American stingless bees. *Psyche* **20**:1–19.

———. 1913*a*. *Ants*. New York: Columbia University Press. 662 pp.

———. 1918. A study of some ant larvae, with a consideration of the origin and meaning of the social habit among insects. *Proc. Amer. Phil. Soc.* **57**: 293–343.

———. 1921. A new case of parabiosis and the "ant gardens" of British Guiana. *Ecology* **2**:89–103.

———. 1923. *Social Life among Insects*. New York: Harcourt Brace. 375 pp.

———. 1926. *Les sociétés d'insectes, leur origine, leur évolution*. Paris: Doin. 469 pp. Reviewed by P. P. Calvert, *Quar. Rev. Biol.* **2**:119–25.

———. 1926*a*. Social habits of some Canary Island spiders. *Psyche* **33**:29–31.

———. 1928. *Insect Societies, Their Origin and Evolution.* New York: Harcourt Brace. 378 pp.

———. 1928*a*. *Emergent Evolution and the Development of Societies.* New York: Norton. 80 pp.

———. 1930. Societal evolution. Chap. vi, pp. 139–55, in *Human Biology and Racial Welfare*, edited by E. V. Cowdry. New York: Hoeber. 612 pp.

WHITEFIELD, R. R. 1882. Description of *Lymnaea* (*Balimnaea*) *megasama*, with an account of changes produced in the offspring by unfavorable conditions of life. *Bull. Amer. Mus. Nat. Hist.* **1**:29–37.

WHITMAN, C. O. 1919. *Posthumous Works of Charles Otis Whitman.* Vol. III, *Behavior in Pigeons.* Edited by H. A. Carr. Carnegie Inst. Publ. No. 257. 161 pp.

WHITNEY, D. D. 1929. Male production in crowded and uncrowded cultures of the rotifer, *Hydatina. Physiol. Zoöl.* **2**:269–74.

WHITTLE, C. L. 1919. Migration of the myrtle warbler. *Auk* **39**:23.

———. 1923. Some aspects of the group habit among birds. *Ibid.* **40**:224–40.

———. 1924. Further observations on the group habit in birds. *Ibid.* **41**:327–33.

WIDMAN, O. 1896. The sociability of birds. *Osprey* **1**:3–5.

———. 1898. Gabbaret Island bird roosts. *Auk* **15**:22–27.

———. 1922. The crows' winter roost at St. Louis. Reprinted in *Trans. Acad. Sci. St. Louis* **24**:17–23.

WILCZYNSKY, J. 1914. Über die Exkretionsvorgänge bei den Männchen von *Bonellia viridis. Bull. Acad. Sci. Cracovie, Classe Math. et Natur.* B, pp. 191–97.

WILDER, B. G. 1875. Bats and their young. *Pop. Sci. Monthly* **7**:641–52.

WILDIERS, E. 1901. Nouvelle substance indispensable au développement de la levure. *La Cellule* **18**:311–33.

WILKINSON, D. S. 1926. The cyprus processionary caterpillar (*Thaumetopoea wilkinsoni*). *Bull. Ent. Res.* **17**:163–83.

WILLEM, V. 1896. Observations sur la respiration cutanée des *Limnées* et son influence sur leur croissance. *Bull. Acad. Roy. des Beaux-Arts de Belgique* **32**:563–77.

WILLER, A., and E. SCHNIGENBERG. 1927. Untersuchungen über das Wachstum bei Fischen: I. Über den Einfluss des Raumfaktors auf das Wachstum der Bachforellenbrut. *Zeitschr. f. Fischerei* **25**:263–90.

WILLETT, G. 1919. Bird notes from southeastern Oregon and northeastern California. *Condor* **21**:194–207.

WILLIAMS, C. B. 1922. Co-ordinated rhythm in insects with a record of sound production in an aphid. *Entomologist* **55**:173–76.

WILLIAMS, ROGER J. 1919. The vitamine requirement of yeast: a simple biological test for vitamine. *Jour. Biol. Chem.* **38**:465–86.

WILLIAMS, S. R. 1920. *Pectinella* colony. *Anat. Rec.* **17**:339–40.

WILLISTON, S. 1908. *Manual of North American Diptera.* 3d ed. New Haven: Hathaway. 405 pp.

WILLMER, E. N. 1928. Tissue culture from the standpoint of general physiology. *Biol. Rev.* **3**:271–302.

WILSON, C. B. 1923. Water beetles and pond fish culture. *U.S. Dept. Agric. Bull. Bur. Fisheries* **39**:231–345.

WILSON, H. V. 1910. Development of sponges from dissociated tissue cells. *Bull. U.S. Bur. Fish.* **30**:1–30.

———. 1911. On the behavior of dissociated cells in hydroids, *Alcyonaria* and *Asterias*. *Jour. Exp. Zoöl.* **11**:281–339.

WISSLER, CLARK. 1930. The integration of the sexes—marriage. From Cowdry's *Human Biology and Racial Welfare,* chap. xii, pp. 266–91. New York: Hoeber.

WOLFE, H. S. 1926. The auximone question. *Bot. Gaz.* **81**:228–31.

WOODRUFF, L. L. 1911. The effect of excretion products of *Paramoecium* on its rate of reproduction. *Jour. Exp. Zoöl.* **10**:551–81.

———. 1912. Observations on the origin and sequence of the protozoan fauna of hay infusions. *Ibid.* **12**:205–64.

———. 1914. The effect of excretion products of Infusoria on the same and on different species, with special reference to the protozoan sequence in infusions. *Ibid.* **14**:575–82.

WOODWORTH, W. M. 1907. The palolo worm, *Eunice viridis* Gray. *Bull. Mus. Comp. Zoöl. Harvard* **51**:3–21.

WRIGHT, A. H. 1911. Early records of the passenger pigeon. *Auk* **28**:346–66, 427–49.

WRIGHT, G. P. 1926. Presence of a growth stimulating substance in the yolk of incubated hen's eggs. *Proc. Soc. Exp. Biol. and Med.* **23**:603–5.

WRIGHT, HOWARD W. 1909. An ornithological trip to Los Coronados Islands, Mexico. *Condor* **11**:96–100.

———. 1913. Birds of San Martin Island, Lower California. *Ibid.* **15**:207–10.

WYLLY, MRS. 1889. Droves of lycaenid caterpillars herded by ants. *Canadian Entomol.* **21**:132–33.

YERKES, R. M. 1916. Mental life of monkeys and apes. *Behavior Mono.* No. III. New York: Holt. 145 pp.

———. 1903. The instincts, habits and reactions of the frog. *Psych. Rev. Mono. Suppl.* **4**:579–97.

———. 1905. The sense of hearing in frogs. *Jour. Comp. Neur. and Psych.* **15** :279–304.

———. 1915. Maternal instinct in a monkey. *Jour. An. Behav.* **5**:403–5.

YERKES, R. M., and A. W. 1929. *The Great Apes. A Study of Anthropoid Life.* New Haven: Yale University Press. 652 pp.

YOCOM, H. B. 1928. The effect of the quantity of culture medium on the division rate of *Oxytricha. Biol. Bull.* **54**:410–17.

YOUNG, W. C. 1929. A study of the function of the epididymus. I. Is the attainment of full spermatozoön maturity attributable to some specific action of the epididymal secretion? *Jour. Morph. and Physiol.* **47**:479–95.

YUNG, E. 1878. Contributions à l'histoire de l'influence des milieus physiques sur les êtres vivants. *Arch. Zoöl. Exp. et Gen.* **7**:251–82.

———. 1885. De l'influence des variations du milieu physico-chimique sur le développement des animaux. *Arch. des Sci. Phys. et Nat.* **14**:502–22.

ZANDER, E. 1921. *Das Leben der Biene.* Stuttgart: Ulmer. 103 pp.

INDEX

INDEX[1]

[1] Author citations in the index do not include reference to bibliography pages.